McGraw-Hill Education
CONQUERING
GRE
Math

THIRD EDITION | Robert E. Moyer, PhD

New York   Chicago   San Francisco   Athens   London   Madrid   Mexico City
Milan   New Delhi   Singapore   Sydney   Toronto

1 2 3 4 5 6 7 8 9 RHR 21 20 19 18 17 16

ISBN: 978-1-259-85950-2
MHID: 1-259-85950-9

e-ISBN: 978-1-259-85951-9
e-MHID: 1-259-85951-7

McGraw-Hill Education books are available at special quantity discounts to use as premiums and sales promotions, or for use in corporate training programs. To contact a representative please visit the Contact Us pages at www.mhprofessional.com.

GRE is a registered trademark of the Educational Testing Service, which was not involved in the production of, and does not endorse, this product.

CONTENTS

Part IV: GRE Math Practice Sections / 293

ABOUT THE AUTHOR

Dr. Robert E. Moyer taught mathematics and mathematics education at Southwest Minnesota State University in Marshall, Minnesota from 2002 to 2009. Before coming to SMSU, he taught at Fort Valley State University in Fort Valley, Georgia, from 1985 to 2000, serving as head of the Department of Mathematics and Physics from 1992 to 1994.

Prior to teaching at the university level, Dr. Moyer spent 7 years as the mathematics consultant for a five-county Regional Educational Service Agency in central Georgia and 12 years as a high school mathematics teacher in Illinois. He has developed and taught numerous in-service courses for mathematics teachers.

He received his Doctor of Philosophy in Mathematics Education from the University of Illinois (Urbana-Champaign) in 1974. He received his Master of Science in 1967 and his Bachelor of Science in 1964, both in Mathematics Education, from Southern Illinois University (Carbondale).

ACKNOWLEDGMENT

The writing of this book has been greatly aided and assisted by my daughter, Michelle Parent-Moyer. She did research on the tests and the mathematics content in them, created the graphics used in the manuscript, and edited the manuscript. Her work also helped ensure consistency of style, chapter format, and overall structure. I owe her a great deal of thanks and appreciation for all of the support she lent to the completion of the manuscript.

PREFACE

Recognizing that the people preparing to take the GRE have widely varying backgrounds and experiences in mathematics, this book provides an orientation to the mathematics content of the tests, an introduction to the formats used for the mathematics test questions, and practice with multiple-choice mathematics questions. There is an explanation of the quantitative comparison questions and data interpretation questions that are on the GRE. Many of the questions on the test are general problem-solving questions in a multiple-choice format with five answer choices.

The mathematics review is quite comprehensive with explanations, example problems, and practice problems covering arithmetic, algebra, and geometry. The mathematics on the GRE is no more advanced than the mathematics taught in high school. The topics are explained in detail and several examples of each concept are provided. After a few concepts have been explained, there is a set of practice problems with solutions. In each of the four mathematics review units, there is at least one multiple-choice test covering the concepts of the unit. The answers and solutions to the questions for each unit test are provided in a separate section following the test. Questions in the GRE formats follow each test. The review materials are structured so that you may select which topics you want to review. The unit tests may also be used to determine what topics you need to review.

There are three practice sections modeled after the GRE mathematics sections. Each section is followed by the answers and solutions for the questions on the section. The recommended time limit for the sections is the same as that on the GRE, 35 minutes. The practice sections are the same length as the actual test, 20 questions. The concepts on the practice sections are similar to those of the actual test and the proportion of questions on each area is also similar to that of the actual test. Information on the most recent changes to the GRE can be found on the www.ets.org website.

Use this book to review your mathematics knowledge, check your understanding of mathematics concepts, and practice demonstrating your math skills in a limited time frame. This will help you become prepared for your actual GRE.

Robert E. Moyer, PhD
Associate Professor of Mathematics (Retired)
Southwest Minnesota State University

PART I
INTRODUCTION

Graduate and professional schools consider a variety of factors when deciding which applicants to admit to their programs. These factors include educational background, work experience, recommendations from faculty, personal essays, and interviews. One factor often considered in admissions decisions is the applicant's performance on a standardized examination. One of the most common graduate school admissions tests is the Graduate Record Examination, generally called the GRE®.

The GRE is developed and administered by Educational Testing Service (ETS). There are nine exams that can be referred to as GRE tests; the GRE General Test and eight subject-specific exams. In this book, the name "GRE" will always refer to the GRE General Test; information on the subject tests is outside the scope of this book.

The GRE General Test consists of three parts: Verbal Reasoning, Quantitative Reasoning, and Analytical Writing. The test does not measure knowledge that comes from the in-depth study of any particular field; instead, it requires skills that are acquired over a period of many years. Many of those skills are developed through the curriculum of the average high school.

ETS revises the GRE General Test from time to time. This book covers the latest version of the test, introduced in Fall 2011. This version includes these features:

➤ A user-friendly testing interface that allows testers to skip questions, go back to previous questions within a question section, change answer choices, and use an on-screen calculator.
➤ Questions that closely reflect graduate-level reasoning skills: data interpretation, real-life scenarios, and questions that may have more than one possible answer.
➤ A scoring system that makes it easy for institutions to compare GRE scores among applicants.

The contents of this book were developed to prepare you for this revised version. The book contains chapters that discuss the various types of questions you will be asked, a review of the mathematics concepts you need, and practice GRE quantitative sections.

For general information about registering for and taking the GRE, visit the ETS website, **www.ets.org**, or the GRE website, **www.gre.org**.

CHAPTER 1

THE GRE QUANTITATIVE REASONING SECTIONS

The GRE is given as a computer-based test in the United States. (In some other countries, a paper-based version is used.) On the computer-based GRE General Test, there are two 35-minute Quantitative Reasoning sections. The test uses a modified computer-adaptive process in which the computer selects the difficulty of your second section based on how you well you scored on the first section. In other words, if you do well on the first section, you will get a harder second section (and a higher score). If you do poorly on the first section, you will get an easier second section (and a lower score). Since you must answer 20 questions in 35 minutes, you need to answer a question approximately every minute and a half. Within a section, you may skip a question and return to it later in order to maximize your efficiency. You need to finish each section in the allotted time. There is an on-screen calculator that you may use to aid in your calculations.

The questions in the Quantitative Reasoning sections assess your ability to solve problems using mathematical and logical reasoning and basic mathematical concepts and skills. The mathematics content on the GRE General Test does not go beyond what is generally taught in high schools. It includes arithmetic, algebra, geometry, and data analysis. The mathematics content, based on GRE sample tests provided by ETS, comes from the following areas:

➤ Number properties: approximately 22%
➤ Arithmetic (often graph-related): approximately 18%
➤ Algebra: approximately 18%
➤ Plane and solid geometry: approximately 14%
➤ Probability and statistics: approximately 8%
➤ Algebra word problems: approximately 6%
➤ Arithmetic ratios: approximately 6%
➤ Coordinate geometry: approximately 4%
➤ Tables: approximately 4%

There is no guarantee that the questions on each Quantitative Reasoning section on a given GRE test will be divided among the content areas according to these exact percentages, but the total Quantitative Reasoning part of the GRE will be spread among the content areas in approximately this way, based on the sample test materials provided by ETS.

In the revised GRE introduced in 2011, two new question formats have been added. These new question types allow some questions to be asked and answered in more natural and complex ways than the older formats permitted. The types of questions in the Quantitative Reasoning sections of the GRE

General Test may now include the following:

➤ Quantitative Comparison
➤ Multiple-Choice
➤ Numeric Entry
➤ Multiple-Response

Quantitative Comparison questions present two mathematical quantities. You must determine whether the first quantity is larger, the second is larger, the two quantities are equal, or if it is impossible to determine the relationship based on the given information.

Multiple-Choice questions are questions for which you are to select a single answer from a list of choices. These are the traditional multiple-choice questions with five possible answers that most test-takers will be familiar with from other standardized examinations.

In **Numeric Entry questions**, you are asked to type in the answer to the problem from the keyboard, rather than choosing from answers provided to you. For example, if the answer to the question is 8.2, you click on the answer box and then type in the number 8.2.

Multiple-Response questions are similar to multiple-choice questions, but you may select more than one of the five choices, if appropriate.

To be successful on the GRE Quantitative Reasoning sections, you need to be familiar with the types of questions you will be asked as well as the relevant mathematical concepts. Later chapters will go into more detail about the different question types and how to approach answering each of them.

CHAPTER 2

THE MATHEMATICS YOU NEED TO REVIEW

The GRE is taken by people with a wide variety of educational backgrounds and undergraduate majors. For that reason, the GRE Quantitative Reasoning sections test mathematical skills and concepts that are assumed to be common for all test-takers. The test questions require you to know arithmetic, algebra, geometry, and basic probability and statistics. You will be expected to apply basic mathematical skills, understand elementary mathematical concepts, reason quantitatively, apply problem-solving skills, recognize what information is relevant to a problem, determine what relationship, if any, exists between two quantities, and interpret tables and graphs.

The GRE does not attempt to assess how much mathematics you know. It seeks to determine whether you can use the mathematics frequently needed by graduate students, and whether you can use quantitative reasoning to solve problems. Specialized or advanced mathematical knowledge is not needed to be successful on the Quantitative Reasoning sections of the GRE. You will **NOT** be expected to know advanced statistics, trigonometry, or calculus, and you will not be required to write a proof.

In general, the mathematical knowledge and skills needed to be successful on the GRE do not extend beyond what is usually covered in the average high school mathematics curriculum. The broad areas of mathematical knowledge needed for success are number properties, arithmetic computation, algebra, and geometry.

Number properties include such concepts as even and odd numbers, prime numbers, divisibility, rounding, and signed (positive and negative) numbers.

In **arithmetic computation**, order of operations, fractions (including computation with fractions), decimals, and averages will be tested. You may also be asked to solve word problems using arithmetic concepts.

The **algebra** needed on the GRE includes linear equations, operations with algebraic expressions, powers and roots, standard deviation, inequalities, quadratic equations, systems of equations, and radicals. Again, algebra concepts may be part of a word problem you are asked to solve.

In **geometry**, concepts tested include the properties of points, lines, planes, and polygons. You may be asked to calculate area, perimeter, and volume, or explore coordinate geometry.

You will be expected to recognize standard symbols for mathematical relationships, such as = (equal), ≠ (not equal), < (less than), > (greater than), ‖ (parallel), and ⊥ (perpendicular). All numbers used will be real numbers. Fractions, decimals, and percentages may be used.

When units of measure are used, they may be in English (or customary) or metric units. If you need to convert between units of measure, the conversion relationship will be given, except for common ones such as converting minutes to hours, inches to feet, or centimeters to meters.

GRE word problems usually focus on doing something or deciding something. The mathematics is only a tool to help you get the necessary result. When answering a question on the GRE, you first need to read the question carefully to see what is being asked. Then, recall the mathematical concepts needed to relate the information you are given in a way that will enable you to solve the problem.

If you have completed the average high school mathematics program, you have been taught the mathematics you need for the GRE. The review of arithmetic, algebra, and geometry provided in this book will help you refresh your memory of the mathematical skills and knowledge you previously learned.

If you are not satisfied with your existing mathematics knowledge in a given area, then review the material provided on that topic in more detail, making sure that you fully understand each section before going on to the next one.

CHAPTER 3

CALCULATORS ON THE GRE QUANTITATIVE REASONING SECTIONS

3.1 OVERVIEW

Calculators are provided for use on the GRE Quantitative Reasoning sections. A handheld calculator is provided for the paper version of the GRE and an online calculator is provided for use during the computer version of the GRE. You are NOT allowed to use any calculator other than the one provided.

Just because a calculator is provided does not mean that one is needed or even helpful for most questions on the Quantitative Reasoning sections. The calculator will aid you in completing computations more quickly and easily; however, most of the questions are based on using reasoning and comparisons, which are not calculator activities. Using reasoning skills, estimation, mental computation, or simple arithmetic will enable you to answer most questions. The calculator can help you answer some questions more quickly or accurately than doing the computations manually. Questions with square roots, long division, or computations with multiple digits are reasonable questions for using a calculator. Also, use a calculator if there is a procedure that has been a frequent source of errors for you in the past.

You need to understand how the calculator will do the computations and display the results. Both the handheld calculator and the online calculator have eight-digit displays, which means only answers of eight digits or fewer can be shown. If the result to a computation is more than eight digits, the calculator will display an error message. For 5555555 times 3, the calculator will display the eight-digit result 16666665, but 55555555 times 3 is a nine-digit answer, so an error message will be displayed.

When the digits that cause the result to have more than eight digits are to the right of the decimal point, the calculator may just drop the extra digits, or it may round the result to eight digits. To see what the calculator you are using will do, you can test the calculator with 2 divided by 3. If the result displayed is 0.6666666, your calculator drops the extra digits. When the result displayed is 0.6666667, your calculator rounds the result to eight digits.

3.2 CALCULATOR FOR THE COMPUTER VERSION OF THE GRE

The computer version of the GRE has an online calculator for use during the Quantitative Reasoning sections. The online calculator has keys for memory storage, parentheses, and square root. A more important aspect of this calculator is that it follows the order of operations from algebra. The calculator will do the operations in parentheses first, then multiplications and divisions in order from left to right, and then additions and subtractions in order from left to right. This calculator will compute $5 + 3 \times 4$ as 17 because it will compute 3×4 to get

12 first and then compute 12 + 5 to get 17. The use of parentheses can get the calculator to do 5 + 3 first. Thus, (5 + 3) × 4 will compute 5 + 3 to get 8 and then compute 8 × 4 to get 32. The way you enter the problem into the calculator can influence the result of the computation. There is a special key, Transfer Display, on the online calculator for use on Numeric Entry questions to record your result in the box for your answer on the computer screen. This will eliminate copying errors when you are recording your answer.

▬ 3.3 CALCULATOR FOR THE PAPER VERSION OF THE GRE

For the paper version of the GRE, a handheld calculator is provided for use on the Quantitative Reasoning sections. The handheld calculator provided has a square root key and memory keys but NO parentheses keys. The calculator uses the rules of arithmetic to perform the operations in the order they are entered into the calculator. Thus, 5 + 3 × 4 will yield a result of 32 because 5 will be added to 3 to get 8, which will then be multiplied by 4 to get 32. If you enter 3 × 4 + 5, the calculator will multiply 3 times 4 to get 12 and then add 5 to get 17. Thus, the order you enter the data on this calculator influences the result.

▬ 3.4 SOME GENERAL GUIDELINES FOR CALCULATOR USAGE

In general, you should do the computations without using a calculator. This keeps you in charge of the work and eliminates one error source, the way the data are entered into the calculator. There are times when using a calculator will yield the result more quickly and easily, however.

- Most questions do not require the use of the calculator because there are no computations required.
- Simple computations are done faster mentally than with a calculator. So do computations like $40 - 295$, $\sqrt{49}$, $256/100$, 90^2, $(6)(800)$, and $56 + 104$ mentally.
- Estimating the result of the computation may let you select the best answer without needing to compute the exact answer.
- When using the calculator, compute the result as a decimal only if the answer choices have decimals or if the answer choices are different enough that the best answer can be matched easily from an approximate result.
- Use the calculator when the computations are complicated such as long division, computations using numbers that have many digits, or square roots.
- Use the calculator for computations in which you are likely to make errors based on your past experience.
- Enter numbers into the calculator carefully so that the numbers entered are correct and the computations will be completed in the order that you want them done.
- Clear the memory on the calculator before you start entering numbers for a new problem, and clear the memory on the calculator after you complete a problem. This will introduce two checks to be sure no left-over data from a previous problem will create errors when doing the current problem.

▆▆ 3.5 ONLINE CALCULATOR EXAMPLES

1. Compute: $\dfrac{(13 + 71)}{12} + (4 \times 6.24)$

   Enter: [(13 + 71) ÷ 12] + (4 × 6.24) = to get the result 31.96.
   or
   Enter: 13 + 71 = ÷12 = to get 7, and then press the M+ key to store the result in memory. Now enter 4 × 6.24 = to get the result 24.96, and then press the M+ key to add this number to the one currently in the memory. Finally, press the MR key to get the answer 31.96. Once you are finished with this result, press the MC key to clear the calculator memory so that the calculator is ready for use on another problem.
   Answer: 31.96

2. Compute: $\sqrt{7^2 + 5^2}$ to the nearest thousandth.
   Enter: (7 × 7 + 5 × 5) √ to get 8.6023253. Now round the result to the nearest thousandth, three decimal places, to get the final result 8.602.
   Answer: 8.602

3. Compute: $-\left[\dfrac{(7 + 15)}{4} + 3\right]$
   Enter: [(7 + 15) ÷ 4] +3 = ± to get −8.5
   Answer: −8.5

4. Compute: $(-11)^3 - 11^3$
   Enter: 11 ± × 11 ± × 11 ± × − 11 × 11 × 11 = to get −2662.
   Answer: −2662

▆▆ 3.6 HANDHELD CALCULATOR EXAMPLES

1. Compute: $\dfrac{(13 + 71)}{12} + (4 \times 6.24)$

   Enter: 13 + 71 ÷12 = to get 7. Now press the M+ key to store this number. Next, enter 4 × 6.24 +, and then press the MR key to add the stored result and get the final result 31.96. Be sure to press the MC key to clear the calculator's memory to prepare the calculator for use on another problem.
   Answer: 31.96

2. Compute: $\sqrt{7^2 + 5^2}$ to the nearest thousandth.
   Enter: 7 × 7 =, then press M+, then enter 5 × 5 +, and then press the MR key = √ to get 8.6023252.
   Now round the result to the nearest thousandth, three decimal places, to get 8.602. Press the MC key to clear the memory.
   Answer: 8.602

3. Compute: $-\left[\dfrac{(7 + 15)}{4} + 3\right]$
   Enter: [(7 + 15) ÷ 4] + 3 = ± to get −8.5.
   Answer: −8.5

4. Compute: $(-11)^3 - 11^3$
   Enter: 11 ± × 11 ± × 11 ± =, then press the M+ key, then enter 11 × 11 × 11 = ± +, and then press the MR key to get −2662. Press the MC key to clear the memory.
   or
   Enter: 11 × 11 × 11 =, then press the M+ key, then enter 11 ± × 11 ± × 11 ± = −, and then press the MR key to get −2662. Press the MC key to clear the memory.
   Answer: −2662.

PART II

TYPES OF GRE MATH QUESTIONS

Prior to 2007, the only types of question in the Quantitative Reasoning sections of the GRE were Quantitative Comparison and Multiple-Choice. Beginning in 2007, ETS developed the Numeric Entry format and began testing it. The revised General Test now includes those three question formats plus a Multiple-Response question format. On the computer-based version, each Quantitative Reasoning section is structured as follows:

Question Type	Number of Questions
Multiple-Choice	Approx. 10
Quantitative Comparison	Approx. 8
Numeric Entry	Approx. 1
Multiple-Response	Approx. 1
Total:	20

A multiple-choice question is simply a question with five answer choices from which you are asked to choose the one best answer. Approximately half of the questions are of this type. Quantitative Comparisons make up slightly less than half of the questions. Quantitative Comparisons always have the same four answer choices: you are asked to compare two quantities (A and B) and choose whether A is greater than B, B is greater than A, A and B are equal, or the relationship between A and B cannot be determined. Only a few questions are of the new types, Numeric Entry and Multiple-Response. In Numeric Entry questions, you are not given answer choices. Instead, you must calculate your own answer and type it into a space provided. Multiple-response questions are like multiple-choice questions, except that more than one of the answer choices may be correct.

While it is important to review the mathematical concepts that will be tested on the GRE, successful test-takers will also familiarize themselves with the ways in which the questions will be asked on the actual exam. The chapters in this section will take a closer look at the question formats used in the Quantitative Reasoning sections of the GRE, give you some strategies for approaching each type of question, and provide you with examples and practice exercises for each type.

CHAPTER 4

GRE QUANTITATIVE COMPARISON QUESTIONS

4.1 QUANTITATIVE COMPARISON ITEM FORMAT

Quantitative Comparison questions are designed to measure your ability to determine the relative sizes of two quantities or to realize that more information is needed to make the comparison. To succeed in answering these questions, you need to make quick decisions about the relative sizes of the two given quantities.

The first quantity appears on the left as "Quantity A." The second quantity appears on the right as "Quantity B." There are only four answer choices for this type of question, and they are always the same:

A. Quantity A is greater.
B. Quantity B is greater.
C. The two quantities are equal.
D. The relationship cannot be determined from the information given.

You are not expected to find precise values for A and B, and in fact, you may not be able to do so. You are merely asked to compare the relative values. If you see that under some conditions A is greater, but under other conditions B is greater, then the relationship cannot be determined, and the correct answer is choice D.

A symbol or other information that appears more than once in a question has the same meaning everywhere in the question. You will sometimes be given general information to be used in determining the relationship; this information will be above and centered between Quantity A and Quantity B.

In some countries, under some circumstances, the GRE may be given in a paper-based format, rather than as a computer-based test. When the test is given in a print format, the answer sheets always has five answer choices: A, B, C, D, and E. For Quantitative Comparison questions, there are only four answer choices: A, B, C, and D. If you are taking the paper-based GRE, never mark E for a Quantitative Comparison question.

4.2 EXAMPLES

Directions: Examples 1--5 each provide two quantities, Quantity A and Quantity B. Compare the two quantities, and then choose one of the following answer choices:

(A) Quantity A is greater.
(B) Quantity B is greater.
(C) The two quantities are equal.
(D) The relationship cannot be determined from the information given.

Example 1:

Quantity A	Quantity B	Answer
n	$\dfrac{1}{n}$	Ⓐ Ⓑ Ⓒ Ⓓ

Solution:

For $n = 2$, $\dfrac{1}{n} = \dfrac{1}{2}$. Because $n > \dfrac{1}{n}$, A is the greater quantity.

For $n = \dfrac{1}{4}$, $\dfrac{1}{n} = 4$. Because $n < \dfrac{1}{n}$, B is the greater quantity.

Thus, the relationship cannot be determined, and the correct answer is choice D.

Example 2:

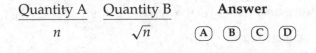

n is a real number greater than 1.

Quantity A	Quantity B	Answer
n	$\sqrt{n}$	Ⓐ Ⓑ Ⓒ Ⓓ

Solution:

By the definition of $\sqrt{n}$, $\sqrt{n} \times \sqrt{n} = n$. If $n > 1$, then $\sqrt{n} > 1$. $\sqrt{n} \times \sqrt{n} > 1 \times \sqrt{n}$, so $n > \sqrt{n}$, and choice A is the correct answer.

Example 3:

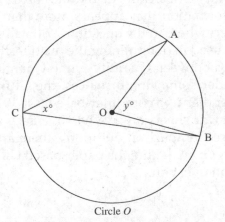

Circle O

Quantity A	Quantity B	Answer
$x°$	**$y°$**	Ⓐ Ⓑ Ⓒ Ⓓ

Solution:

In circle O, the central angle, $\angle AOB$, is equal in degrees to arc AB. In circle O, the inscribed angle, $\angle ACB$, is equal in degrees to $\frac{1}{2}$(arc AB). Thus, m$\angle AOB >$ $\angle ACB$, so $y° > x°$, and the correct answer is choice B.

Example 4:

Quantity A	Quantity B	Answer
The greatest prime factor of 20	The greatest prime factor of 15	Ⓐ Ⓑ Ⓒ Ⓓ

Solution:

The prime factorization of $20 = 2 \times 2 \times 5$, so the greatest prime factor is 5. The prime factorization of $15 = 3 \times 5$, so the greatest prime factor is 5. The greatest prime factor of each number is 5; thus, the correct answer is choice C.

Example 5:

Quantity A	Quantity B	Answer
The area of a rectangle with a perimeter of 60 m	200 m²	Ⓐ Ⓑ Ⓒ Ⓓ

Solution:

Because the perimeter of the rectangle is 60 m, the sum of the length and width is 30 m, since $P = 2l + 2w$. For any given fixed perimeter, the rectangle shape with the greatest area is a square. So if length and width are equal and have a sum of 30 m, each is 15 m, and the area is 225 m². In this case, Quantity A is greater than Quantity B.

However, there is no guarantee that the rectangle is a square, and you must investigate other pairs of numbers that add up to 30 as possible measures for the length and width. If the length is 10 m and the width is 20 m, the area of the rectangle is 200 m², and the two Quantities are equal. If the length is 29 m and the width is 1 m, the area of the rectangle is 29 m², and Quantity B is greater.

Therefore, the relationship cannot be determined from the given information, and the correct answer is choice D.

▰▰ 4.3 SOLUTION STRATEGIES

1. **If quantities A and B do not contain any variables, then the relationship between the quantities can always be determined: one quantity will be greater than the other, or they will be equal. Never choose D if there are no unknown quantities.**
   Example 4 shows this situation.

2. **If quantities A and B each have a fixed relationship to a third quantity, you can use those relationships to compare A and B.**
   Example 3 shows that both A and B have fixed relationships to the measure of arc AB, with Quantity A equal to half of it and Quantity B equal to it. Since arc AB has a positive measure, Quantity B is greater.

3. **Pay attention to restrictions put on the variable.**
   In Example 2, $n > 1$, so $n > \sqrt{n}$, and Quantity A is greater. However, if the restriction were $n > 0$, then for $n = \frac{1}{4}$, $\sqrt{n} = \frac{1}{2}$, and B is greater than A. Because there are now cases in which A is greater and cases in which B is greater, the relationship cannot be determined.

4. **In some cases, all possible values for the quantities are in a fixed interval. You should be sure to try numbers at the beginning, the middle, and the end of the interval.**
   In Example 5, because the perimeter is 60, the length plus the width of the rectangle is 30, so you should test cases in which l is equal to 1, 15, and 9. If $l = 1$, then $w = 30 - 1 = 29$, and the area is $1 \times 29 = 29$ m^2. If $l = 15$, then $w = 15$, and the area is $15 \times 15 = 225$ m^2. Because you can already see that there are cases in which the area of the rectangle could be greater than 200 m^2 and cases in which it could be less than 200 m^2, you do not need to test any further possible values of l. You already know that the given information is not enough to allow you to determine the relationship between Quantities A and B.

5. **Consider all possible numbers allowed.**
   When they are allowed, be sure to consider zero and negative numbers in your testing values. Also consider numbers between zero and one if those are allowed. In Example 1, as long as you use whole numbers greater than 1, A is always greater than B. However, if $n = \frac{1}{4}$, then $\frac{1}{n} = 4$, and B is greater. If $n = -2$, then $\frac{1}{n} = -\frac{1}{2}$, and B is greater. If $n = 1$, then $\frac{1}{n} = 1$, and A = B. Thus, choosing values of n only > 1 does not give a full picture of the relationship between A and B and will lead you to answer the question incorrectly.

6. **If the problem includes a figure (either provided for you or described in the question), try to visualize parts of the figure that are variable while the given information is still true.**
   If the size and shape of the figure can change while the given information remains true, the relationship between Quantities A and B can probably not be determined.
   In Example 5, the rectangle can have many different shapes while still having a perimeter of 60 m. For a rectangle with length 16 m and width 14 m, the area is 224 m^2. With length 20 m and width 10 m, the area is 200 m^2, and with length 25 m and width 5 m, the area is 125 m^2.
   Thus, by changing the shape of the rectangle, you can produce rectangles of areas greater than, equal to, and less than 200 m^2 while the perimeter remains fixed at 60 m.

4.4 EXERCISES

Directions: Exercises 1 – 5 each provide two quantities, Quantity A and Quantity B. Compare the two quantities, and then choose one of the following answer choices:

(A) Quantity A is greater.
(B) Quantity B is greater.
(C) The two quantities are equal.
(D) The relationship cannot be determined from the information given.

	Quantity A	Quantity B	Answer
1.	1.352×10^3	$135{,}620 \times 10^{-3}$	Ⓐ Ⓑ Ⓒ Ⓓ
2.	Sum of the solutions of $x^2-3x-10=0$	Sum of the solutions of $x^2-4x+3=0$	Ⓐ Ⓑ Ⓒ Ⓓ
3.	x	y	Ⓐ Ⓑ Ⓒ Ⓓ

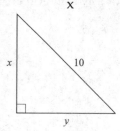

4.	x^2	y^2	Ⓐ Ⓑ Ⓒ Ⓓ
	$1 > x > y > 0$		
5.	Sum of the exterior angles in a regular hexagon	$720°$	Ⓐ Ⓑ Ⓒ Ⓓ

▰ 4.5 SOLUTIONS

1. **A** Because the values for A and B are numbers, you know immediately that answer choice D is not correct. You can compute A and B to see the relationship.

   $A = 1.352 \times 10^3 = 1{,}352$

   $B = 135{,}620 \times 10^{-3} = 135.62$

   Thus, A > B. The correct answer is choice A.

2. **B** Solve the equations to find the sum of the solutions, or use the fact that if $ax^2 + bx + c = 0$, then the sum of the solutions is $\dfrac{-b}{a}$.
   A: $x^2 - 3x - 10 = 0$; $(x-5)(x+2) = 0$; $x = 5$, $x = -2$; $5 + (-2) = 3$, **or** $\dfrac{-b}{a} = \dfrac{-(-3)}{1} = 3$.

   B: $x^2 - 4x + 3 = 0$; $(x-3)(x-1) = 0$; $x = 3$, $x = 1$; $3 + 1 = 4$, **or** $\dfrac{-b}{a} = \dfrac{-(-4)}{1} = 4$.

   Because 4 > 3, Quantity B is greater than Quantity A. The correct answer is choice B.

3. **D** The Pythagorean theorem states that $x^2 + y^2 = 10^2$. x^2 and y^2 can be any two numbers whose sum is 100. Setting $x = \sqrt{10}$ and $y = \sqrt{90}$ yields a short but wide triangle; if $x = \sqrt{50}$ and $y = \sqrt{50}$, the triangle is isosceles; and if $x = \sqrt{70}$ and $y = \sqrt{30}$, the triangle is tall but not very wide. Because many right triangles are possible, there is no way to determine the relative sizes of x and y. Thus, the correct answer is choice D.

4. **A** Because there is an interval, try any x and y in the interval. $1 > x > y > 0$. With $x = \dfrac{3}{4}$ and $y = \dfrac{1}{4}$, $x^2 = \dfrac{3}{4} \cdot \dfrac{3}{4} = \dfrac{9}{16}$, and $y^2 = \dfrac{1}{4} \cdot \dfrac{1}{4} = \dfrac{1}{16}$. Because each

factor of x^2 is greater than each factor of y^2, $x^2 > y^2$. Because x and y are always positive, x^2 will be greater than y^2 each time.

Also, since $x > y > 0$, multiplying by x you get $x^2 > xy > 0$. Because $x > y > 0$, multiplying by y gives you $xy > y^2 > 0$. Thus, $x^2 > xy > y^2$ and $x^2 > y^2$. Because A > B, the correct answer is choice A.

5. **B** The sum of the exterior angles for any polygon is 360°. So A < B, and choice B is the correct answer.

CHAPTER 5

GRE MULTIPLE-CHOICE QUESTIONS

5.1 MULTIPLE-CHOICE ITEM FORMAT

Multiple-choice questions are the typical standardized test questions most test-takers are familiar with. About 50% of the questions on the Quantitative Reasoning sections of the GRE are of this type; on the computer-based test, out of 20 questions in each section, approximately 10 will be multiple-choice questions. These questions have five answer choices, only one of which is correct. They focus on general problem-solving skills. You are to use the given information and your reasoning skills to select the best answer.

Unless you are told differently, you can assume all numbers are real numbers. Operations on real numbers are also assumed.

Figures provided for these problems show general relationships such as straight lines, collinear points, and adjacent angles. In general, you cannot determine the measures of angles or the lengths of segments from the figure alone. When a figure is NOT drawn to scale, it will be clearly identified as such. From a figure that is drawn to scale, you may estimate the lengths of segments and measurements of angles.

5.2 EXAMPLES

Example 1:

If $\dfrac{b}{a+b} = \dfrac{7}{12}$, then what does $\dfrac{a}{b}$ equal?

(A) $\dfrac{7}{19}$

(B) $\dfrac{5}{12}$

(C) $\dfrac{5}{7}$

(D) $\dfrac{7}{5}$

(E) $\dfrac{19}{12}$

Solution:

Because $\dfrac{b}{a+b} = \dfrac{7}{12}$ is a proportion, you can use two properties to transform it. First, use the reciprocal property to get $\dfrac{a+b}{b} = \dfrac{12}{7}$; then use the subtraction property to get $\dfrac{a+b-b}{b} = \dfrac{12-7}{7}$. So $\dfrac{a+b-b}{b} = \dfrac{a}{b} = \dfrac{5}{7}$, and answer choice C is correct.

Example 2:

In circle P, the two chords intersect at point X, with the lengths as indicated in the figure. Which could NOT be the sum of the lengths *a* and *b*, if *a* and *b* are integers?

(A) 30
(B) 26
(C) 19
(D) 16
(E) 14

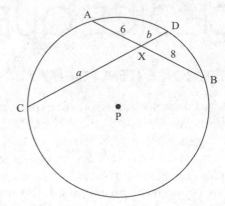

Solution:

When two chords intersect within a circle, the product of the segments on one chord is equal to the product of the segments on the other chord. Because the segments of the first chord are 6 and 8, the product of the lengths is 48. Thus, the product of the lengths *a* and *b* must be 48. Possible lengths are 48 and 1, 24 and 2, 16 and 3, 12 and 4, and 8 and 6, so possible values for *a* + *b* are 49, 26, 19, 16, and 14. The correct answer is choice A because 30 is not the sum of two integer factors of 48.

Example 3:

In one can of mixed nuts, 30% of the mixture is peanuts. In another can of mixed nuts that is one-half the size of the first one, 40% is peanuts. If both cans are emptied into the same bowl, what percent of the mixed nuts in the bowl is peanuts?

(A) 16²/₃%
(B) 20%
(C) 25%
(D) 33¹/₃%
(E) 40%

Solution:

Let the first can contain 16 ounces of nuts, which means that the second can contains 8 ounces of nuts. Thirty percent of 16 ounces is 4.8 ounces of peanuts, and forty percent of 8 ounces is 3.2 ounces of peanuts. The bowl contains (4.8 + 3.2) ounces of peanuts in the (16 + 8) ounces bowl, and $\dfrac{4.8 + 3.2}{16 + 8} = \dfrac{8}{24} = \dfrac{1}{3} = 33\dfrac{1}{3}\%$. So $33\dfrac{1}{3}\%$ of the nuts are peanuts, and choice D is the correct answer.

Example 4:

What is the sum of the prime numbers between $\frac{1}{2}$ and $9\frac{1}{5}$?

(A) 15
(B) 16
(C) 17
(D) 18
(E) 25

Solution:

The prime numbers between $\frac{1}{2}$ and $9\frac{1}{5}$ are 2, 3, 5, and 7. The sum of these prime numbers is 17, so the answer is choice C.

Example 5:

A paint store mixes $\frac{3}{4}$ pint of red paint and $\frac{2}{3}$ pint of blue paint to make a new paint color called Perfectly Purple. How many pints of red paint would be needed to make 34 pints of Perfectly Purple paint?

(A) 1.42
(B) 18.0
(C) 36.13
(D) 47.5
(E) 48.12

Solution:

First, determine how much paint the recipe for Perfectly Purple will make. $\frac{3}{4}$ pint $+\frac{2}{3}$ pint $= \frac{9}{12}$ pint $+\frac{8}{12}$ pint $= \frac{17}{12}$ pints, or $1\frac{5}{12}$ pints. The ratio of red paint in the recipe is the same as it will be in the 34 pints of paint. Let N be the number of pints of red paint needed.

$$\frac{\frac{3}{4}}{1\frac{5}{12}} = \frac{N}{34}$$

$$\frac{3}{4}(34) = 1\frac{5}{12}(N)$$

$$\frac{102}{4} = \frac{17}{12}N$$

$$\frac{102}{4} \div \frac{17}{12} = N$$

$$18 = N$$

To make 34 pints of Perfectly Purple paint, 18 pints of red paint are needed. The correct answer is choice B.

5.3 SOLUTION STRATEGIES

1. **Apply a general rule or formula to answer the question.**
   In Example 2, you can apply a property from geometry that says that when two chords intersect inside a circle, the segments formed have lengths such that the product of the segment lengths is the same number for each chord.
2. **Apply basic properties of numbers.**
   In Example 4, you need to use the definition of a prime number so that you do not include 1, but do include 2.
3. **Eliminate as many answers as possible so that you are selecting from a smaller set of answer choices.**
   In Example 3, you can eliminate some of the answers by noting that because each can of mixed nuts is at least 30% peanuts, the mixture of the two cans will be least 30% peanuts. Thus, before doing any computation, you can see that answers A, B, and C are wrong. Therefore, you only need to select from two answer choices.
4. **Substitute answer choices into the given expression to see which one produces the correct result.**
   In Example 1, you are given $\dfrac{b}{a+b} = \dfrac{7}{12}$, and you want the value of $\dfrac{a}{b}$. You can divide the numerator and denominator of $\dfrac{b}{a+b}$ by b to get $\dfrac{1}{\frac{a}{b}+1}$. Now substitute the answer choices into the expression to see which answer produces a value of $\dfrac{7}{12}$. Choice A produces $\dfrac{19}{26}$, so it is wrong. Choice C produces $\dfrac{7}{12}$, so it is correct. Because this type of question only has one correct answer, the correct answer must be choice C. You do not have to test the rest of the answer choices.

   This strategy works on only a few questions, so only use it when you can see a way to quickly test the answer choices.
5. **Break the situation into individual steps.**
   In Exercise 1 below, there is an everyday situation of a discount sale. Step 1 in problems of this type is often to represent or find the amount of discount. Step 2 is to subtract the discount amount from the regular price to find the sale price. Similar situations would be a sale with commission or a meal with sales tax and tip.

5.4 EXERCISES

1. **Bella's Baubles wants to sell a necklace for $179.95 next week at a 60% off sale. How much should the price of the necklace be this week?**

   (A) $71.98
   (B) $251.93
   (C) $287.92
   (D) $299.92
   (E) $449.88

2. **What is the median for this set of data: 9, 2, 5, 7, 10, 9, 2, 8, 11, 10?**

   (A) 8
   (B) 8.5
   (C) 9
   (D) 9.5
   (E) 10

3. **Which number is divisible by 3, 4, 5, and 6?**

   (A) 30
   (B) 48
   (C) 75
   (D) 120
   (E) 160

4. **The length of a rectangle is 4 cm longer than the width, and the perimeter is 96 cm. The area of the rectangle is how many square centimeters?**

   (A) 22
   (B) 26
   (C) 48
   (D) 192
   (E) 572

5. **Which quadratic equation has roots of 4 and $\frac{1}{2}$?**

   (A) $4x^2 - 9x + 2 = 0$
   (B) $2x^2 - 9x + 4 = 0$
   (C) $2x^3 - 9x^2 + 4x = 0$
   (D) $2x^2 + 9x + 4 = 0$
   (E) $x^2 - 2x + \frac{1}{2} = 0$

5.5 SOLUTIONS

1. **E** Break the problem into steps to solve it. To find the original price when you know the sale price, you must subtract the discount amount from the original price to get the sale price. Let P be the original price. Then $P - 0.6P = 179.95$. Then solve for P to find the original price.

   $0.4P = 179.95$

   $P = 179.95 \div 0.4$

   $P = 449.875$

   Because the answer is a sum of money, round it to $449.88. The correct answer is choice E.

2. **B** Apply the definition of the median as the middle value of an ordered sequence of values. To find the median, you need to arrange the data in order

from lowest to highest: 2, 2, 5, 7, 8, 9, 9, 10, 10, 11. Because there is an even number of values, average the two middle values to get the median.

$$Md = (8 + 9) \div 2$$
$$Md = 8.5$$

3. **D** Apply the divisibility rules for 3, 4, and 5. Since any number divisible by 3 and 4 is divisible by 6, there is no need to check separately for divisibility by 6. When a number is divisible by 5, its units digit must be either 0 or 5. If a number is divisible by 3, then the sum of the digits must be divisible by 3. To be divisible by 4, the last two digits must form a number divisible by 4.

   Because you want an answer that ends in 0 or 5, choice B can be eliminated. The sum of the digits must be divisible by 3, so choice E can be eliminated. Finally, the last two digits of the number must be divisible by 4, so choices A and C can be eliminated. The correct answer is choice D.

4. **E** Apply the formulas for perimeter and area of a rectangle. The perimeter of a rectangle is given by the formula $P = 2l + 2w$, and the area is given by the formula $A = lw$. Let w equal the width of the given rectangle. The length can then be represented as $w + 4$.

$$2w + 2(w + 4) = 96$$
$$2w + 2w + 8 = 96$$
$$4w + 8 = 96$$
$$4w = 88$$
$$w = 22$$
$$w + 4 = 26$$
$$A = lw = 22 \times 26 = 572$$

5. **B** Apply the factoring procedure and then find the solution for each factor. Note that choice C is not a quadratic equation and can be eliminated immediately. If the roots of a quadratic equation are 4 and $\frac{1}{2}$, then $x = 4$ and $x = \frac{1}{2}$ will yield $x - 4 = 0$ and $2x - 1 = 0$. The quadratic equation is therefore $(x - 4)(2x - 1) = 0$, which is $2x^2 - 9x + 4 = 0$. The correct answer is choice B.

CHAPTER 6

OTHER GRE MATH QUESTION FORMATS

6.1 NUMERIC ENTRY ITEM FORMAT

Numeric entry questions require you to compute the answer to a question and then enter that answer into a box or boxes provided. These questions are problem-solving and reasoning situations similar to GRE multiple-choice questions, except that no answer choices are provided. The correct answer to a numeric entry question is a decimal or integer with up to eight digits; it may be positive, zero, or negative. The negative sign or decimal point must be included if necessary.

For single-box answers, type the answer directly into the box, but take care in your typing. A typo will count as a wrong answer.

When the answer is a fraction, you will be given two boxes, one above a fraction bar for the numerator, and the other below the bar for the denominator. You cannot have a decimal point in the numerator or denominator of a fraction. Fractions do **NOT** have to be in lowest terms for numeric entry questions.

6.2 EXAMPLES

Example 1:

A coat was sold at a 30% discount sale for $140. What was the original price of the coat?

Solution:

The original price of the coat is 100% of P, and the discount amount is 30% of P, so the sale price is 70% of the original price.

$$0.70P = 140$$

$$P = 140 \div 0.70$$

$$P = 200$$

On the real test, you will answer the question by clicking on the box and entering 200. It would also be correct to enter 200.00, but it is not necessary to do so.

Example 2:

George earns $15 an hour for each regular-time hour he works; he earns time and one-half for each overtime hour. Overtime hours start after he has

worked 32 hours in a week. What is George's pay this week if he works 45 hours?

$

Solution:

George works $45 - 32 = 13$ overtime hours. Because overtime hours are paid at a rate of time and one-half, the pay rate is $1.5 \times \$15 = \$22.50$ per hour.

George's pay is his regular pay plus his overtime pay. $32 \times \$15 + 13 \times \$22.50 = \$480 + \$292.50 = \$772.50$. Enter 772.50 into the box. The decimal point must be entered from the keyboard, too. If you enter 772.5, that is also correct.

6.3 SOLUTION STRATEGIES

1. Read the problem carefully to make sure your solution answers the question that is asked.
2. Make sure you enter a complete answer, including any negative signs or decimal points.
3. There are no answer choices to check your solution against, so check to see that it satisfies all of the conditions in the problem.

6.4 EXERCISES

1. Calls made using a particular calling card cost $0.50 for the first 2 minutes and $0.15 for each minute after the first 2 minutes. What would be the cost of a 10-minute call?

$

2. A room is 12 ft wide, 18 ft long, and 9 ft high. How many square yards of carpet would it take to cover the floor of the room?

yards^2

3. What is the sum of the positive factors of 75?

6.5 SOLUTIONS

1. **1.70** Break the problem into parts: the cost of the first 2 minutes and the cost of the remaining 8 minutes. The cost of the first 2 minutes is $0.50. The remaining 8 minutes cost $0.15 each.

$$\$0.50 + 8(\$0.15) = \$0.50 + \$1.20 = \$1.70.$$

The call costs $1.70, so enter 1.70 or 1.7.

2. **24** The floor is 12 ft wide and 18 ft long. (Note that you were also given the height of the room. On the GRE, you may be given more information than you need to solve a problem.) Change the dimensions to yards; 3 ft equals 1 yard, and the length and width are divisible by 3. The room is 4 yards wide and 6 yards long, so the area of the room is 4×6 square yards, or 24 square yards. Enter 24.

3. **124** $75 = 1 \times 75, 75 = 3 \times 25, 75 = 5 \times 15$. Thus, the positive factors of 75 are 1, 3, 5, 15, 25, and 75. $1 + 3 + 5 + 15 + 25 + 75 = 124$. The sum of the positive factors of 75 is 124, so enter 124.

6.6 MULTIPLE-RESPONSE ITEM FORMAT

Multiple-response questions are similar to multiple-choice questions, except that you are allowed to select more than one correct answer. You will probably only encounter one of these questions on a section. These questions have five answer choices, but more than one may be correct. Like multiple-choice questions, they focus on general problem-solving skills. You are to use the given information and your reasoning skills to select the best answer. Unless you are told differently, you can assume all numbers are real numbers. Operations on real numbers are also assumed.

Figures provided for these problems show general relationships such as straight lines, collinear points, and adjacent angles. In general, you cannot determine the measures of angles or the lengths of segments from the figure alone. When a figure is NOT drawn to scale, it will be clearly identified as such. From a figure that is drawn to scale, you may estimate the lengths of segments and measurements of angles.

6.7 EXAMPLES

Example 1:

Which of the following is equivalent to $12\frac{1}{2}\%$? Select all that apply.

- A. $\frac{1}{8}$
- B. 0.125
- C. $\frac{125}{100}$
- D. 12.5
- E. $\frac{25}{200}$

Solution:

Because percent means "per hundred," $12\frac{1}{2}\% = \frac{12\frac{1}{2}}{100}$. You can multiply both the

numerator and the denominator by 2 to simplify the fraction to $\frac{25}{200}$, choice E.

That fraction can be simplified to $\frac{1}{8}$, choice A. You can also convert $\frac{12\frac{1}{2}}{100}$ to a

decimal to get 0.125, choice B. You should click on each correct answer to the question, so here you should select three choices: A, B, and E.

Example 2:

Which of the following are prime factors of 24? Select all that apply.

A. 1
B. 2
C. 3
D. 4
E. 6

Solution:

$24 = 4 \times 6 = 2 \times 2 \times 2 \times 3$. The prime factors of 24 are 2 and 3. Select choices B and C only.

■■■ 6.8 SOLUTION STRATEGIES

1. **Read the problem carefully to make sure you understand the parameters of the question.**

2. **Do not simply choose the first correct answer that you see and move on.** Carefully consider EACH answer choice to see whether it could be correct. Remember, these questions may have up to five correct answers!

3. **Consider alternative ways in which the same answer could be expressed.** In Example 1, both fraction and decimal representations of the same value are expressed.

■■■ 6.9 EXERCISES

1. **Which of the following are properties of zero? Select all that apply.**

   A. Negative
   B. Positive
   C. Even
   D. Odd
   E. Integer

2. In which of the following figures do the diagonals always bisect each other? Select all that apply.

   - [A] Quadrilateral
   - [B] Rhombus
   - [C] Rectangle
   - [D] Parallelogram
   - [E] Trapezoid

3. The numbers x and y are integers. If the ratio of x to y is 5 to 3, which of the following are possible values for $x + y$? Select all that apply.

   - [A] 2
   - [B] 8
   - [C] 15
   - [D] 40
   - [E] 72

6.10 SOLUTIONS

1. **C** and **E** Zero is neither positive nor negative. It is an even number, since it can be divided by 2. Zero is also an integer, so you should select both choice C and choice E.

2. **B, C,** and **D** The diagonals of a parallelogram always bisect each other. Since a rectangle and a rhombus are also parallelograms, you should select choices B, C, and D. Trapezoids and quadrilaterals do not necessarily have bisecting diagonals.

3. **B, D,** and **E** If $x : y = 5 : 3$, then 8 could be a value of $x + y$. Increase x and y by a factor of 5, and 40 could be a value of $x + y$. Increase x and y by a factor of 9, and 72 could also be a value of $x + y$. You should select choices B, D, and E.

CHAPTER 7

GRE DATA INTERPRETATION QUESTIONS

7.1 DATA INTERPRETATION ITEM FORMAT

Data interpretation questions involve information presented in a table or a graph. The questions may involve simply reading the graph, drawing conclusions from the data shown, making inferences based on the given data, and making comparisons using the data shown. One thing to note about the table or graph provided for a question is that it will always contain a great deal of information not needed for answering your question.

For the GRE, about one-fifth of the 20 questions in a quantitative section will be data interpretation questions.

The data presented in charts and tables are classified by both row and column. The units will always be stated, and the table will frequently involve data that have been rounded. Occasionally, the data will be expressed as percentages. The questions related to a table may ask you to merely read one particular item from the table, to combine values from the table, or to compare values found in the table.

In questions involving graphs, the graph will have a name that indicates what information is being presented. The type of graph used may also aid you in the interpretation of the data.

Line graphs and bar graphs have a scale that may or may not start at zero. The horizontal and vertical axes will each be labeled to show both what is being graphed and the scale along that dimension. Line graphs show the change in a quantity over time. A bar graph can also show the change in a quantity over time, but it can show a comparison of quantities, as well. In a bar graph, the width of the bars is generally uniform, and only the height or length varies.

A circle graph or pie chart is used to show a whole divided up into sectors that represent a percentage of the whole. The size of the sector is proportional to the percentage of the whole that it represents. The sum of all of the sectors is always 100%.

When working with graphs and tables, the goal is to use the visual representation to help you find the answer, rather than doing extensive computation. In a line graph that has values for 1980 and 1990, if you are asked to find the value for 1985, the easiest way is to locate the midpoint of the segment between 1980 and 1990 and read the answer from the axis representing the requested value. If you are asked how many times the number of accidents exceeded 40,000, then hold your pencil at the correct level to form a line at 40,000 and count the number of bars going past the line. In circle graphs, it is easier to make comparisons using the percentages given than to use the actual numbers. Whenever possible, use the visual image to answer the question; that is why the data are presented in graphical form.

7.2 EXAMPLES

Example 1:

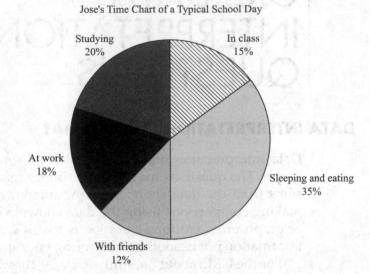

Jose's Time Chart of a Typical School Day

Studying 20%
In class 15%
At work 18%
Sleeping and eating 35%
With friends 12%

What is the ratio of the time Jose spent in class to the time he spent at work?

(A) 3% more

(B) 33% together

(C) $\dfrac{6}{5}$

(D) $\dfrac{3}{4}$

(E) $\dfrac{5}{6}$

Solution:

The circle graph shows that Jose spent 15% of his time in class and 18% of his time at work. There is no need to change the percentages to hours to get the ratio of time spent in class to time spent at work. The ratio is $\dfrac{15\%}{18\%} = \dfrac{0.15}{0.18} = \dfrac{15}{18} = \dfrac{5}{6}$

The answer is E.

Example 2:

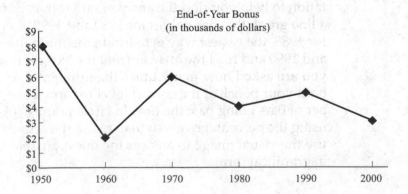

End-of-Year Bonus
(in thousands of dollars)

What is the best estimate of the end-of-year bonus for 1965?

(A) $3
(B) $4
(C) $3,000
(D) $4,000
(E) $5,000

Solution:

On the line graph, if you place an imaginary dot in the middle of the segment between 1960 and 1970, then follow that point across the graph to the scale, it hits at about the $4 mark. Since the scale is in thousands of dollars, the bonus for 1965 is $4,000. The answer is D.

Example 3:

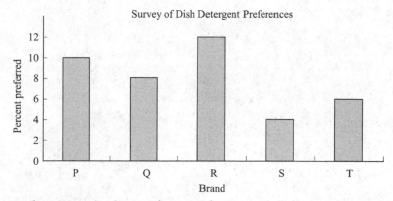

Survey of Dish Detergent Preferences

How much greater is the preference for Brand Q over Brand S?

(A) 4%
(B) 3%
(C) 2%
(D) 1%
(E) 12%

Solution:

The bar graph shows that Brand Q is preferred by 8% and Brand S is preferred by 4%. Thus, 4% more people preferred Brand Q over Brand S. The answer is A.

Example 4:

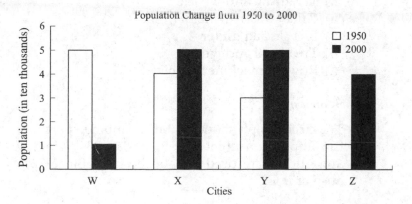

Population Change from 1950 to 2000

What is the least amount of growth for cities that grew between 1950 and 2000?

(A) 10,000
(B) 20,000
(C) 30,000
(D) 40,000
(E) 50,000

Solution:

The bar graph shows that the population of city W declined by 40,000, city X increased by 10,000, city Y increased by 20,000, and city Z increased by 30,000. Of the cities that had a population increase, city X had the smallest increase. The answer is A.

Example 5:

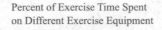

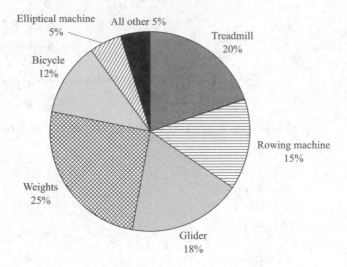

Percent of Exercise Time Spent on Different Exercise Equipment

What two pieces of equipment together are used the same amount of time as weights and the elliptical machine?

(A) Treadmill and bicycle
(B) Glider and rowing machine
(C) Bicycle and glider
(D) Treadmill and weights
(E) Rowing machine and bicycle

Solution:

The circle graph shows that the combined percent of time spent on weights and the elliptical machine is 25% + 5% = 30%. Bicycles are used 12% of the time, and gliders are used 18% of the time. Since 12% + 18% = 30%, the correct answer is C.

7.3 SOLUTION STRATEGIES

1. **Determine whether you can just read the graph to answer the question.**
   In Example 4, you just need to read the graph to see which black bar exceeds the corresponding white bar by the least amount and that amount is the answer.

2. **Determine whether you need to combine two values on the graph by addition or subtraction.**
   In Example 3, you have to subtract the percentage for Brand S from the percentage for Brand Q.

3. **Determine whether you need to compare values from the graph.**
   In Example 1, you need to get the ratio, in fraction form, for time spent in class to time spent at work.
   In Example 5, you have to find the total for the two given categories and then see which of the answers has two categories with the same total.

4. **Interpret the data in the graph to get the value at another point not stated.**
   In Example 2, you are asked for the value half way between two given years. You must average the values for the two given years to get the best estimate of the values requested.

7.4 EXERCISES

Questions 1 and 2 refer to the following graph.

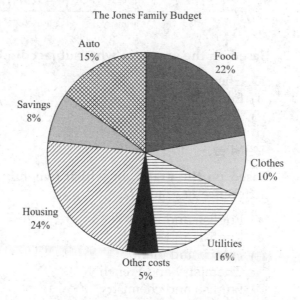

The Jones Family Budget

1. **Based on the graph, what is the ratio of housing costs to savings?**
   (A) 1 : 2
   (B) 2 : 1
   (C) 1 : 3
   (D) 3 : 1
   (E) 3 : 2

2. Based on the graph, what is the fractional part of the Jones family income budgeted for auto costs?

(A) $\frac{1}{7}$

(B) $\frac{3}{20}$

(C) $\frac{3}{10}$

(D) $\frac{4}{7}$

(E) $\frac{5}{8}$

Questions 3 to 5 refer to the following graph.

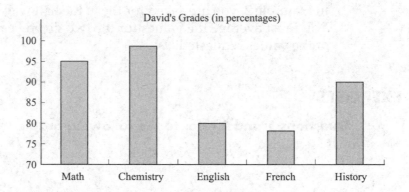

David's Grades (in percentages)

3. Based on the graph, in what subject did David score a 78%?

(A) Chemistry
(B) English
(C) French
(D) History
(E) Math

4. Based on the graph, in which two subjects did David have the greatest difference in grades?

(A) English and math
(B) English and French
(C) French and math
(D) Chemistry and French
(E) English and chemistry

5. Based on the graph, in which subject did David get his fourth highest grade?

(A) Chemistry
(B) English
(C) French
(D) History
(E) Math

▬ 7.5 SOLUTIONS

1. **D** Using Strategy 3, read from the graph to see that housing is 24% and savings is 8%. The ratio of housing to savings is 24:8 or 3:1.

2. **B** Using Strategy 1, read from the graph that auto costs are budgeted at 15%.
$15\% = \dfrac{15}{100} = \dfrac{3}{20}$.

3. **C** Using Strategy 1, when you read the graph there is only one bar that ends in the 75 to 80 range, so that is the correct answer. David's grade in French is 78%.

4. **D** Using Strategy 2, read the graph to see that David's highest grade, in chemistry, is 98%, and his lowest grade, in French, is 78%. Since these are the greatest value and the least value, they have the greatest difference between them.

5. **B** Using Strategy 2, read the graph to see that chemistry has the highest bar, math has the second highest bar, history has the third highest bar, and English has the fourth highest bar.

PART III

GRE MATHEMATICS REVIEW

The mathematics sections of the GRE require a knowledge of mathematics that is acquired over a period of years. This section is to review topics in arithmetic, algebra, and geometry that could form the content of the mathematics questions on the test.

Each chapter includes definitions of key concepts, worked-out examples, and practice exercises with explanations. Within each chapter there are summary tasks, solved problems in the various GRE question formats, and practice problems in these formats. The longer chapters will have multiple sets of these activities. The examples let you review the concepts and recall information you had learned previously. The practice exercises enable you to demonstrate your understanding of the concepts. Finally, you can apply your knowledge to questions in a format similar to test questions.

Chapter 8, Number Properties, reviews the number line, the types of real numbers, rounding numbers, computing with signed numbers, and the properties of numbers. The properties of the special numbers 0, 1, and −1 are reviewed, as well as those of even and odd numbers. Number theory properties such as prime and composite numbers, multiples, factors, and divisors are covered as well as least common multiple, greatest common divisor, and prime factorization.

In Chapter 9, Arithmetic Computation, the order of operations, properties of operations, fractions, decimals, ratios, proportions, percents, averaging, powers, and roots are discussed. The associative, commutative, and distributive properties are reviewed and applied. Converting among fractions, decimals, and percents shows the relationships among the various forms of rational numbers. Word problems will allow you to apply information you know about numbers.

Chapter 10, Algebra, reviews the basic concepts of algebra, including evaluating expressions, solving equations, and solving inequalities. Computation with algebraic expressions will include monomials, binomials, and polynomials. Graphs of points and linear equations will be discussed. The solution of quadratic equations by factoring and by the quadratic formula will be demonstrated. Algebraic word problems will include consecutive integer problems, age problems, mixture problems, and motion problems.

Chapter 11, Geometry, reviews the fundamental concepts of geometry. Proofs will not be part of the GRE, so they will not be considered. Angles, angle relationships, the relationship between lines, and types of polygons will be reviewed. Properties of triangles, parallelograms, rectangles, squares, and circles will be discussed. Formulas for area, perimeter, and volume of common geometric figures will be used in problems.

CHAPTER 8

NUMBER PROPERTIES

8.1 THE NUMBER LINE

A number line is a line with a scale for locating numbers indicated. The scale is determined when two numbers have their location indicated on the line. Usually, 0 and 1 are located first, but that is not a requirement.

The **origin** of a number line is the location of 0 on that number line.

The **coordinate** of a point on a number line is the number associated with that point.

Figure 8.1 is an example of a number line. The space between two consecutive numbers is equal everywhere on the number line. On this number line, the integers from -5 to $+5$ are shown, but all integers can be located by using the scale shown.

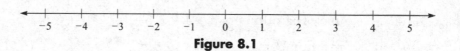

Figure 8.1

Numbers not shown on the number line can be represented by estimating their location between two consecutive numbers.

Example:

Locate $A = \dfrac{1}{2}$, $B = \sqrt{2}$, and $C = 3.1$ on a number line, using Figure 8.1.

Solution:

See Figure 8.2.

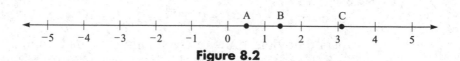

Figure 8.2

Since $A = 1/2$, it is located midway between 0 and 1. Since $B = \sqrt{2}$, you have to approximate $\sqrt{2}$ to get $\sqrt{2} \approx 1.414$. Locate B at about 1.4, which is a little less than halfway between 1 and 2. For $C = 3.1$, Locate C just to the right of 3 on the number line.

In working out the above example, you used a common practice of having the numbers increase as you go from left to right. Unless you are told otherwise, you may assume this is true for all number lines you encounter in this book or on the GRE. Also, there are an infinite number of points between any two of the indicated coordinates. In general, attempt to locate the number by thinking of the interval between two numbers as divided into halves. Then in each half, locate the number by deciding if is closer to the left end of the space, in the middle of the space, or closer to the right end of the space.

If greater accuracy is needed on the number line, the space between two numbers can be enlarged and subdivided into tenths of a unit. See Figure 8.3.

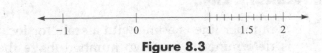

Figure 8.3

For even greater accuracy in locating points, the space between 1.2 and 1.3 could be enlarged and subdivided into 10 parts, or hundredths of a unit.

8.2 THE REAL NUMBERS

A **real number** is any number that can be the coordinates of a point on a number line. The numbers 3, -4, 2.3, $5\frac{1}{3}$, -7.21, $-\frac{2}{3}$, $\sqrt{2}$, 5, and $\frac{5}{7}$ are all examples of real numbers.

The set of **counting numbers** is the set of numbers 1, 2, 3, 4, ... Counting numbers are evenly spaced on the number line. Each number is one more than the previous number, except for 1, which is the smallest counting number. The counting numbers are also known as the **natural numbers**.

The **whole numbers** are the counting numbers plus zero.

The **integers** are made up of the counting numbers, zero, and the negatives of the counting numbers. The integers are $\{\ldots, -4, -3, -2, -1, 0, 1, 2, 3, 4, \ldots\}$.

The **rational numbers** are all the numbers that can be written as the ratio of two integers a and b, when b is not zero. The rational numbers include all the integers, since you can let a be any integer and b be 1. You also get fractions such as $\frac{2}{3}$, $-\frac{1}{4}$, $\frac{6}{5}$, and $-\frac{11}{7}$. These fractions can also be written as decimals through division. $\frac{1}{4} = 1 \div 4 = 0.25$; $\frac{1}{3} = 1 \div 3 = 0.333\ldots$. Thus, you have rational numbers that yield finite or terminating decimals, and rational numbers that yield infinite repeating decimals. Every rational number can be written in a decimal form.

The **irrational numbers** are the real numbers that are not rational. The irrational numbers you are most familiar with include $\sqrt{2}$, $\sqrt{3}$, π, and e. Irrational numbers can be written as infinite, nonrepeating decimals. There are decimal approximations for $\sqrt{2} \approx 1.41421\ldots$, $\sqrt{3} \approx 1.73205\ldots$, $\pi \approx 3.14159\ldots$, and $e \approx 2.71828\ldots$. However, these approximations do not clearly show that the decimals are nonrepeating. To show a decimal that is both infinite and nonrepeating, you need a pattern that is not based on repetition. One such pattern is to start with 1, then 01, then 001, then 0001, with each step in the pattern adding another zero. You get $0.101001000100001\ldots$. Another pattern yields $0.1121231234\ldots$.

The real numbers are made up of the combination of all rational numbers and all irrational numbers. Thus, the set of real numbers is the set of all decimals, finite and infinite.

8.3 ROUNDING NUMBERS

When numbers are approximated, the results need to be rounded to maintain the accuracy of the data. For example, frequently, when you work on problems with money, you get results that contain a fractional part of a cent. In these cases, you round the result to the nearest cent. In general, the accuracy is specified in the problem.

To round a number, first locate the digit that has the accuracy wanted; then examine the digit to the right. If the digit to the right is 5 or more, round up by increasing the accuracy digit by one and dropping all digits to the right of it. If the digit to the right of the accuracy digit is 4 or less, you round down by leaving the accuracy digit the same and dropping all digits to the right of it.

If the digits dropped are to the left of the decimal point, they are replaced with zeros.

Example:

Consider the number 2,643.718.

A. Round the number to the nearest tenth.
B. Round the number to the nearest hundredth.
C. Round the number to the nearest thousand.
D. Round the number to the nearest unit.
E. Round the number to the nearest hundred.

Solution:

A. There is a 7 in the tenths digit, and the digit to the right of the tenths digit is a 1. Thus, the tenths digit stays the same, and the digits to the right of it are dropped. When 2,643.718 is rounded to the nearest tenth, you get 2,643.7.
B. There is a 1 in the hundredths digit and an 8 in the place to the right. Thus, you increase the 1 by 1 and drop all digits to the right. When rounded to the nearest hundredth, 2,643.718 becomes 2,643.72.
C. There is a 2 in the thousands place, and the digit to the right is a 6, so you add 1 to the 2 and drop all digits to the right. Since the thousands place is to the left of the decimal point, you have to fill in zeros for the dropped digits between the thousands digit and the decimal point. Thus, when you round 2,643.718 to the nearest thousand, you get 3,000.
D. When rounding 2,643.718 to the nearest unit, you have a 3 in the units place and a 7 in the place to the right. The result of the rounding is 2,644.
E. When rounding to the nearest hundred, note that there is a 6 in the hundreds place and a 4 in the place to the right. Thus, you leave the 6 unchanged and drop the digits to the right of the hundreds place. Since the hundreds place is to the left of the decimal point, you must fill in zeros for the dropped digits between the hundreds place and the decimal point. The answer after rounding to hundreds is 2,600.

■ 8.4 EXPANDED NOTATION

Place value is used to write numbers in expanded notation. Each digit in the number is multiplied by the place value of that digit.

$300 = 3 \times 100$, or $300 = 3 \times 10^2$ since $10^2 = 100$

$50 = 5 \times 10$, or $50 = 5 \times 10^1$ since $10^1 = 10$

$7 = 7 \times 1$, or $7 = 7 \times 10^0$ since $10^0 = 1$

$357 = 300 + 50 + 7 = 3 \times 100 + 5 \times 10 + 7 \times 1$, or

$357 = 300 + 50 + 7 = 3 \times 10^2 + 5 \times 10^1 + 7 \times 10^0$

Example:

Write 568 in expanded notation.

$568 = 500 + 60 + 8 = 5 \times 100 + 6 \times 10 + 8 \times 1$, or

$568 = 500 + 60 + 8 = 5 \times 10^2 + 6 \times 10^1 + 8 \times 10^0$.

Example:

Write 21,000 in expanded notation.

$21,000 = 20,000 + 1,000 = 2 \times 10,000 + 1 \times 1,000$, or

$21,000 = 20,000 + 1,000 = 2 \times 10^4 + 1 \times 10^3$.

Expanded notation also applies to decimals. Fractions or 10 with negative exponents are used to indicate the place value to the right of the decimal point.

$0.1 = \dfrac{1}{10} = 10^{-1}$

$0.01 = \dfrac{1}{100} = 10^{-2}$

$0.001 = \dfrac{1}{1,000} = 10^{-3}$

$0.352 = 0.3 + 0.05 + 0.002 = 3 \times 0.1 + 5 \times 0.01 + 2 \times 0.001$

$\qquad = 3 \times \dfrac{1}{10} + 5 \times \dfrac{1}{100} + 2 \times \dfrac{1}{1,000}$, or

$0.352 = 0.3 + 0.05 + 0.002 = 3 \times 0.1 + 5 \times 0.01 + 2 \times 0.001$

$\qquad = 3 \times 10^{-1} + 5 \times 10^{-2} + 2 \times 10^{-3}$.

Example:

Write 0.804 in expanded notation.

$$0.804 = 0.8 + 0.004 = 8 \times \frac{1}{10} + 4 \times \frac{1}{1,000}, \text{ or}$$

$$0.804 = 0.8 + 0.004 = 8 \times 10^{-1} + 4 \times 10^{-3}.$$

A number that has both a whole number and a decimal point can also be written in expanded notation.

Example:

Write 147.23 in expanded notation.

$$147.23 = 100 + 40 + 7 \mid 0.2 + 0.03$$

$$= 1 \times 100 + 4 \times 10 + 7 \times 1 + 2 \times \frac{1}{10} + 3 \times \frac{1}{100}, \text{ or}$$

$$147.23 = 100 + 40 + 7 + 0.2 + 0.03$$

$$= 1 \times 10^2 + 4 \times 10^1 + 7 \times 10^0 + 2 \times 10^{-1} + 3 \times 10^{-2}.$$

■■■ 8.5   PRACTICE PROBLEMS

1. Graph the points $A = 2\frac{1}{2}$, $B = -4$, $C = 0$, $D = \frac{3}{4}$, $E = 5$ on a number line.

2. What are the coordinates of points F, G, H, I, and J in Figure 8.4?

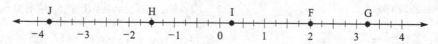

Figure 8.4

3. Which of these numbers are counting numbers?

$$3, 0, -5, \frac{2}{3}, 0.4, 1\frac{1}{2}, 8, -\frac{1}{2}, \sqrt{2}$$

4. Which of these numbers are whole numbers?

$$-2, 4, \frac{3}{4}, 1\frac{1}{2}, 6, \sqrt{5}, -\frac{1}{3}, 0, 0.8$$

5. Which of these numbers are integers?

$$-4, 3, \frac{1}{4}, \sqrt{8}, 0, -\frac{1}{3}, 2, 1\frac{3}{5}, \frac{5}{4}$$

6. Which of these numbers are rational?

$$2, 1\frac{1}{3}, \sqrt{6}, -5, 0.333\ldots, 0.51,$$

$$0.5152253335\ldots$$

7. Which of these numbers are irrational?

$$0, 2\frac{1}{5}, \sqrt{15}, -2, 0.525252\ldots,$$

$$0.010010001\ldots, 0.86, 9$$

8. Which types of number contain 18?

9. Which types of number contain −15?

10. Which types of number contain $\frac{7}{3}$?

11. Which types of number contain 0?

12. Which types of number contain 0.64?

13. Which types of number contain 0.672672672 . . . ?

14. Round 26,854 to the nearest thousand.

15. Round 0.6384 to the nearest tenth.

16. Round $112.465 to the nearest cent.

17. Round 435.982 to the nearest unit.

18. Round 264.382 to the nearest hundredth.

19. Round 468.451 to the nearest ten.

20. Round 751.364 to the nearest hundred.

21. Write 3,471 in expanded notation.

22. Write 258,000,000 in expanded notation.

23. Write 37.42 in expanded notation.

24. Write 516.4 in expanded notation.

25. Write 0.8724 in expanded notation.

8.6 SOLUTIONS

1. See Figure 8.5.

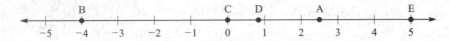

Figure 8.5

2. $F = 2, G = 3\frac{1}{4}, H = -1\frac{1}{2}, I = \frac{1}{4}, J = -3\frac{3}{4}.$

3. 3 and 8 are counting numbers.

4. 0, 4, and 6 are whole numbers.

5. −4, 0, 2, and 3 are integers.

6. $-5, 0.333\ldots, 0.51, 1\frac{1}{3}$, and 2 are rational numbers.

7. $0.010010001\ldots$ and $\sqrt{15}$ are irrational numbers.

8. 18 is a whole number, a counting number, an integer, a rational number, and a real number.

9. −15 is an integer, a rational number, and a real number.

10. $\frac{7}{3}$ is a rational number and a real number.

11. 0 is a whole number, an integer, a rational number, and a real number.

12. 0.64 is a rational number and a real number.

13. 0.672672672 . . . is a rational number and a real number.

14. To the nearest thousand, 26,854 is 27,000.

15. To the nearest tenth, 0.6384 is 0.6.

16. To the nearest cent, $112.465 is $112.47.

17. To the nearest unit, 435.982 is 436.

18. To the nearest hundredth, 264.382 is 264.38.

19. To the nearest ten, 468.451 is 470.

20. To the nearest hundred, 751.364 is 800.

21. $3,471 = 3,000 + 400 + 70 + 1 = 3 \times 1,000 + 4 \times 100 + 7 \times 10 + 1 \times 1$, or

    $3,471 = 3,000 + 400 + 70 + 1 = 3 \times 10^3 + 4 \times 10^2 + 7 \times 10^1 + 1 \times 10^0.$

22. $258,000,000 = 200,000,000 + 50,000,000 + 8,000,000 = 2 \times 100,000,000 + 5 \times 10,000,000 + 8 \times 1,000,000$, or

    $258,000,000 = 200,000,000 + 50,000,000 + 8,000,000 = 2 \times 10^8 + 5 \times 10^7 + 8 \times 10^6.$

23. $37.42 = 30 + 7 + 0.4 + 0.02 = 3 \times 10 + 7 \times 1 + 4 \times \dfrac{1}{10} + 2 \times \dfrac{1}{100}$, or

$37.42 = 30 + 7 + 0.4 + 0.02 = 3 \times 10^1 + 7 \times 10^0 + 4 \times 10^{-1} + 2 \times 10^{-2}$.

24. $516.4 = 500 + 10 + 6 + 0.4 = 5 \times 100 + 1 \times 10 + 6 \times 1 + 4 \times \dfrac{1}{10}$, or

$516.4 = 500 + 10 + 6 + 0.4 = 5 \times 10^2 + 1 \times 10^1 + 6 \times 10^0 + 4 \times 10^{-1}$.

25. $0.8724 = 0.8 + 0.07 + 0.002 + 0.0004 = 8 \times \dfrac{1}{10} + 7 \times \dfrac{1}{100} + 2 \times \dfrac{1}{1,000} + 4 \times \dfrac{1}{10,000}$, or

$0.8724 = 0.8 + 0.07 + 0.002 + 0.0004 = 8 \times 10^{-1} + 7 \times 10^{-2} + 2 \times 10^{-3} + 4 \times 10^{-4}$.

Properties of Zero:

➤ For any real number a, $a \times 0 = 0$.
➤ For any real number a, $a + 0 = a$.
➤ For any real number a, $a - 0 = a$ and $0 - a = -a$.
➤ For any nonzero real number a, $0 \div a = 0$.
➤ For real numbers a and b, if $a \times b = 0$, then $a = 0$ or $b = 0$.

Properties of 1 and -1:

➤ $1 \times 1 = 1$
➤ $(-1) \times (-1) = 1$
➤ $(-1) \times (1) = -1$
➤ For all real numbers a, $a \times 1 = a$.
➤ For all real numbers a, $a \times (-1) = -a$.
➤ For all real numbers a and b, $-1(a \times b) = (-a) \times b = a \times (-b)$.
➤ For all real numbers a, $-1 \times (-a) = a$.
➤ For all real numbers a, $a \div 1 = a$.
➤ For all real numbers a, $a \div (-1) = -a$.
➤ For any integer n, $n + 1$ is the next larger integer.
➤ For any integer n, $n + (-1)$ is the next smaller integer.
➤ The smallest counting number is 1.
➤ $1 + (-1) = 0$.
➤ For any real number a, $(1 \times a) + (-1 \times a) = a + (-a) = 0$.

▰ 8.7 ODD AND EVEN NUMBERS

An **even number** is an integer that is divisible by 2.

An **odd number** is an integer that has a remainder of 1 when divided by 2.

An even number can be written in the form $2n$, where n is an integer. The even numbers are $\ldots, -6, -4, -2, 0, 2, 4, 6, \ldots$. To get the positive even numbers, restrict n to the counting numbers.

An odd number can be written in the form $2k + 1$, where k is an integer. The odd numbers are $\ldots, -5, -3, -1, 1, 3, 5, \ldots$.

Every integer is either even or odd. No integer can be both even and odd. Zero is an even number. The smallest positive odd number is 1. The smallest positive even number is 2.

An even number added to another even number always yields a third even number. $2 + 6 = 8$, $0 + (-4) = -4$, $16 + (-4) = 12$.

An even number subtracted from another even number always yields a third even number. $4 - 12 = -8$, $16 - 2 = 14$, $0 - 8 = -8$, $10 - (-8) = 18$.

When an even number is multiplied by an even number, the result is an even number. $4 \times 6 = 24$, $8 \times 0 = 0$, $-4 \times 2 = -8$, $-6 \times -10 = 60$.

However, when you divide an even number by another even number, you do not always get an even number, and may not get an integer at all. $18 \div (-2) = -9$, $24 \div 6 = 4$, $6 \div 2 = 3$, $4 \div 8 = \dfrac{1}{2}$, $6 \div 0 =$ undefined.

Summary: $even + even = even$
               $even - even = even$
               $even \times even = even$

An odd number added to an odd number is an even number. $3 + 7 = 10$, $-5 + 11 = 6$, $-3 + 3 = 0$, $1 + 1 = 2$.

When an odd number is subtracted from an odd number, the answer is an even number. $7 - 13 = -6$, $5 - 3 = 2$, $7 - 7 = 0$, $11 - 3 = 8$.

When two odd numbers are multiplied together, the product will be a third odd number. $3 \times 5 = 15$, $7 \times -5 = -35$, $7 \times (-9) = -63$.

When two odd numbers are divided, the result is always defined, but may be either odd or even, or may not be an integer at all. $21 \div 3 = 7$, $15 \div (-5) = -3$, $11 \div 5 = \dfrac{11}{5}$, $5 \div 7 = \dfrac{5}{7}$.

Summary: $odd + odd = even$
               $odd - odd = even$
               $odd \times odd = odd$

Adding an even number and an odd number always gives an odd number for the answer. $3 + 4 = 7$, $8 + (-5) = 3$, $11 + 12 = 23$, $15 + 0 = 15$, $8 + 5 = 13$.

When subtracting an even number and an odd number, the result is an odd number. $5 - 2 = 3$, $6 - 1 = 5$, $-2 - (-3) = 1$, $8 + (-3) = 5$, $-7 + 6 = -1$.

If an even number and an odd number are multiplied, the answer is an even number. $3 \times 6 = 18$, $-7 \times 2 = -14$, $0 \times 3 = 0$, $5 \times 8 = 40$, $6 \times 11 = 66$.

When dividing an even number and an odd number, the answer **will not** be an integer when the divisor is an even number, and **may not** be an integer when the divisor is an odd number. $7 \div 0 =$ undefined, $5 \div 2 = \dfrac{5}{2}$, $11 \div 14 = \dfrac{11}{14}$, $18 \div 3 = 6$, $10 \div 7 = \dfrac{10}{7}$.

Summary: $odd + even = odd$
               $even + odd = odd$
               $odd - even = odd$
               $even - odd = odd$
               $odd \times even = even$
               $even \times odd = even$

▬▬ 8.8 NUMBER PROPERTIES TEST 1

Use the following test to assess how well you have mastered the material in this chapter. Mark your answers by blackening the corresponding answer oval in each question. An answer key and solutions are provided at the end of the test.

1. What is the coordinate of point P?

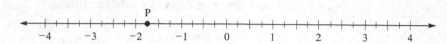

(A) $\dfrac{1}{4}$

(B) $-\dfrac{1}{4}$

(C) $-2\dfrac{1}{4}$

(D) $-1\dfrac{3}{4}$

(E) $-\dfrac{3}{4}$

2. Which point has a coordinate of $3\dfrac{2}{5}$?

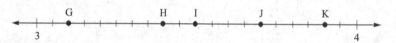

(A) Point G
(B) Point H
(C) Point I
(D) Point J
(E) Point K

3. Which point on the number line represents the smallest number?

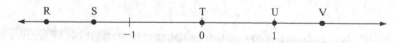

(A) Point R
(B) Point S
(C) Point T
(D) Point U
(E) Point V

4. Which of these numbers is a counting number?

   (A) 5
   (B) −4
   (C) 0
   (D) $\sqrt{3}$
   (E) 0.223344 . . .

5. Which of these numbers is a whole number?

   (A) $\frac{1}{2}$
   (B) 0.121212 . . .
   (C) 0
   (D) $\sqrt{10}$
   (E) −3

6. Which is a FALSE statement?

   (A) A whole number is an integer.
   (B) An irrational number is a real number.
   (C) A whole number is a rational number.
   (D) A counting number is a whole number.
   (E) An integer is an irrational number.

7. Which shows 3.15 rounded to tenths?

   (A) 3.05
   (B) 3.1
   (C) 3.15
   (D) 3.2
   (E) 3.25

8. In which number would the hundredths digit round up?

   (A) 3.1724
   (B) 5.6043
   (C) 5.6149
   (D) 5.6351
   (E) 5.8134

9. In which number would the thousandths digit NOT round up?

   (A) 2.34647
   (B) 2.48762
   (C) 2.76251
   (D) 3.10093
   (E) 5.92185

10. Which number shows 286,435 rounded to the ten-thousands place?

    (A) 286,440
    (B) 286,430
    (C) 290,000
    (D) 280,000
    (E) 270,000

11. Which is the expanded notation for 2,065?

    (A) $2 \times 10^5 + 6 \times 10^3 + 5 \times 10^1$
    (B) $2 \times 10^4 + 6 \times 10^3 + 5 \times 10^2$
    (C) $2 \times 10^3 + 6 \times 10^1 + 5 \times 10^0$
    (D) $2 \times 10^2 + 6 \times 10^1 + 5 \times 10^0$
    (E) $2 \times 10^3 + 6 \times 10^2 + 5 \times 10^1$

12. Which is the standard number for $3 \times 10^5 + 5 \times 10^3 + 7 \times 10^1$?

    (A) 357
    (B) 3,057
    (C) 3,507
    (D) 35,070
    (E) 305,070

13. Which is the expanded notation for 0.2841?

    (A) $2 \times 10^{-4} + 8 \times 10^{-3} + 4 \times 10^{-2} + 1 \times 10^{-1}$
    (B) $2 \times 10^0 + 8 \times 10^{-1} + 4 \times 10^{-2} + 1 \times 10^{-3}$
    (C) $2 \times 10^3 + 8 \times 10^2 + 4 \times 10^1 + 1 \times 10^0$
    (D) $2 \times 10^{-1} + 8 \times 10^{-2} + 4 \times 10^{-3} + 1 \times 10^{-4}$
    (E) $2 \times 10^{-3} + 8 \times 10^{-2} + 4 \times 10^{-1} + 1 \times 10^0$

14. Which of these is an odd number?

    (A) even + even
    (B) odd + odd
    (C) even × odd
    (D) odd × odd
    (E) odd − odd

15. Which of these is an even number?

    (A) $7 - 4$
    (B) $0 - 5$
    (C) $6 + 3$
    (D) $5 - 11$
    (E) 11×5

NUMBER PROPERTIES TEST 1

Answer Key

1. D	6. E	11. C
2. B	7. D	12. E
3. A	8. D	13. D
4. A	9. A	14. D
5. C	10. C	15. D

■■■ 8.9 SOLUTIONS

1. **D** $-1\frac{3}{4}$
   The point P is between -1 and -2. The units are divided into quarter units, so P is at $-1\frac{3}{4}$.

2. **B** Point H
   The unit between 3 and 4 is divided into twentieths, with darker lines at each fifth. Thus, H is located at $3\frac{2}{5}$.

3. **A** Point R
   The point R is farthest left on the number line, so it represents the smallest value of the labeled points.

4. **A** 5
   The counting numbers are 1, 2, 3, 4, ... The only counting number listed is 5.

5. **C** 0
   The whole numbers are 0, 1, 2, 3, ... The only whole number listed is 0.

6. **E** An integer is an irrational number.
   Since the integers are rational numbers, they cannot also be irrational numbers.

7. **D** 3.2
   There is a 1 in the tenths place, and the digit to the right is 5. Increase the tenths digit by 1 to get 2, and drop the digits to the right of the tenths place.

8. **D** 5.6351
   To round up, the digit to the right of the hundredths place must be 5 or greater, and this is only true in 5.6351.

9. **A** 2.34647
   The thousandths digit will not round up if the digit to the right of the thousandths place is 4 or less. This is only true for 2.34647.

10. **C** 290,000

    The digit in the ten-thousands place is an 8, and the digit to the right is a 6. Add one to the 8 to get 9, and drop the digits to the right of the ten-thousands. You must replace the dropped digits between the ten-thousands digit and the decimal point with zeros.

11. **C** $2 \times 10^3 + 6 \times 10^1 + 5 \times 10^0$

    $2,065 = 2,000 + 60 + 5 = 2 \times 10^3 + 6 \times 10^1 + 5 \times 10^0$

12. **E** 305,070

    $$3 \times 10^5 + 5 \times 10^3 + 7 \times 10^1 = 3 \times 100,000 + 5 \times 1,000 + 7 \times 10$$
    $$= 300,000 + 5,000 + 70 = 305,070$$

13. **D** $2 \times 10^{-1} + 8 \times 10^{-2} + 4 \times 10^{-3} + 1 \times 10^{-4}$

    $0.2841 = 0.2 + 0.08 + 0.004 + 0.0001$
    $$= 2 \times 10^{-1} + 8 \times 10^{-2} + 4 \times 10^{-3} + 1 \times 10^{-4}$$

14. **D** odd $\times$ odd

    even + even = even, odd + odd = even, even $\times$ odd = even,

    odd $\times$ odd = odd, odd − odd = even

15. **D** $5 - 11$

    $7 - 4 = 3$, $0 - 5 = -5$, $6 + 3 = 9$, $5 - 11 = -6$, $11 \times 5 = 55$

    -6 is even, so the answer is $5 - 11$.

■ 8.10 SOLVED GRE PROBLEMS

For each question, select the best answer unless otherwise instructed.

Quantity A	Quantity B

1. $5 \times 10^2 + 3 \times 10^{-2} + 4 \times 10^{-3}$ $9 \times 10^0 + 2 \times 10^{-1} + 4 \times 10^{-2} + 8 \times 10^{-3}$

    (A) Quantity A is greater.
    (B) Quantity B is greater.
    (C) The two quantities are equal.
    (D) The relationship cannot be determined from the given information.

2. Which is 728.359 rounded to the nearest tenth?

    (A) 730
    (B) 728.4
    (C) 728.36
    (D) 720
    (E) 73

For this question, enter your answer in the boxes.

3. What fraction is equal to $-\sqrt{\dfrac{16}{25}}$?

$$\boxed{} \atop \boxed{}$$

4. **Which of the following are even numbers? Select all that apply.**

- [A] 93,956
- [B] 4,249
- [C] 27
- [D] 0
- [E] −388

Quantity A	Quantity B
$(-1)^5$	$\left(\dfrac{2}{3}\right)^2$

5.

- (A) Quantity A is greater.
- (B) Quantity B is greater.
- (C) The two quantities are equal.
- (D) The relationship cannot be determined from the given information.

■ 8.11 SOLUTIONS

1. **A** Quantity A is greater.

$$5 \times 10^3 + 3 \times 10^{-2} + 4 \times 10^{-3} = 500 + 0.03 + 0.004 = 500.034$$
$$9 \times 10^0 + 2 \times 10^{-1} + 4 \times 10^{-2} + 8 \times 10^{-3} = 9 + 0.2 + 0.04 + 0.008 = 9.248$$

2. **B** In 728.359, the tenths digit is 3. The digit to the right of the tenths digit is a 5 which is greater than or equal to 5, so add 1 to the 3 and drop the digits to the right of the tenths digits. Thus, 728.359 rounded to the nearest tenth is **728.4.**

3. $\dfrac{-4}{5}$

$$-\sqrt{\dfrac{16}{25}} = -\dfrac{\sqrt{16}}{\sqrt{25}} = -\dfrac{4}{5} = \dfrac{-4}{5}$$

Enter the negative sign and the 4 in the numerator box and the 5 in the denominator box.

4. **A, D,** and **E** Even numbers are divisible by 2 and must have a 0, 2, 4, 6, or 8 as the units digit. A, D, and E are all correct.

5. **B** Quantity B is greater.

$$(-1)^5 = -1 \times -1 \times -1 \times -1 \times -1 = -1$$
$$\left(\dfrac{2}{3}\right)^2 = \dfrac{2}{3} \times \dfrac{2}{3} = \dfrac{4}{9}$$

8.12 GRE PRACTICE PROBLEMS

For each question, select the best answer unless otherwise instructed.

1. Which of these numbers is an irrational number?

   Ⓐ $\sqrt{9}$
   Ⓑ 0.15135135 . . .
   Ⓒ 0.1223334444 . . .
   Ⓓ 0
   Ⓔ $-\dfrac{2}{5}$

2. Which number is 1,674 rounded to the nearest hundred?

   Ⓐ 16
   Ⓑ 17
   Ⓒ 1,600
   Ⓓ 1,700
   Ⓔ 2,000

Quantity A	Quantity B
$4 \times 10^3 + 3 \times 10^2 + 5 \times 10^1$	400030050

3.

   Ⓐ Quantity A is greater.
   Ⓑ Quantity B is greater.
   Ⓒ The two quantities are equal.
   Ⓓ The relationship cannot be determined from the given information.

Quantity A	Quantity B
$(-4)^6$	$(16)^3$

4.

   Ⓐ Quantity A is greater.
   Ⓑ Quantity B is greater.
   Ⓒ The two quantities are equal.
   Ⓓ The relationship cannot be determined from the given information.

Quantity A	Quantity B
5. The smallest counting number.	The smallest whole number.

   Ⓐ Quantity A is greater.
   Ⓑ Quantity B is greater.
   Ⓒ The two quantities are equal.
   Ⓓ The relationship cannot be determined from the given information.

6. Which of the following statements are true? Select all that apply.

   ☐A Odd × odd = odd
   ☐B Even + odd = even
   ☐C Odd + odd = odd
   ☐D Even × odd = odd
   ☐E Odd − odd = odd

ANSWER KEY

> 1. C
> 2. D
> 3. B
> 4. C
> 5. A
> 6. A only

8.13 PRIMES, MULTIPLES, AND DIVISORS

A counting number a is a **multiple** of a counting number b if there is a counting number c such that $b \times c = a$.

To find the multiples of 3, multiply 3 by each counting number in turn. $3 \times 1 = 3$, $3 \times 2 = 6$, $3 \times 3 = 9$, $3 \times 4 = 12$... therefore, the multiples of 3 are 3, 6, 9, 12, 15....

The **least common multiple** of two or more counting numbers is the smallest positive number that is a multiple of each of the numbers.

The least common multiple of 6 and 10 is found by listing the multiples of each number and finding the smallest number that is in both lists. The multiples of 6 are 6, 12, 18, 24, 30, 36, 42, 48... and the multiples of 10 are 10, 20, 30, 40, 50, 60.... Since 30 is the smallest number that is in both lists, the least common multiple (LCM) of 6 and 10 is 30. You can write this as LCM $(6, 10) = 30$.

Example:

Find the first five multiples of each number.

A. 7   B. 8   C. 9   D. 10   E. 15

Solution:

A. $7 \times 1 = 7$, $7 \times 2 = 14$, $7 \times 3 = 21$, $7 \times 4 = 28$, $7 \times 5 = 35$ 7, 14, 21, 28, 35
B. $8 \times 1 = 8$, $8 \times 2 = 16$, $8 \times 3 = 24$, $8 \times 4 = 32$, $8 \times 5 = 40$ 8, 16, 24, 32, 40
C. $9 \times 1 = 9$, $9 \times 2 = 18$, $9 \times 3 = 27$, $9 \times 4 = 36$, $9 \times 5 = 45$ 9, 18, 27, 36, 45
D. $10 \times 1 = 10$, $10 \times 2 = 20$, $10 \times 3 = 30$, $10 \times 4 = 40$,
   $10 \times 5 = 50$ 10, 20, 30, 40, 50
E. $15 \times 1 = 15$, $15 \times 2 = 30$, $15 \times 3 = 45$, $15 \times 4 = 60$,
   $15 \times 5 = 75$ 15, 30, 45, 60, 75

Example:

Find the least common multiple of each pair of numbers.

A. 4 and 6   B. 30 and 24   C. 8 and 15

Solution:

A. The multiples of 4 are 4, 8, 12, 16, 20, 24, 28, 32, 36, ...
   The multiples of 6 are 6, 12, 18, 24, 30, 36, 42, ...
   12 is the smallest number that appears in both lists. LCM $(4, 6) = 12$.
B. The multiples of 30 are 30, 60, 90, 120, 150, 180, 210, ...
   The multiples of 24 are 24, 48, 72, 96, 120, 144, 168, 192, ...
   120 is the smallest number common to both lists. LCM $(30, 24) = 120$.

C. The multiples of 8 are 8, 16, 24, 32, 40, 48, 56, 64, 72, 80, 88, 96, 104, 112, 120, 128, ...

The multiples of 15 are 15, 30, 45, 60, 75, 90, 105, 120, 135, 150, ...

The smallest number in both lists is 120. LCM (8, 15) = 120.

Notes:

➤ If there is not a multiple in common in your lists, extend your list of multiples until one is reached.

➤ For counting numbers a and b, LCM $(a, b) \le a \times b$. This means that the least common multiple of two numbers is less than or equal to the product of the two numbers.

Example:

Find the least common multiple for these sets of three numbers.

A. 5, 12, 8   B. 10, 25, 15

Solution:

A. The multiples of 5 are 5, 10, 15, 20, 25, 30, 35, 40, 45, 50, 55, 60, 65, 70, 75, 80, 85, 90, 95, 100, 105, 110, 115, 120, 125, ...

The multiples of 12 are 12, 24, 36, 48, 60, 72, 84, 96, 108, 120, 132, 144, ...

The multiples of 8 are 8, 16, 24, 32, 40, 48, 56, 64, 72, 80, 88, 96, 104, 112, 120, 128, 136, ...

The smallest number common to all three lists is 120. LCM (5, 12, 8) = 120.

B. The multiples of 10 are 10, 20, 30, 40, 50, 60, 70, 80, 90, 100, 110, 120, 130, 140, 150, 160, 170, 180, ...

The multiples of 25 are 25, 50, 75, 100, 125, 150, 175, 200, ...

The multiples of 15 are 15, 30, 45, 60, 75, 90, 105, 120, 135, 150, 165, 180, ...

The smallest number common to all three lists is 150. LCM (10, 25, 15) = 150.

Factors and Greatest Common Divisors

A counting number a is a divisor of a counting number b if there is a counting number c such that $a \times c = b$.

Since $a \times c = b$, a is a **factor** of b.

In most cases, the terms **factor** and **divisor** are used as though they are the same; however, there are times when there is a slight difference in their meanings. Since the GRE uses counting numbers, this difference is not relevant.

The **greatest common divisor** for two or more counting numbers is the largest counting number that is a divisor of each of the counting numbers.

To find the divisors of a counting number such as 12, write down the factors, which come in pairs. However, when both factors are the same, list it just once. The divisors of 12 are: 1 and 12, 2 and 6, and 3 and 4. You usually list the divisors in numerical order. Thus, the divisors of 12 are: 1, 2, 3, 4, 6, and 12. Since 12 is equal to the number itself, 12 is an **improper divisor** of 12, and 1, 2, 3, 4, and 6 are the **proper divisors** of 12.

Example:

List all the divisors of the given numbers.

A. 16   B. 20   C. 21   D. 23   E. 36

Solution:

A. $1 \times 16 = 16$, $2 \times 8 = 16$, $4 \times 4 = 16$
The divisors of 16 are 1, 2, 4, 8, 16.

B. $1 \times 20 = 20$, $2 \times 10 = 20$, $4 \times 5 = 20$
The divisors of 20 are 1, 2, 4, 5, 10, 20.

C. $1 \times 21 = 21$, $3 \times 7 = 21$
The divisors of 21 are 1, 3, 7, 21.

D. $1 \times 23 = 23$
The divisors of 23 are 1, 23.

E. $1 \times 36 = 36$, $2 \times 18 = 36$, $3 \times 12 = 36$, $4 \times 9 = 36$, $6 \times 6 = 36$
The divisors of 36 are 1, 2, 3, 4, 6, 9, 12, 18, 36.

Example:

Find the greatest common divisor of each pair of numbers.
A. 8 and 12   B. 15 and 40   C. 30 and 24   D. 8 and 15   E. 7 and 14

Solution:

A. $1 \times 8 = 8$, $2 \times 4 = 8$ The divisors of 8 are 1, 2, 4, 8.
$1 \times 12 = 12$, $2 \times 6 = 12$, $3 \times 4 = 12$ The divisors of 12 are 1, 2, 3, 4, 6, 12.
1, 2, and 4 are the common divisors of 8 and 12. The greatest common divisor of 8 and 12 is 4. You can also write this as GCD (8, 12) = 4.

B. $1 \times 15 = 15$, $3 \times 5 = 15$ The divisors of 15 are 1, 3, 5, 15.
$1 \times 40 = 40$, $2 \times 20 = 40$, $4 \times 10 = 40$, $5 \times 8 = 40$ The divisors of 40 are 1, 2, 4, 5, 8, 10, 20, 40. The common divisors of 15 and 40 are 1 and 5. GCD (15, 40) = 5.

C. $1 \times 30 = 30$, $2 \times 15 = 30$, $3 \times 10 = 30$, $5 \times 6 = 30$ The divisors of 30 are 1, 2, 3, 5, 6, 10, 15, 30.
$1 \times 24 = 24$, $2 \times 12 = 24$, $3 \times 8 = 24$, $4 \times 6 = 24$ The divisors of 24 are 1, 2, 3, 4, 6, 8, 12, 24.
The common divisors of 30 and 24 are 1, 2, 3, and 6. GCD (30, 24) = 6.

D. $1 \times 8 = 8$, $2 \times 4 = 8$ The divisors of 8 are 1, 2, 4, 8.
$1 \times 15 = 15$, $3 \times 5 = 15$ The divisors of 15 are 1, 3, 5, 15.
The only common divisor of 8 and 15 is 1. GCD (8, 15) = 1.

E. $1 \times 7 = 7$ The divisors of 7 are 1, 7.
$1 \times 14 = 14$, $2 \times 7 = 14$ The divisors of 14 are 1, 2, 7, 14.
The common divisors of 7 and 14 are 1 and 7. GCD (7, 14) = 7.

The greatest common divisor of two counting numbers is at least 1 and never larger than the smaller of the two numbers.

Example:

Find the greatest common divisor for each set of three numbers.
A. 8, 10, and 12   B. 39, 65, and 91   C. 250, 375, and 625

Solution:

A. $1 \times 8 = 8$, $2 \times 4 = 8$ The divisors of 8 are 1, 2, 4, 8.
$1 \times 10 = 10$, $2 \times 5 = 10$ The divisors of 10 are 1, 2, 5, 10.
$1 \times 12 = 12$, $2 \times 6 = 12$, $3 \times 4 = 12$ The divisors of 12 are 1, 2, 3, 4, 6, 12.
The common divisors of 8, 10, and 12 are 1 and 2. GCD (8, 10, 12) = 2.

B. $1 \times 39 = 39$, $3 \times 13 = 39$ The divisors of 39 are 1, 3, 13, 39.
   $1 \times 65 = 65$, $5 \times 13 = 65$ The divisors of 65 are 1, 5, 13, 65.
   $1 \times 91 = 91$, $7 \times 13 = 91$ The divisors of 91 are 1, 7, 13, 91.
   The common divisors of 39, 65, and 91 are 1 and 13. GCD $(39, 65, 91) = 13$.
C. $1 \times 250 = 250$, $2 \times 125 = 250$, $5 \times 50 = 250$, $10 \times 25 = 250$ The divisors of 250 are 1, 2, 5, 10, 25, 50, 125, 250.
   $1 \times 375 = 375$, $3 \times 125 = 375$, $5 \times 75 = 375$, $15 \times 25 = 375$ The divisors of 375 are 1, 3, 5, 15, 25, 75, 125, 375.
   $1 \times 625 = 625$, $5 \times 125 = 625$, $25 \times 25 = 625$ The divisors of 625 are 1, 5, 25, 125, 625.
   The common divisors of 250, 375, and 625 are 1, 5, 25, and 125. GCD $(250, 375, 625) = 125$.

When a number is a divisor of a second number, it means that there is a counting number such that the product of it and the divisor equals the second number. However, when the first number is not a divisor of the second, you cannot find a counting number that will yield the second. In this case, you try to get as close as you can and stay less than the second number. To make them equal, you add the missing amount, called the *remainder*.

If a is a divisor of b, then there is a counting number c such that $a \times c = b$. If a is not a divisor of b, then there exist counting numbers q and r such that $a \times q + r = b$, where $0 < r < a$.

Prime Numbers

A **prime number** is a counting number greater than 1 such that its only counting number divisors are 1 and itself. Since a prime number is greater than 1, 0 and 1 are not prime numbers.

A **composite number** is a counting number greater than 1 such that it has at least three divisors. A counting number greater than 1 is either prime or composite. No counting number is both prime and composite.

The **prime factorization** of a number is a set of prime numbers such that the product of these prime factors will yield the given number. A prime factor may be used more than once in the product. The prime factorization of 6 is $6 = 2 \times 3$, and the prime factorization of 8 is $8 = 2 \times 2 \times 2$, or $8 = 2^3$.

Example:

Find the prime factorization of each number.
A. 24   B. 50   C. 72   D. 23   E. 39

Solution:

A. $24 = 2 \times 12 = 2 \times 2 \times 6 = 2 \times 2 \times 2 \times 3 = 2^3 \times 3$
B. $50 = 2 \times 25 = 2 \times 5 \times 5 = 2 \times 5^2$
C. $72 = 2 \times 36 = 2 \times 2 \times 18 = 2 \times 2 \times 2 \times 9 = 2 \times 2 \times 2 \times 3 \times 3 = 2^3 \times 3^2$
D. $23 = 23$
E. $39 = 3 \times 13$

To find the prime factorization of a counting number, look for divisors (or factors) of the number that are prime. Thus, it is helpful to know some prime numbers. The prime numbers less than 100 are 2, 3, 5, 7, 11, 13, 17, 19, 23, 29, 31, 37, 41, 43, 47, 53, 59, 61, 67, 71, 73, 79, 83, 89, and 97. It is important to notice that 2 is the only even number that is prime, since for a number to be even means that it is divisible by 2.

Example:

Find the prime factorization of each number.

A. 54   B. 150   C. 168   D. 500   E. 144

Solution:

A. $54 = 2 \times 27 = 2 \times 3 \times 9 = 2 \times 3 \times 3 \times 3 = 2 \times 3^3$
B. $150 = 2 \times 75 = 2 \times 3 \times 25 = 2 \times 3 \times 5 \times 5 = 2 \times 3 \times 5^2$
C. $168 = 2 \times 84 = 2 \times 2 \times 42 = 2 \times 2 \times 2 \times 21 = 2 \times 2 \times 2 \times 3 \times 7 = 2^3 \times 3 \times 7$
D. $500 = 2 \times 250 = 2 \times 2 \times 125 = 2 \times 2 \times 5 \times 25 = 2 \times 2 \times 5 \times 5 \times 5 = 2^2 \times 5^3$
E. $144 = 2 \times 72 = 2 \times 2 \times 36 = 2 \times 2 \times 2 \times 18 = 2 \times 2 \times 2 \times 2 \times 9 = 2 \times 2 \times 2 \times 2 \times 3 \times 3 = 2^4 \times 3^2$

Finding the prime factorization of a number requires you to divide the number by primes until you find one that is a divisor; then use that same prime to see if it is a divisor of the second factor. Continue to work on the factor that is not known to be a prime until you have reduced it to a prime. Check your work by multiplying all of the prime factors together to get the original number.

■■■ 8.14 GCD AND LCM REVISITED

A second way to find the GCD and LCM of a number is to use prime factorization. When you are finding the divisors of numbers with many factors, it is easy to overlook a factor that may be a common divisor. Similarly, in listing the multiples of a number, any error in finding one could cause other numbers to be wrong. It could also take a long time to find the common multiple if it is large.

To find the least common multiple (LCM) using the prime factorization, you must first find the prime factorization of each number. Then you need to find the greatest power each different factor of the numbers has. The least common multiple will be the product of each different factor to the greatest power it occurs.

Example:

Find the least common multiple of these sets of numbers.

A. 150 and 225   B. 63 and 84   C. 24, 60, and 96

Solution:

A. $150 = 2 \cdot 75 = 2 \cdot 3 \cdot 25 = 2 \cdot 3 \cdot 5 \cdot 5 = 2 \cdot 3 \cdot 5^2$
   $225 = 3 \cdot 75 = 3 \cdot 3 \cdot 25 = 3 \cdot 3 \cdot 5 \cdot 5 = 3^2 \cdot 5^2$
   The different prime factors are 2, 3, and 5. The greatest power of 2 is 1, the greatest power of 3 is 2, and the greatest power of 5 is 2.
   LCM $(150, 225) = 2^1 \cdot 3^2 \cdot 5^2 = 2 \cdot 9 \cdot 25 = 450$
B. $63 = 3 \cdot 21 = 3 \cdot 3 \cdot 7 = 3^2 \cdot 7$
   $84 = 2 \cdot 42 = 2 \cdot 2 \cdot 21 = 2 \cdot 2 \cdot 3 \cdot 7 = 2^2 \cdot 3 \cdot 7$
   The greatest power of 2 is 2, the greatest power of 3 is 2, and the greatest power of 7 is 1.
   LCM $(63, 84) = 2^2 \cdot 3^2 \cdot 7 = 4 \cdot 9 \cdot 7 = 252$
C. $24 = 2 \cdot 12 = 2 \cdot 2 \cdot 6 = 2 \cdot 2 \cdot 2 \cdot 3 = 2^3 \cdot 3$
   $60 = 2 \cdot 30 = 2 \cdot 2 \cdot 15 = 2 \cdot 2 \cdot 3 \cdot 5 = 2^2 \cdot 3 \cdot 5$
   $96 = 2 \cdot 48 = 2 \cdot 2 \cdot 24 = 2 \cdot 2 \cdot 2 \cdot 12 = 2 \cdot 2 \cdot 2 \cdot 2 \cdot 6 = 2 \cdot 2 \cdot 2 \cdot 2 \cdot 2 \cdot 3 = 2^5 \cdot 3$
   The greatest occurring power of 2 is 5, the greatest power of 3 is 1, and the greatest power of 5 is 1.
   LCM $(24, 60, 96) = 2^5 \cdot 3 \cdot 5 = 32 \cdot 3 \cdot 5 = 480$

To find the greatest common divisor (GCD) of a set of numbers, you must first find the prime factorization of each number. You next identify the primes that are common to all the numbers and then the greatest common power that these primes have. The greatest common divisor of the numbers is the product of these common divisors to their greatest common powers.

Example:

Find the greatest common divisor for these sets of numbers.

A. 32 and 104 B. 120 and 216 C. 225, 375, and 825

Solution:

A. $32 = 2 \cdot 16 = 2 \cdot 2 \cdot 8 = 2 \cdot 2 \cdot 2 \cdot 4 = 2 \cdot 2 \cdot 2 \cdot 2 \cdot 2 = 2^5$
$104 = 2 \cdot 52 = 2 \cdot 2 \cdot 26 = 2 \cdot 2 \cdot 2 \cdot 13 = 2^3 \cdot 13$
2 is the only common prime factor for the numbers, and 3 is the greatest common power of 2.
$\text{GCD}(32, 104) = 2^3 = 8$

B. $120 = 2 \cdot 60 = 2 \cdot 2 \cdot 30 = 2 \cdot 2 \cdot 2 \cdot 15 = 2 \cdot 2 \cdot 2 \cdot 3 \cdot 5 = 2^3 \cdot 3 \cdot 5$
$216 = 2 \cdot 108 = 2 \cdot 2 \cdot 54 = 2 \cdot 2 \cdot 2 \cdot 27 = 2 \cdot 2 \cdot 2 \cdot 3 \cdot 9 = 2 \cdot 2 \cdot 2 \cdot 3 \cdot 3 \cdot 3 = 2^3 \cdot 3^3$
The common prime divisors are 2 and 3. The greatest common power of 2 is 3, and the greatest common power of 3 is 1.
$\text{GCD}(120, 216) = 2^3 \cdot 3^1 = 8 \cdot 3 = 24$

C. $225 = 3 \cdot 75 = 3 \cdot 3 \cdot 25 = 3 \cdot 3 \cdot 5 \cdot 5 = 3^2 \cdot 5^2$
$375 = 3 \cdot 125 = 3 \cdot 5 \cdot 25 = 3 \cdot 5 \cdot 5 \cdot 5 = 3 \cdot 5^3$
$825 = 3 \cdot 275 = 3 \cdot 5 \cdot 55 = 3 \cdot 5 \cdot 5 \cdot 11 = 3 \cdot 5^2 \cdot 11$
The common divisors are 3 and 5. The greatest common power of 3 is 1, and of 5 is 2.
$\text{GCD}(225, 375, 825) = 3^1 \cdot 5^2 = 3 \cdot 25 = 75$

8.15 PRACTICE PROBLEMS

1. Find the least common multiple.
   A. 20 and 24 B. 15 and 10
   C. 8 and 48 D. 14 and 10

2. Find the greatest common divisor.
   A. 20 and 24 B. 15 and 28
   C. 8 and 48 D. 14 and 10

3. Which of these numbers are prime?
   A. 14 B. 23 C. 79
   D. 51 E. 117

4. Which of these numbers are composite?
   A. 81 B. 18 C. 43
   D. 19 E. 91

5. Find the prime factorization of each number.
   A. 400 B. 98 C. 54
   D. 96 E. 150

6. Find the least common multiple.
   A. 252 and 588 B. 54 and 144 C. 56, 140, and 168

7. Find the greatest common divisor.
   A. 252 and 588 B. 54 and 144 C. 56, 140, and 168

▰▰▰ 8.16 SOLUTIONS

1. A. The multiples of 20 are 20, 40, 60, 80, 100, 120, 140, 160, ...
   The multiples of 24 are 24, 48, 72, 96, 120, 144, 180, ...
   LCM (20, 24) = 120
   B. The multiples of 15 are 15, 30, 45, 60, 75, 90, 105, ...
   The multiples of 10 are 10, 20, 30, 40, 50, ...
   LCM (15, 10) = 30
   C. The multiples of 8 are 8, 16, 24, 32, 40, 48, 56, ...
   The multiples of 48 are 48, 96, 144, ...
   LCM (8, 48) = 48
   D. The multiples of 14 are 14, 28, 42, 56, 70, 84, ...
   The multiples of 10 are 10, 20, 30, 40, 50, 60, 70, ...
   LCM (14, 10) = 70

2. A. $1 \times 20 = 20$, $2 \times 10 = 20$, $4 \times 5 = 20$
   The divisors of 20 are 1, 2, 4, 5, 10, 20.
   $1 \times 24 = 24$, $2 \times 12 = 24$, $3 \times 8 = 24$, $4 \times 6 = 24$
   The divisors of 24 are 1, 2, 3, 4, 6, 8, 12, 24.
   GCD (20, 24) = 4
   B. $1 \times 15 = 15$, $3 \times 5 = 15$ The divisors of 15 are 1, 3, 5, 15.
   $1 \times 28 = 28$, $2 \times 14 = 28$, $4 \times 7 = 28$
   The divisors of 28 are 1, 2, 4, 7, 14, 28.
   GCD (15, 28) = 1
   C. $1 \times 8 = 8$, $2 \times 4 = 8$ The divisors of 8 are 1, 2, 4, 8.
   $1 \times 48 = 48$, $2 \times 24 = 48$, $3 \times 16 = 48$, $4 \times 12 = 48$, $6 \times 8 = 48$ The divisors of 48 are 1, 2, 3, 4, 6, 8, 12, 16, 24, 48.
   GCD (8, 48) = 8
   D. $1 \times 14 = 14$, $2 \times 7 = 14$ The divisors of 14 are 1, 2, 7, 14.
   $1 \times 10 = 10$, $2 \times 5 = 10$ The divisors of 10 are 1, 2, 5, 10.
   GCD (14, 10) = 2

3. A. $14 = 2 \cdot 7$ 14 is NOT prime
   B. $23 = 23$ 23 is prime
   C. $79 = 79$ 79 is prime
   D. $51 = 3 \cdot 17$ 51 is NOT prime
   E. $117 = 3 \cdot 39$ 117 is NOT prime

4. A. $81 = 3 \cdot 27$ 81 is composite
   B. $18 = 2 \cdot 9$ 18 is composite
   C. $43 = 43$ 43 is NOT composite
   D. $19 = 19$ 19 is NOT composite
   E. $91 = 7 \cdot 13$ 91 is composite

5. A. $400 = 2 \cdot 200 = 2 \cdot 2 \cdot 100 = 2 \cdot 2 \cdot 2 \cdot 50 = 2 \cdot 2 \cdot 2 \cdot 2 \cdot 25 = 2 \cdot 2 \cdot 2 \cdot 2 \cdot 5 \cdot 5 = 2^4 \cdot 5^2$
   B. $98 = 2 \cdot 49 = 2 \cdot 7 \cdot 7 = 2^1 \cdot 7^2$
   C. $54 = 2 \cdot 27 = 2 \cdot 3 \cdot 9 = 2 \cdot 3 \cdot 3 \cdot 3 = 2^1 \cdot 3^3$
   D. $96 = 2 \cdot 48 = 2 \cdot 2 \cdot 24 = 2 \cdot 2 \cdot 2 \cdot 12 = 2 \cdot 2 \cdot 2 \cdot 2 \cdot 6 = 2 \cdot 2 \cdot 2 \cdot 2 \cdot 2 \cdot 3 = 2^5 \cdot 3^1$
   E. $150 = 2 \cdot 75 = 2 \cdot 3 \cdot 25 = 2 \cdot 3 \cdot 5 \cdot 5 = 2^1 \cdot 3^1 \cdot 5^2$

6. A. $252 = 2 \cdot 126 = 2 \cdot 2 \cdot 63 = 2 \cdot 2 \cdot 3 \cdot 21 = 2 \cdot 2 \cdot 3 \cdot 3 \cdot 7 = 2^2 \cdot 3^2 \cdot 7^1$
   $588 = 2 \cdot 294 = 2 \cdot 2 \cdot 147 = 2 \cdot 2 \cdot 3 \cdot 49 = 2 \cdot 2 \cdot 3 \cdot 7 \cdot 7 = 2^2 \cdot 3^1 \cdot 7^2$
   LCM (252, 588) = $2^2 \cdot 3^2 \cdot 7^2 = 4 \cdot 9 \cdot 49 = 1,764$
   B. $54 = 2 \cdot 27 = 2 \cdot 3 \cdot 9 = 2 \cdot 3 \cdot 3 \cdot 3 = 2^1 \cdot 3^3$
   $144 = 2 \cdot 72 = 2 \cdot 2 \cdot 36 = 2 \cdot 2 \cdot 2 \cdot 18 = 2 \cdot 2 \cdot 2 \cdot 2 \cdot 9 = 2 \cdot 2 \cdot 2 \cdot 2 \cdot 3 \cdot 3 = 2^4 \cdot 3^2$
   LCM (54, 144) = $2^4 \cdot 3^3 = 16 \cdot 27 = 432$
   C. $56 = 2 \cdot 28 = 2 \cdot 2 \cdot 14 = 2 \cdot 2 \cdot 2 \cdot 7 = 2^3 \cdot 7^1$
   $140 = 2 \cdot 70 = 2 \cdot 2 \cdot 35 = 2 \cdot 2 \cdot 5 \cdot 7 = 2^2 \cdot 5^1 \cdot 7^1$
   $168 = 2 \cdot 84 = 2 \cdot 2 \cdot 42 = 2 \cdot 2 \cdot 2 \cdot 21 = 2 \cdot 2 \cdot 2 \cdot 3 \cdot 7 = 2^3 \cdot 3^1 \cdot 7^1$
   LCM (56, 140, 168) = $2^3 \cdot 3^1 \cdot 5^1 \cdot 7^1 = 8 \cdot 3 \cdot 5 \cdot 7 = 840$

7. The numbers in these exercises are the same as in the above exercise, so the prime factorization from that problem can be used here.
   A. $252 = 2^2 \cdot 3^2 \cdot 7^1$, $588 = 2^2 \cdot 3^1 \cdot 7^2$
   GCD (252, 588) = $2^2 \cdot 3^1 \cdot 7^1 = 4 \cdot 3 \cdot 7 = 84$
   B. $54 = 2^1 \cdot 3^3$, $144 = 2^4 \cdot 3^2$
   GCD (54, 144) = $2^1 \cdot 3^2 = 2 \cdot 9 = 18$
   C. $56 = 2^3 \cdot 7^1$, $140 = 2^2 \cdot 5^1 \cdot 7^1$, $168 = 2^3 \cdot 3^1 \cdot 7^1$
   GCD (56, 140, 168) = $2^2 \cdot 7^1 = 4 \cdot 7 = 28$

8.17 NUMBER PROPERTIES TEST 2

Use the following test to assess how well you have mastered the material in this chapter. Mark your answers by blackening the corresponding answer oval in each question. An answer key and solutions are provided at the end of the test.

1. Which is NOT a multiple of 7?

   (A) 140
   (B) 84
   (C) 56
   (D) 49
   (E) 1

2. Which is a divisor of 96?

   (A) 36
   (B) 32
   (C) 18
   (D) 9
   (E) 5

3. Which is the least common multiple of 15 and 10?

   (A) 300
   (B) 150
   (C) 90
   (D) 30
   (E) 5

4. Which is the least common multiple of 24 and 36?

   (A) 72
   (B) 48
   (C) 12
   (D) 6
   (E) 2

5. Which is the greatest common divisor of 8 and 40?

   (A) 2
   (B) 5
   (C) 8
   (D) 40
   (E) 320

6. Which is the greatest common divisor of 80 and 144?

   (A) 11,520
   (B) 720
   (C) 16
   (D) 8
   (E) 1

7. Which is the greatest common divisor of 8 and 15?

(A) 120
(B) 60
(C) 40
(D) 1
(E) 0

8. Which is a prime number?

(A) 9
(B) 4
(C) 2
(D) 1
(E) 0

9. Which is a prime number?

(A) 17
(B) 57
(C) 63
(D) 82
(E) 91

10. Which is a composite number?

(A) 0
(B) 1
(C) 2
(D) 29
(E) 51

11. Which is the prime factorization of 180?

(A) $2 \cdot 3 \cdot 5$
(B) $12 \cdot 15$
(C) $10 \cdot 18$
(D) $2^2 \cdot 3^2 \cdot 5$
(E) $2^2 \cdot 3^2 \cdot 5^2$

12. Which is the prime factorization of 588?

(A) $2 \cdot 3 \cdot 7$
(B) $2^2 \cdot 3^2 \cdot 7^2$
(C) $2^2 \cdot 3 \cdot 7^2$
(D) $2^2 \cdot 3 \cdot 49$
(E) $2^2 \cdot 147$

13. What is the prime factorization of 75?

    (A) $1 \times 3 \times 5$
    (B) 3×5
    (C) 3×25
    (D) 5×15
    (E) 3×5^2

14. What is the greatest prime factor of 130?

    (A) 5
    (B) 13
    (C) 26
    (D) 65
    (E) 130

15. What is the least common multiple of 12 and 75?

    (A) 3
    (B) 300
    (C) 450
    (D) 600
    (E) 900

16. What is the prime factorization of 288?

    (A) 2×3
    (B) 9×32
    (C) $6^2 \times 8$
    (D) $2^5 \times 3^2$
    (E) $2 \times 3 \times 48$

17. What is the great common factor for 35 and 66?

    (A) 1
    (B) 5
    (C) 7
    (D) 11
    (E) 2,310

18. What is the smallest prime number?

    (A) 0
    (B) 1
    (C) 2
    (D) 3
    (E) 5

NUMBER PROPERTIES TEST 2

Answer Key

1. E	7. D	13. E
2. B	8. C	14. B
3. D	9. A	15. B
4. A	10. E	16. D
5. C	11. D	17. A
6. C	12. C	18. C

8.18 SOLUTIONS

1. **E** 1

   The multiples of 7 are 7, 14, 21, 28, 35, 42, 49, 56, 63, 70, 77, 84, 91, 98, 105, 112, 119, 126, 133, 140, ...

   84, 49, 140, and 56 are all in the list of multiples of 7. 1 is not in the list, so it is not a multiple of 7.

2. **B** 32

   $1 \times 96 = 96$, $2 \times 48 = 96$, $3 \times 32 = 96$, $4 \times 24 = 96$, $6 \times 16 = 96$, $8 \times 12 = 96$. The divisors of 96 are 1, 2, 3, 4, 6, 8, 12, 16, 24, 32, 48, and 96. Only 32 is listed in the divisors of 96.

3. **D** 30

   The multiples of 15 are 15, 30, 45, 60, ...
   The multiples of 10 are 10, 20, 30, 40, 50, ...
   LCM (10, 15) = 30

4. **A** 72

   $24 = 2 \cdot 12 = 2 \cdot 2 \cdot 6 = 2 \cdot 2 \cdot 2 \cdot 3 = 2^3 \cdot 3$
   $36 = 2 \cdot 18 = 2 \cdot 2 \cdot 9 = 2 \cdot 2 \cdot 3 \cdot 3 = 2^2 \cdot 3^2$
   LCM (24, 36) $= 2^3 \cdot 3^2 = 8 \cdot 9 = 72$

5. **C** 8

   $1 \times 8 = 8$, $2 \times 4 = 8$. The divisors of 8 are 1, 2, 4, 8.
   $1 \times 40 = 40$, $2 \times 20 = 40$, $4 \times 10 = 40$, $5 \times 8 = 40$. The divisors of 40 are 1, 2, 4, 5, 8, 10, 20, 40.
   GCD (8, 40) = 8

6. **C** 16

   $80 = 2 \cdot 40 = 2 \cdot 2 \cdot 20 = 2 \cdot 2 \cdot 2 \cdot 10 = 2 \cdot 2 \cdot 2 \cdot 2 \cdot 5 = 2^4 \cdot 5$
   $144 = 2 \cdot 72 = 2 \cdot 2 \cdot 36 = 2 \cdot 2 \cdot 2 \cdot 18 = 2 \cdot 2 \cdot 2 \cdot 2 \cdot 9 = 2 \cdot 2 \cdot 2 \cdot 2 \cdot 3 \cdot 3 = 2^4 \cdot 3^2$.
   GCD (80, 144) $= 2^4 = 16$

7. **D** 1

   $8 = 2 \cdot 4 = 2 \cdot 2 \cdot 2 = 2^3$
   $15 = 3 \cdot 5$
   Since they have no common prime factors, their only common factor is 1, which is a divisor of all counting numbers. GCD (8, 15) = 1.

8. **C** 2

   $9 = 3 \cdot 3$, and $4 = 2 \cdot 2$. 2 is a prime. 0 and 1 are not greater than 1, so are not primes.

9. **A** 17

   $17 = 1 \cdot 17$, $82 = 2 \cdot 41$, $63 = 3 \cdot 21$, $57 = 3 \cdot 19$, $91 = 7 \cdot 13$

   All the choices but 17 have a counting number divisor that is greater than 1 but less than the number.

10. **E** 51

    0 and 1 are not greater than 1, so they are not composite. 2 and 29 are primes. $51 = 3 \cdot 17$, so 51 is a composite number.

11. **D** $2^2 \cdot 3^2 \cdot 5$

    $180 = 2 \cdot 90 = 2 \cdot 2 \cdot 45 = 2 \cdot 2 \cdot 3 \cdot 15 = 2 \cdot 2 \cdot 3 \cdot 3 \cdot 5 = 2^2 \cdot 3^2 \cdot 5$

12. **C** $2^2 \cdot 3 \cdot 7^2$

    $588 = 2 \cdot 294 = 2 \cdot 2 \cdot 147 = 2 \cdot 2 \cdot 3 \cdot 49 = 2 \cdot 2 \cdot 3 \cdot 7 \cdot 7 = 2^2 \cdot 3 \cdot 7^2$

13. **E** Since $3 \times 5^2 = 3 \times 25 = 75$ and 3 and 5 are prime numbers, 3×5^2 is the prime factorization of 75.

14. **B** $130 = 2 \cdot 65 = 2 \cdot 5 \cdot 13$ Thus, the greatest prime number divisor of 130 is 13.

15. **B** $12 = 2 \cdot 6 = 2^2 \cdot 3$ and $75 = 3 \cdot 25 = 3 \cdot 5^2$ The least common multiple of 12 and 75 is $2^2 \cdot 3 \cdot 5^2$, which is equal to 300.

16. **D** $288 = 2 \cdot 144 = 2 \cdot 12 \cdot 12 = 2 \cdot 2 \cdot 2 \cdot 3 \cdot 2 \cdot 2 \cdot 3 = 2^5 \cdot 3^2$

17. **A** The divisors of 35 are 1, 5, 7, and 35. The divisors of 66 are 1, 2, 3, 6, 11, 22, 33, and 66. The only common divisor for 35 and 66 is 1.

18. **C** By definition, prime numbers are whole numbers greater than 1, so 0 and 1 are not prime numbers. However 2, 3, and 5 are prime numbers. Thus, 2 is the smallest prime number.

8.19 SOLVED GRE PROBLEMS

For each question, select the best answer unless otherwise instructed.

1. **What is the sum of the prime factors of 770?**

   (A) 25
   (B) 81
   (C) 88
   (D) 117
   (E) 159

	Quantity A	Quantity B
2.	LCM(4, 12)	GCD(36, 84)

(A) Quantity A is greater.
(B) Quantity B is greater.
(C) The two quantities are equal.
(D) The relationship cannot be determined from the given information.

	Quantity A	Quantity B
3.	3	Smallest prime number

(A) Quantity A is greater.
(B) Quantity B is greater.
(C) The two quantities are equal.
(D) The relationship cannot be determined from the given information.

For this question, enter your answer in the box.

4. **What is the product of the prime numbers less than 10?**

5. **Which is a prime number divisor of 3,542?**

(A) 3
(B) 17
(C) 22
(D) 23
(E) 161

8.20 SOLUTIONS

1. **A** $770 = 77 \cdot 10 = 7 \cdot 11 \cdot 2 \cdot 5$ $\quad 7 + 11 + 2 + 5 = 25$
2. **C** LCM(4, 12) = 12 $\quad$ GCD(36, 84) = 12
3. **A** The smallest prime number is 2 and 3 > 2.
4. 210
   The prime numbers less than 10 are 2, 3, 5, and 7. Their product is $2 \times 3 \times 5 \times 7 = 210$.
5. **D** The prime number divisors of 3,542 are 2, 7, 11, and 23.

8.21 GRE PRACTICE PROBLEMS

For each question, select the best answer unless otherwise instructed.

	Quantity A	Quantity B
1.	GCD (25, 40)	GCD (80, 125)

(A) Quantity A is greater.
(B) Quantity B is greater.
(C) The two quantities are equal.
(D) The relationship cannot be determined from the given information.

2. **What is the sum of the prime factors of 4,030?**

   (A) 51
   (B) 80
   (C) 98
   (D) 185
   (E) 413

3. **What is the prime factorization of 405?**

   (A) (5)(81)
   (B) (9)(45)
   (C) (15)(27)
   (D) $(3^4)(5)$
   (E) $(3^2)(5^2)$

	Quantity A	**Quantity B**
4.	LCM (12, 30)	LCM (4, 90)

   (A) Quantity A is greater.
   (B) Quantity B is greater.
   (C) The two quantities are equal.
   (D) The relationship cannot be determined from the given information.

For this question, enter your answer in the box.

5. **What prime number is between 45 and 50?**

6. **What is the prime factorization of 135?**

   (A) 5^3
   (B) $(5)(3^3)$
   (C) (15)(9)
   (D) (27)(5)
   (E) (1)(3)(5)

	Quantity A	**Quantity B**
7.	6	Sum of the proper divisors of 6

   (A) Quantity A is greater.
   (B) Quantity B is greater.
   (C) The two quantities are equal.
   (D) The relationship cannot be determined from the given information.

ANSWER KEY

1. C
2. A
3. D
4. B
5. 47
6. B
7. C

CHAPTER 9
ARITHMETIC COMPUTATION

9.1 SYMBOLS

$=$	equals
$>$	is greater than
$<$	is less than
$\geq$	is greater than or equal to
$\leq$	is less than or equal to
$\neq$	is not equal to
x^2	x squared
x^3	x cubed
$\sqrt{x}$	square root of x
$|x|$	absolute value of x
π	pi (about 3.14)
$+$	addition, plus
$-$	subtraction, minus
$a \times b$	a times b, multiplication
$a \cdot b$	a times b, multiplication
$a * b$	a times b, multiplication
ab	a times b, multiplication
$a \div b$	a divided by b, division
$\dfrac{a}{b}$	a divided by b, division
$a : b$	the ratio of a to b
$\%$	percent

9.2 ORDER OF OPERATIONS

When there is more than one operation in an expression, you have to use the standard order of operations to do the computation in order to get a consistent, correct result. You use the order PEMDAS, which stands for **P**arentheses, **E**xponents, **M**ultiplication and **D**ivision, and **A**ddition and **S**ubtraction.

The first step is to work within each set of parentheses or brackets. A fraction bar means that the numerator and denominator are worked separately as though each is enclosed by parentheses. The work inside parentheses follows these rules also.

The second level is exponents, which means that you simplify all powers and roots before trying to compute using them.

The next level is multiplication and division. These operations are done in the order that they occur from left to right in the problem. In $6 \times 5 \div 2 \times 4$, multiply 6×5 first to get $30 \div 2 \times 4$; then divide 30 by 2 to get 15×4. Finally, multiply 15 by 4 to get 60. Multiplication and division are of equal rank in the order of operations.

71

The last step is to do the additions and subtractions in the order in which they occur from left to right. In $15 - 7 + 4$, subtract first, then add: $15 - 7 + 4 = 8 + 4 = 12$. Addition and subtraction are of equal rank in the order of operations.

To help you remember the order of operations, PEMDAS, you can create a sentence based on these letters: *Please Excuse Me, Dear Aunt Sally*. The memory device helps you keep the letters in the correct sequence.

Example:

Evaluate each expression.

A. $7^2 - (-5) + 12$ B. $[5(-7) - 8] \times 2$ C. $-2 * 5 + 15 \div 5$
D. $6(-2) - (-2)(-5)^3$ E. $-(-5)^2 - (-5)^3$ F. $(8 - 2^4)(-3 + \sqrt{49})$

Solution:

A. $7^2 - (-5) + 12 = 49 - (-5) + 12 = 54 + 12 = 66$
B. $[5(-7) - 8] \times 2 = [-35 - 8] \times 2 = [-43] \times 2 = -86$
C. $-2 * 5 + 15 \div 5 = -10 + 15 \div 5 = -10 + 3 = -7$
D. $6(-2) - (-2)(-5)^3 = 6(-2) - (-2)(-125) = -12 - (250) = -262$
E. $-(-5)^2 - (-5)^3 = -(25) - (-125) = -25 + 125 = 100$
F. $(8 - 2^4)(-3 + \sqrt{49}) = (8 - 16)(-3 + 7) = (-8)(4) = -32$

9.3   PROPERTIES OF OPERATIONS

Commutative Properties

If a and b are real numbers, then $a + b = b + a$ and $ab = ba$. The commutative properties say that $4 + 6$ and $6 + 4$ have the same answer. Also, $-3 * 12$ and $12 * (-3)$ have the same answer.

It is important to note that the commutative properties are not valid for subtraction and division. $4 - 2 = 2$, but $2 - 4 = -2$ and $2 \neq -2$. Also, $8 \div 2 = 4$, but $2 \div 8 = \frac{1}{4}$ and $4 \neq -\frac{1}{4}$.

(A single example can show a statement is false, but one example does not show that a statement is true.)

Associative Properties

If a, b, and c are real numbers, then $a + (b + c) = (a + b) + c$ and $a(bc) = (ab)c$. The associative properties say that $a + b + c$ and abc have meaning, and the value is the same whichever way you compute the values. $5 + 2 + (-3)$ can be evaluated as $[5 + 2] + (-3)$ or by $5 + [2 + (-3)]$. $[5 + 2] + (-3) = 7 + (-3) = 4$, and $5 + [2 + (-3)] = 5 + (-1) = 4$, so you say $5 + 2 + (-3) = 4$. Similarly, $(-3) * 2 * (-2)$ can be evaluated as $[(-3) * 2] * (-2)$ or as $(-3) * [2 * (-2)]$. $[(-3) * 2] * (-2) = (-6) * (-2) = 12$, and $(-3) * [2 * (-2)] = (-3) * (-4) = 12$, so you can say $(-3) * 2 * (-2) = 12$.

Distributive Property

If a, b, and c are real numbers, then $a * (b + c) = ab + ac$. The distributive property allows you to change the order of the operations from addition and then multiplication to multiplication and then addition, or the reverse. $3 * (-7 + 5) = 3 * (-7) + 3 * 5 = -21 + 15 = -6$, and $(-8) * 5 + (-8) * 2 = -8(5 + 2)$. Since subtraction is the same as addition of the opposite, you can convert $a * (b - c) = a * (b + (-c)) = a * b + a * (-c) = ab - ac$.

Identity Properties

For any real number a, $a + 0 = 0 + a = a$, and $a * 1 = 1 * a = a$. The identity for addition is 0 and the identity for multiplication is 1.

Inverse Properties

For any real number a, there is a real number that is the opposite of a, $-a$, such that $a + (-a) = (-a) + a = 0$. For any nonzero real number b, there is a real number that is the reciprocal of b, $\frac{1}{b}$, such that $b \times \frac{1}{b} = \frac{1}{b} \times b = 1$.

For 6, the opposite is -6, and the reciprocal is $\frac{1}{6}$. Thus $6 + (-6) = (-6) + 6 = 0$, and $6 \times \frac{1}{6} = \frac{1}{6} \times 6 = 1$. The additive inverse of 6 is -6, and the multiplicative inverse of 6 is $\frac{1}{6}$.

Example:

State the property illustrated by each statement.

A. $5 + (6 + 8) = (5 + 6) + 8$
B. $-7 + 0 = -7$
C. $3 * 1 = 3$
D. $4 * (3 * 9) = (4 * 3) * 9$
E. $(8 + 5) * 3 = 3 * (8 + 5)$
F. $-5 + 5 = 0$

Solution:

A. Since the numbers are in the same order and only the grouping has changed, the property is the associative property for addition.
B. Since 0 was added to the number, the property is the addition identity property.
C. Since 1 is multiplied by the number, the property is the multiplication identity property.
D. Since the order of the numbers is unchanged and only the grouping has changed, the property is the associative property for multiplication.
E. Since the order of the factors has been changed, the property is the commutative property of multiplication. You treat $(8 + 5)$ as a single number here.
F. Since you have added a number and its opposite together, this is the addition inverse property.

▬ 9.4 PRACTICE PROBLEMS

1. Use the order of operations to evaluate each expression.

   A. $6^2 - 4(9 - 1)$

   B. $(-3)(-2) - [7 + (8 - 12)]$

   C. $\dfrac{(-8+5) - (4+7)}{15 - 17}$

   D. $(-4 - 1)(-3 - 5) - 3^2$

   E. $\dfrac{8(-3) - 2^3(-3)^2}{-3[4 - (-8)]}$

2. Evaluate each expression when $x = -1$, $y = -2$, and $z = 4$.

   A. $-z(2x - 5y)$

   B. $(x - 2) \div 5y + z$

   C. $9x + 2y - 5z$

   D. $-7x + 2y + 3z$

3. Find the addition inverse of each number.

   A. 6   B. 3   C. −7   D. 0   E. −3   F. $\dfrac{1}{2}$

4. Find the multiplication inverse of each number.

   A. 5   B. −3   C. $\dfrac{1}{2}$ D. $-\dfrac{2}{3}$ E. 1   F. −1

5. Which property is illustrated by each statement?

   A. $(9 + 3) + 0 = 9 + 3$

   B. $5 * (6 * -2) = (5 * 6) * (-2)$

   C. $5 * (3 + 8) = (3 + 8) * 5$

   D. $5 * -7 + 5 * 4 = 5 * (-7 + 4)$

   E. $-3 + 3 = 0$

   F. $3 + (4 + 2) = (3 + 4) + 2$

   G. $-2 * (8 + (-3)) = -2 * 8 + (-2) * (-3)$

   H. $-1/4 * (-4) = 1$

   I. $7 * 1 = 7$

   J. $0 * 6 = 0$

   K. $-5 * (-6) = -6 * (-5)$

   L. $5 + 2 = 2 + 5$

▬ 9.5 SOLUTIONS

1. A. $6^2 - 4(9-1) = 6^2 - 4(8) = 36 - 4*8 = 36 - 32 = 4$

   B. $(-3)(-2) - [7 + (8 - 12)] = (-3)(-2) - [7 + (-4)] = 6 - [+3] = 6 - 3 = 3$

   C. $\dfrac{(-8+5) - (4+7)}{15 - 17} = \dfrac{(-3) - 11}{-2} = \dfrac{-14}{-2} = 7$

   D. $(-4 - 1)(-3 - 5) - 3^2 = (-5)(-8) - 9 = 40 - 9 = 31$

   E. $\dfrac{8(-3) - 2^3(-3)^2}{-3[4 - (-8)]} = \dfrac{-24 - 8(9)}{-3[4 + 8]} =$

   $\dfrac{-24 - 72}{-3(12)} = \dfrac{-96}{-36} = +\dfrac{16}{6} = \dfrac{8}{3} \text{ or } 2\dfrac{2}{3}$

2. A. $-z(2x - 5y) = -4(2(-1) - 5(-2)) = -4(-2 + 10) = -4(8) = -32$

   B. $(x - 2) \div 5y + z = (-1 - 2) \div 5(-2) + 4 = -3 \div 5(-2) + 4 = -\dfrac{3}{5} * -2 + 4 = \dfrac{6}{5} + 4 =$

   $\dfrac{6}{5} + \dfrac{20}{5} = \dfrac{26}{5} = 5\dfrac{1}{5}$

   C. $9x + 2y - 5z = 9(-1) + 2(-2) - 5(4) = -9 - 4 - 20 = -33$

   D. $-7x + 2y + 3z = -7(-1) + 2(-2) + 3(4) = 7 - 4 + 12 = 3 + 12 = 15$

3. A. Because $6 + (-6) = 0$ and $-6 + 6 = 0$, -6 is the addition inverse of 6.

   B. Because $+3 + (-3) = 0$ and $-3 + (+3) = 0$, -3 is the addition inverse of $+3$.

   C. Because $-7 + 7 = 0$ and $7 + (-7) = 0$, 7 is the addition inverse of -7.

   D. Because $0 + 0 = 0$, 0 is the addition inverse of 0.

   E. Because $-3 + (+3) = 0$ and $+3 + (-3) = 0$, $+3$ is the addition inverse of -3.

   F. Because $\dfrac{1}{2} + \left(-\dfrac{1}{2}\right) = 0$ and $-\dfrac{1}{2} + \dfrac{1}{2} = 0$, $-\dfrac{1}{2}$ is the addition inverse of $\dfrac{1}{2}$.

4. A. Because $5 * \dfrac{1}{5} = 1$ and $\dfrac{1}{5} * 5 = 1$, $\dfrac{1}{5}$ is the multiplication inverse of 5.

   B. Because $-3 * \left(-\dfrac{1}{3}\right) = 1$ and $-\dfrac{1}{3} * (-3) = 1$, $-\dfrac{1}{3}$ is the multiplication inverse of -3.

C. Because $\frac{1}{2} * 2 = 1$ and $2 * \frac{1}{2} = 1$, 2 is the multiplication inverse of $\frac{1}{2}$.

D. Because $-\frac{2}{3} * \left(-\frac{3}{2}\right) = 1$ and $-\frac{3}{2} * \left(-\frac{2}{3}\right) = 1$, $-\frac{3}{2}$ is the multiplication inverse of $-\frac{2}{3}$.

E. Because $1 * 1 = 1$, 1 is the multiplication inverse of 1.

F. Because $-1*(-1) = 1$, -1 is the multiplication inverse of -1.

5. A. Because 0 was added to $(9 + 3)$, this shows the addition identity property.

B. Because you just changed the grouping, this shows the associative property for multiplication.

C. Because the order of the factors 5 and $(3 + 8)$ is all that has been changed, the property is commutative for multiplication.

D. The common factor of 5 has been taken out of the two terms, so it is the distributive property.

E. Opposites are being added, so this is the addition inverse property.

F. The grouping is changed, so this is the associative property for addition.

G. Since the -2 has been distributed over the sum, this is the distributive property.

H. You are multiplying reciprocals, so this is the multiplication inverse.

I. You are multiplying by 1, so this is the multiplication identity.

J. You are multiplying by 0, so this is the zero multiplication property.

K. The order of the factors has changed, so this is the commutative property of multiplication.

L. The order of the addends has changed, so this is the commutative property of addition.

9.6 FRACTIONS

A fraction is made up of three parts: the **numerator,** the **fraction bar,** and the **denominator.** The numerator tells you how many parts you have, and the denominator tells you how many parts the whole is divided into. The fraction bar is read as "out of." Otherwise, it is used as a grouping indicator to remind you to simplify the numerator and denominator separately before doing the division.

A fraction can tell you what part of a whole unit that you have, and even that you have more parts than are needed for one whole.

The fraction $\frac{3}{4}$ means that you have 3 of the 4 parts that the whole is divided into, while $\frac{5}{4}$ means that you have all 4 parts of the one whole, plus 1 part of another whole that is divided into 4 equal parts.

In a fraction $\frac{a}{b}$ where a and b are whole numbers and b is not zero, you say that $\frac{a}{b}$ is a **proper fraction** when $a < b$. If $a = b$ or $a > b$, you say that it is an **improper fraction.**

You can use fractions to express the ratio of two quantities. If there are units with the numbers, then they need to be the same units. When the units are different, convert to a common unit.

A fraction can be used to indicate division; $a \div b = \frac{a}{b}$. This use of fractions is especially helpful when you are dividing a smaller counting number by a larger one.

When you have an improper fraction such as $\frac{4}{4}$, $\frac{7}{5}$, or $\frac{11}{2}$, you can change it so that the number of whole units involved is clear. For $\frac{4}{4}$, you have all 4 parts for 1 unit, and $\frac{4}{4} = 1$. With $\frac{7}{5}$, you have 2 more parts than needed to make a whole unit, so $\frac{7}{5} = 1\frac{2}{5}$. It is understood that $1\frac{2}{5} = 1 + \frac{2}{5}$. With $\frac{11}{2}$, you see that $11 \div 2$ yields 5 with a remainder of 1. The remainder is 1 of 2 parts needed to make the next whole, so it represents $\frac{1}{2}$. Thus, $\frac{11}{2} = 5\frac{1}{2}$. It is very helpful to be able to convert between improper fractions and mixed numbers.

Example:

Convert these improper fractions to mixed numbers.

A. $\frac{7}{3}$ B. $\frac{5}{2}$ C. $\frac{8}{5}$ D. $\frac{14}{5}$ E. $\frac{13}{8}$

Solution:

A. $\frac{7}{3} = 7 \div 3 = 2$ with remainder 1 $\frac{7}{3} = 2\frac{1}{3}$

B. $\frac{5}{2} = 5 \div 2 = 2$ with remainder 1 $\frac{5}{2} = 2\frac{1}{2}$

C. $\frac{8}{5} = 8 \div 5 = 1$ with remainder 3 $\frac{8}{5} = 1\frac{3}{5}$

D. $\frac{14}{5} = 14 \div 5 = 2$ with remainder 4 $\frac{14}{5} = 2\frac{4}{5}$

E. $\frac{13}{8} = 13 \div 8 = 1$ with remainder 5 $\frac{13}{8} = 1\frac{5}{8}$

Note that in each case, the remainder tells you how many parts of the next unit you have, but the denominator is needed to show how many parts are needed to make the whole.

A **complex fraction** has fractions in the numerator, the denominator, or both. Examples of complex fractions are $\frac{1/2}{5}$, $\frac{3}{5/6}$, $\frac{2/3}{1/5}$, and $\frac{1 + (3/4)}{3/10}$.

One way to simplify complex fractions is to find the least common multiple of all of the fractions in the numerator and denominator and then multiply the numerator and denominator by this number.

Example:

Simplify each complex fraction.

A. $\frac{1/2}{5}$ B. $\frac{3}{5/6}$ C. $\frac{2/3}{1/5}$ D. $\frac{1 + (3/4)}{3/10}$

Solution:

A. The only fraction is $\frac{1}{2}$, so you need to multiply by 2. $\dfrac{1/2}{5} = \dfrac{1/2 * 2}{5 * 2} = \dfrac{1}{10}$

B. The only fraction is $\frac{5}{6}$, so you need to multiply by 6. $\dfrac{3}{5/6} = \dfrac{3 * 6}{5/6 * 6} = \dfrac{18}{5}$

C. The fractions are $\frac{2}{3}$ and $\frac{1}{5}$. You find LCM(3, 5), which is 15.

$$\frac{2/3}{1/5} = \frac{2/3 * 15}{1/5 * 15} = \frac{2 * 5}{1 * 3} = \frac{10}{3}$$

D. The fractions are $\frac{3}{4}$ and $\frac{3}{10}$. You find LCM(4,10), which is 20.

$$\frac{1 + 3/4}{3/10} = \frac{(1 + (3/4))\,20}{3/10 * 20} = \frac{1 * 20 + 3/4 * 20}{3 * 2} = \frac{20 + 3 * 5}{6} = \frac{20 + 15}{6} = \frac{35}{6}$$

Equivalent fractions are fractions that are different representations of the same number. For example, in this set, there are six objects, and two are @s: @@####. So the fraction that shows the ratio of @s to all objects in the set is $\frac{2}{6}$. In the set @@@@########, there are 4 @s and 12 objects, so the fraction representing @s to objects is $\frac{4}{12}$. However in each case, there is one @ out of every three objects. The ratio of objects in each set is $\frac{1}{3} \cdot \frac{2}{6} = \frac{2 * 1}{2 * 3} = \frac{1}{3}$, and $\frac{4}{12} = \frac{4 * 1}{4 * 3} = \frac{1}{3}$. Thus, you say that $\frac{1}{3}$, $\frac{2}{6}$, and $\frac{4}{12}$ are equivalent fractions.

You use equivalent fractions when you put a fraction in lowest terms, and when you want to add or subtract fractions. You can also use equivalent fractions when you compare fractions, but other ways may be faster.

Example:

Write each fraction as an equivalent fraction whose denominator is the number given.

A. $\frac{2}{3}$, 30   B. $\frac{5}{7}$, 28   C. $\frac{11}{12}$, 24   D. $\frac{3}{5}$, 40

Solution:

A. To write $\frac{2}{3}$ as an equivalent fraction with a denominator of 30, you need to find what number times 3 equals 30. $30 \div 3 = 10$, so $3 \times 10 = 30$.

$$\frac{2}{3} = \frac{2}{3} * 1 = \frac{2}{3} * \frac{10}{10} = \frac{2 * 10}{3 * 10} = \frac{20}{30}$$

B. $28 \div 7 = 4$, so $7 \times 4 = 28$.

$$\frac{5}{7} = \frac{5}{7} * 1 = \frac{5}{7} * \frac{4}{4} = \frac{5 * 4}{7 * 4} = \frac{20}{28}$$

C. $24 \div 12 = 2$, so $12 \times 2 = 24$.

$$\frac{11}{12} = \frac{11}{12} * 1 = \frac{11}{12} * \frac{2}{2} = \frac{11 * 2}{12 * 2} = \frac{22}{24}$$

D. $40 \div 5 = 8$, so $5 \times 8 = 40$.

$$\frac{3}{5} = \frac{3}{5} * 1 = \frac{3}{5} * \frac{8}{8} = \frac{3 * 8}{5 * 8} = \frac{24}{40}$$

To reduce a fraction to its lowest terms, find the greatest common divisor between numerator and denominator and divide out that common factor. $\frac{20}{36}$ can be reduced to lowest terms because GCD(20, 36) = 4. $\frac{20}{36} = \frac{5 * 4}{9 * 4} = \frac{5}{9} * \frac{4}{4} = \frac{5}{9} * 1 = \frac{5}{9}$ or $\frac{20}{36} = \frac{20 \div 4}{36 \div 4} = \frac{5}{9}$.

Example:

Reduce each fraction to lowest terms.

A. $\frac{14}{16}$ B. $\frac{28}{400}$ C. $\frac{21}{28}$ D. $\frac{18}{24}$ E. $\frac{42}{60}$

Solution:

A. GCD(14, 16) = 2

$$\frac{14}{16} = \frac{7 * 2}{8 * 2} = \frac{7}{8} * \frac{2}{2} = \frac{7}{8} * 1 = \frac{7}{8} \text{ or } \frac{14}{16} = \frac{14 \div 2}{16 \div 2} = \frac{7}{8}$$

B. GCD(28, 400) = 4

$$\frac{28}{400} = \frac{28 \div 4}{400 \div 4} = \frac{7}{100}$$

C. GCD(21, 28) = 7

$$\frac{21}{28} = \frac{21 \div 7}{28 \div 7} = \frac{3}{4}$$

D. GCD(18, 24) = 6

$$\frac{18}{24} = \frac{18 \div 6}{24 \div 6} = \frac{3}{4}$$

E. GCD(42, 60) = 6

$$\frac{42}{60} = \frac{42 \div 6}{60 \div 6} = \frac{7}{10}$$

Comparing fractions allows you to determine whether the two fractions are equal or whether one of the fractions is greater than the other. There are many ways to make the comparisons, and you need to decide which procedure works best for you and use that one.

To compare fractions when the denominators are equal, the fraction with the greater numerator is the greater fraction.

When comparing $\frac{11}{25}$ and $\frac{15}{25}$, for example, since $15 > 11$, $\frac{15}{25} > \frac{11}{25}$.

This procedure can be used to compare $\frac{3}{8}$ and $\frac{4}{7}$ if you change them to equivalent fractions. The easiest denominator to use is the product of the given denominators (even though the least common multiple of them might be much smaller). $7 \times 8 = 56$, so change each fraction to an equivalent one with a denominator of 56. $\frac{3}{8} = \frac{3*7}{8*7} = \frac{21}{56}$, and $\frac{4}{7} = \frac{4*8}{7*8} = \frac{32}{56}$. Since $32 > 21$, $\frac{32}{56} > \frac{21}{56}$; so $\frac{4}{7} > \frac{3}{8}$. You can also convert the fractions to decimals with the on-screen calculator.

Occasionally, the fractions you want to compare will have equal numerators. In this case, the fraction with the smallest denominator will be the greater fraction. Compare $\frac{11}{7}$ and $\frac{11}{5}$. Since with sevenths, the whole is divided into 7 parts, and with fifths, the whole is divided into 5 parts, the fifths are larger pieces than the sevenths, so 11 fifths is greater than 11 sevenths. $\frac{11}{5} > \frac{11}{7}$.

Another procedure that is helpful is to make a mental comparison to another number. For example, $\frac{1}{3}$ is less than $\frac{1}{2}$, and $\frac{3}{5}$ is more than $\frac{1}{2}$, so $\frac{3}{5} > \frac{1}{2}$. The comparison value needs to be such that one fraction is greater than this value and the other is less than it.

The cross-multiplication procedure is a relatively easy method to use and may be faster than converting to decimals. To compare $\frac{a}{b}$ and $\frac{c}{d}$, cross-multiply to get ad and bc. If $ad > bc$, then $\frac{a}{b} > \frac{c}{d}$. If $ad = bc$, then $\frac{a}{b} = \frac{c}{d}$, and if $ad < bc$, then $\frac{a}{b} < \frac{c}{d}$. To compare $\frac{3}{8}$ and $\frac{4}{7}$, cross-multiply to get 21 and 32. $21 < 32$, so $\frac{3}{8} < \frac{4}{7}$.

Example:

Compare each pair of fractions.

A. $\frac{7}{8}, \frac{7}{10}$ B. $\frac{3}{17}, \frac{5}{17}$ C. $\frac{4}{5}, \frac{3}{4}$ D. $\frac{8}{9}, \frac{3}{4}$ E. $\frac{3}{5}, \frac{5}{12}$

Solution:

A. Since the numerators are the same, the fraction with the smaller denominator is the greater fraction. Since $8 < 10$, $\dfrac{7}{8} > \dfrac{7}{10}$.

B. Since the fractions have the same denominators, the fraction with the greater numerator is the greater fraction. Since $5 > 3$, $\dfrac{5}{17} > \dfrac{3}{17}$.

C. Cross-multiply the fractions. $16 > 15$, so $\dfrac{4}{5} > \dfrac{3}{4}$.

D. Cross-multiply the fractions. $32 > 27$, so $\dfrac{8}{9} > \dfrac{3}{4}$.

E. $\dfrac{3}{5}$ is more than $\dfrac{1}{2}$, and $\dfrac{5}{12}$ is less than $\dfrac{1}{2}$, so $\dfrac{3}{5} > \dfrac{5}{12}$.

▰▰ 9.7 PRACTICE PROBLEMS

1. Write these mixed numbers as improper fractions.

   A. $1\dfrac{2}{3}$ B. $7\dfrac{1}{3}$ C. $5\dfrac{3}{4}$ D. $2\dfrac{1}{4}$ E. $3\dfrac{1}{7}$

2. Write these improper fractions as mixed numbers.

   A. $\dfrac{11}{6}$ B. $\dfrac{17}{5}$ C. $\dfrac{32}{7}$ D. $\dfrac{15}{8}$ E. $\dfrac{17}{10}$

3. Simplify each of these complex fractions.

   A. $\dfrac{5/6}{8}$ B. $\dfrac{5}{2/3}$ C. $\dfrac{4/5}{5/8}$

   D. $\dfrac{2/3 + 1/4}{3/5}$ E. $\dfrac{1/2 + 5/6}{4}$

4. Write each fraction as an equivalent fraction with the given denominator.

   A. $\dfrac{3}{5}, 35$ B. $\dfrac{2}{7}, 42$ C. $\dfrac{3}{8}, 40$ D. $\dfrac{2}{9}, 18$ E. $\dfrac{3}{4}, 24$

5. Reduce each fraction to lowest terms.

   A. $\dfrac{16}{20}$ B. $\dfrac{15}{18}$ C. $\dfrac{48}{60}$ D. $\dfrac{9}{12}$ E. $\dfrac{15}{60}$

6. Compare these fractions.

   A. $\dfrac{12}{16}, \dfrac{7}{8}$ B. $\dfrac{5}{9}, \dfrac{5}{7}$ C. $\dfrac{10}{15}, \dfrac{14}{21}$ D. $\dfrac{3}{4}, \dfrac{7}{12}$ E. $\dfrac{2}{3}, \dfrac{7}{8}$

▰▰ 9.8 SOLUTIONS

1. A. $1\dfrac{2}{3} = 1 + \dfrac{2}{3} = \dfrac{3}{3} + \dfrac{2}{3} = \dfrac{5}{3}$

   B. $7\dfrac{1}{3} = 7 + \dfrac{1}{3} = \dfrac{21}{3} + \dfrac{1}{3} = \dfrac{22}{3}$

   C. $5\dfrac{3}{4} = 5 + \dfrac{3}{4} = \dfrac{20}{4} + \dfrac{3}{4} = \dfrac{23}{4}$

   D. $2\dfrac{1}{4} = 2 + \dfrac{1}{4} = \dfrac{8}{4} + \dfrac{1}{4} = \dfrac{9}{4}$

   E. $3\dfrac{1}{7} = 3 + \dfrac{1}{7} = \dfrac{21}{7} + \dfrac{1}{7} = \dfrac{22}{7}$

2. A. $\dfrac{11}{6} = 11 \div 6 = 1$ with remainder $5 = 1\dfrac{5}{6}$

   B. $\dfrac{17}{5} = 17 \div 5 = 3$ with remainder $2 = 3\dfrac{2}{5}$

   C. $\dfrac{32}{7} = 32 \div 7 = 4$ with remainder $4 = 4\dfrac{4}{7}$

   D. $\dfrac{15}{8} = 15 \div 8 = 1$ with remainder $7 = 1\dfrac{7}{8}$

   E. $\dfrac{17}{10} = 17 \div 10 = 1$ with remainder $7 = 1\dfrac{7}{10}$

3. A. The only fraction has a denominator of 6.

$$\frac{5/6}{8} = \frac{5/6 * 6}{8 * 6} = \frac{5}{48}$$

B. The only fraction has a denominator of 3.

$$\frac{5}{2/3} = \frac{5 * 3}{2/3 * 3} = \frac{15}{2}$$

C. The denominators are 5 and 8.
LCM(5, 8) = 40.

$$\frac{4/5}{5/8} = \frac{4/5 * 40}{5/8 * 40} = \frac{4 * 8}{5 * 5} = \frac{32}{25}$$

D. The denominators are 3, 4, and 5.
LCM(3, 4, 5) = 60.

$$\frac{2/3+1/4}{3/5} = \frac{(2/3+1/4)*60}{3/5*60} = \frac{2/3*60+1/4*60}{3*12} -$$

$$\frac{2*20+1*15}{36} = \frac{40+15}{36} = \frac{55}{36}$$

E. The denominators are 2 and 6.
LCM(2, 6) = 6.

$$\frac{1/2 + 5/6}{4} = \frac{(1/2 + 5/6) * 6}{4 * 6} = \frac{1/2 * 6 + 5/6 * 6}{24} =$$

$$\frac{1 * 3 + 5 * 1}{24} = \frac{3 + 5}{24} = \frac{8}{24} = \frac{8 * 1}{8 * 3} = \frac{1}{3}$$

4. A. $35 \div 5 = 7, \dfrac{3}{5} = \dfrac{3 * 7}{5 * 7} = \dfrac{21}{35}$

B. $42 \div 7 = 6, \dfrac{2}{7} = \dfrac{2 * 6}{7 * 6} = \dfrac{12}{42}$

C. $40 \div 8 = 5, \dfrac{3}{8} = \dfrac{3 * 5}{8 * 5} = \dfrac{15}{40}$

D. $18 \div 9 = 2, \dfrac{2}{9} = \dfrac{2 * 2}{9 * 2} = \dfrac{4}{18}$

E. $24 \div 4 = 6, \dfrac{3}{4} = \dfrac{3 * 6}{4 * 6} = \dfrac{18}{24}$

5. A. $GCD(16, 20) = 4, \dfrac{16}{20} = \dfrac{16 \div 4}{20 \div 4} = \dfrac{4}{5}$

B. $GCD(15, 18) = 3, \dfrac{15}{18} = \dfrac{15 \div 3}{18 \div 3} = \dfrac{5}{6}$

C. $GCD(48, 60) = 12, \dfrac{48}{60} = \dfrac{48 \div 12}{60 \div 12} = \dfrac{4}{5}$

D. $GCD(9, 12) = 3, \dfrac{9}{12} = \dfrac{9 \div 3}{12 \div 3} = \dfrac{3}{4}$

E. $GCD(15, 60) = 15, \dfrac{15}{60} = \dfrac{15 \div 15}{60 \div 15} = \dfrac{1}{4}$

6. A. Cross-multiply. $8 * 12 = 96; 16 * 7 = 112.$ Since $96 < 112, \dfrac{12}{16} < \dfrac{7}{8}.$

B. The numerators are the same, so the one with the smaller denominator is bigger.
$\dfrac{5}{9} < \dfrac{5}{7}.$

C. Cross-multiply. $10 * 21 = 210, 15 * 14 = 210.$ Since $210 = 210, \dfrac{10}{15} = \dfrac{14}{21}.$

D. Cross-multiply. $3 * 12 = 36, 4 * 7 = 28.$ Since $36 > 28, \dfrac{3}{4} > \dfrac{7}{12}.$

E. Cross-multiply. $2 * 8 = 16, 3 * 7 = 21.$ Since $16 < 21, \dfrac{2}{3} < \dfrac{7}{8}.$

▬▬ 9.9 OPERATIONS WITH FRACTIONS

Adding Fractions

To add two fractions, the fractions must have a common denominator. Add the numerators to get the numerator of the sum, and the denominator of the sum is the common denominator.

$$\frac{3}{5} + \frac{1}{5} = \text{three fifths} + \text{one fifth} = \text{four fifths} = \frac{4}{5}$$

$$\frac{7}{11} + \frac{2}{11} = \frac{7 + 2}{11} = \frac{9}{11}$$

If the fractions do not have a common denominator, create one by finding the least common multiple of the denominators, or use the product of the denominators. The product of the denominators is quicker to find, but it can be much larger than the LCM, which could make the computation more difficult.

Example:

A. $\dfrac{1}{2}+\dfrac{1}{4}$ B. $\dfrac{2}{9}+\dfrac{5}{9}$ C. $\dfrac{2}{5}+\dfrac{5}{6}$ D. $\dfrac{5}{8}+\dfrac{5}{12}$ E. $\dfrac{3}{10}+\dfrac{1}{6}$

Solution:

A. $2*4=8$ $\dfrac{1}{2}+\dfrac{1}{4}=\dfrac{1*4}{2*4}+\dfrac{1*2}{4*2}=\dfrac{4}{8}+\dfrac{2}{8}=\dfrac{6}{8}=\dfrac{6\div2}{8\div2}=\dfrac{3}{4}$

or LCM$(2,4)=4$ $\dfrac{1}{2}+\dfrac{1}{4}=\dfrac{1*2}{2*2}+\dfrac{1}{4}=\dfrac{2}{4}+\dfrac{1}{4}=\dfrac{3}{4}$

B. $\dfrac{2}{9}+\dfrac{5}{9}=\dfrac{2+5}{9}=\dfrac{7}{9}$

C. $5*6=30$ $\dfrac{2}{5}+\dfrac{5}{6}=\dfrac{2*6}{5*6}+\dfrac{5*5}{6*5}=\dfrac{12}{30}+\dfrac{25}{30}=\dfrac{12+25}{30}=\dfrac{37}{30}$

or LCM$(5,6)=30$

D. $8*12=96$ $\dfrac{5}{8}+\dfrac{5}{12}=\dfrac{5*12}{8*12}+\dfrac{5*8}{12*8}=\dfrac{60}{96}+\dfrac{40}{96}=\dfrac{100}{96}=\dfrac{100\div4}{96\div4}=\dfrac{25}{24}$

or LCM$(8,12)=24$ $\dfrac{5}{8}+\dfrac{5}{12}=\dfrac{5*3}{8*3}+\dfrac{5*2}{12*2}=\dfrac{15}{24}+\dfrac{10}{24}=\dfrac{25}{24}$

E. $10*6=60$ $\dfrac{3}{10}+\dfrac{1}{6}=\dfrac{3*6}{10*6}+\dfrac{1*10}{6*10}=\dfrac{18}{60}+\dfrac{10}{60}=\dfrac{28}{60}=\dfrac{28\div4}{60\div4}=\dfrac{7}{15}$

or LCM$(10,6)=30$ $\dfrac{3}{10}+\dfrac{1}{6}=\dfrac{3*3}{10*3}+\dfrac{1*5}{6*5}=\dfrac{9}{30}+\dfrac{5}{30}=\dfrac{14}{30}=\dfrac{14\div2}{30\div2}=\dfrac{7}{15}$

Subtracting Fractions

Subtraction of fractions also requires that the fractions have a common denominator. Once the fractions have a common denominator, subtract the numerators, and the result goes over the common denominator.

Order is important in subtraction. If it is reversed, you get the opposite value.

$$\dfrac{3}{4}-\dfrac{3}{5}=\dfrac{3*5}{4*5}-\dfrac{3*4}{5*4}=\dfrac{15}{20}-\dfrac{12}{20}=\dfrac{15-12}{20}=\dfrac{3}{20}$$

$$\dfrac{3}{5}-\dfrac{3}{4}=\dfrac{3*4}{5*4}-\dfrac{3*5}{4*5}=\dfrac{12}{20}-\dfrac{15}{20}=\dfrac{12-15}{20}=\dfrac{-3}{20}$$

You can write $\dfrac{-3}{20}$ as $-\dfrac{3}{20}$ with the fraction being negative rather than the numerator being negative, since the two fractions are equal.

Example:

A. $\dfrac{5}{8} - \dfrac{1}{4}$ B. $\dfrac{1}{2} - \dfrac{1}{3}$ C. $\dfrac{7}{10} - \dfrac{3}{10}$ D. $\dfrac{3}{5} - \dfrac{7}{20}$ E. $\dfrac{5}{8} - \dfrac{1}{12}$

Solution:

A. $\dfrac{5}{8} - \dfrac{1}{4} = \dfrac{5}{8} - \dfrac{1*2}{4*2} = \dfrac{5}{8} - \dfrac{2}{8} = \dfrac{3}{8}$

B. $\dfrac{1}{2} - \dfrac{1}{3} = \dfrac{1*3}{2*3} - \dfrac{1*2}{3*2} = \dfrac{3}{6} - \dfrac{2}{6} = \dfrac{1}{6}$

C. $\dfrac{7}{10} - \dfrac{3}{10} = \dfrac{4}{10} = \dfrac{4 \div 2}{10 \div 2} = \dfrac{2}{5}$

D. $\dfrac{3}{5} - \dfrac{7}{20} = \dfrac{3*4}{5*4} - \dfrac{7}{20} = \dfrac{12}{20} - \dfrac{7}{20} = \dfrac{5}{20} = \dfrac{5 \div 5}{20 \div 5} = \dfrac{1}{4}$

E. $\dfrac{5}{8} - \dfrac{1}{12} = \dfrac{5*3}{8*3} - \dfrac{1*2}{12*2} = \dfrac{15}{24} - \dfrac{2}{24} = \dfrac{13}{24}$

Addition and subtraction of mixed numbers is similar to the procedures for fractions.

$$5\dfrac{2}{3} = \quad 5\dfrac{8}{12} \qquad 5\dfrac{3}{5} = \quad 5\dfrac{12}{20}$$

$$+1\dfrac{1}{4} = +1\dfrac{3}{12} \qquad +2\dfrac{11}{20} = +2\dfrac{11}{20}$$

$$6\dfrac{11}{12} \qquad\qquad 7\dfrac{23}{20} = 7 + \dfrac{23}{20} = 7 + 1 + \dfrac{3}{20} = 8\dfrac{3}{20}$$

Example:

Add these mixed numbers.

A. $\begin{array}{r}\dfrac{3}{4}\\[4pt]+5\dfrac{2}{5}\\\hline\end{array}$ B. $\begin{array}{r}2\\[4pt]+3\dfrac{1}{5}\\\hline\end{array}$ C. $\begin{array}{r}10\dfrac{1}{2}\\[4pt]+9\dfrac{3}{8}\\\hline\end{array}$ D. $\begin{array}{r}6\dfrac{3}{4}\\[4pt]+16\dfrac{7}{8}\\\hline\end{array}$

Solution:

A. $\begin{array}{r}\dfrac{3}{4} = \dfrac{15}{20}\\[6pt]+5\dfrac{2}{5} = 5\dfrac{8}{20}\\\hline\end{array}$

$$5\dfrac{23}{20} = 5 + 1 + \dfrac{3}{20} = 6\dfrac{3}{20}$$

B.
$$\begin{array}{r} 2 \\ +3\frac{1}{5} \\ \hline 5\frac{1}{5} \end{array}$$

C.
$$\begin{array}{r} 10\frac{1}{2} = 10\frac{4}{8} \\ + 9\frac{3}{8} = +9\frac{3}{8} \\ \hline 19\frac{7}{8} \end{array}$$

D.
$$\begin{array}{r} 6\frac{3}{4} = 6\frac{6}{8} \\ +16\frac{7}{8} = +16\frac{7}{8} \\ \hline 22\frac{13}{8} = 22 + 1 + \frac{5}{8} = 23\frac{5}{8} \end{array}$$

When you subtract mixed numbers, you may have a fraction to subtract that is the greater fraction.

$$\begin{array}{r} 12\frac{1}{2} = 12\frac{4}{8} = 11\frac{12}{8} \\ -9\frac{7}{8} = -9\frac{7}{8} = -9\frac{7}{8} \\ \hline 2\frac{5}{8} \end{array}$$

Example:

Subtract these mixed numbers.

A.
$$\begin{array}{r} 8\frac{3}{4} \\ -5\frac{1}{6} \\ \hline \end{array}$$

B.
$$\begin{array}{r} 3\frac{7}{12} \\ -1\frac{3}{8} \\ \hline \end{array}$$

C.
$$\begin{array}{r} 9\frac{1}{6} \\ -2\frac{3}{5} \\ \hline \end{array}$$

D.
$$\begin{array}{r} 13 \\ - 9\frac{3}{4} \\ \hline \end{array}$$

Solution:

A.
$$\begin{array}{r} 8\frac{3}{4} = 8\frac{9}{12} \\ -5\frac{1}{6} = -5\frac{2}{12} \\ \hline 3\frac{7}{12} \end{array}$$

$$3\frac{7}{12} = 3\frac{14}{24}$$

B. $-1\frac{3}{8} = -1\frac{9}{24}$

$$2\frac{5}{24}$$

C.
$$9\frac{1}{6} = 9\frac{5}{30} = 8 + \frac{30}{30} + \frac{5}{30} = 8\frac{35}{30}$$
$$-2\frac{3}{5} = -2\frac{18}{30} = -2\frac{18}{30} \qquad = -2\frac{18}{30}$$
$$6\frac{17}{30}$$

D.
$$13 = 12 + 1 = 12\frac{4}{4}$$
$$-9\frac{3}{4} = -9\frac{3}{4} = -9\frac{3}{4}$$
$$3\frac{1}{4}$$

Multiplying Fractions

Multiplication of fractions does not require the fractions to have common denominators.

$$\frac{2}{3} * \frac{3}{4} = \frac{2*3}{3*4} = \frac{6}{12} = \frac{6 \div 6}{12 \div 6} = \frac{1}{2}$$

You can multiply the numerators together and place the product over the product of the denominators. The next step is to reduce the fraction to lowest terms.

A second way is to divide a numerator and a denominator by a common factor. The second approach is most helpful when the numbers are large or when several fractions are being multiplied.

$$\frac{2}{3} * \frac{3}{4} = \frac{{}^1\cancel{2}}{{}_1\cancel{3}} * \frac{\cancel{3}^1}{\cancel{4}_2} = \frac{1*1}{1*2} = \frac{1}{2}$$

Exercise:

Multiply these fractions.

A. $\frac{7}{8} * \frac{4}{5}$ B. $\frac{4}{5} * \frac{15}{16}$ C. $\frac{9}{16} * \frac{5}{6}$ D. $\frac{3}{10} * \frac{5}{9}$

Solution:

A. $\frac{7}{8} * \frac{4}{5} = \frac{7*4}{8*5} = \frac{28}{40} = \frac{28 \div 4}{40 \div 4} = \frac{7}{10}$, or

$$\frac{7}{8} * \frac{4}{5} = \frac{7}{{}_2\cancel{8}} * \frac{\cancel{4}^1}{5} = \frac{7*1}{2*5} = \frac{7}{10}$$

B. $\dfrac{4}{5} * \dfrac{15}{16} = \dfrac{4*15}{5*16} = \dfrac{60}{80} = \dfrac{60 \div 20}{80 \div 20} = \dfrac{3}{4}$, or

$\dfrac{4}{5} * \dfrac{15}{16} = \dfrac{{}^{1}\cancel{4}}{{}_{1}\cancel{5}} * \dfrac{\cancel{15}^{3}}{\cancel{16}_{4}} = \dfrac{1*3}{1*4} = \dfrac{3}{4}$

C. $\dfrac{9}{16} * \dfrac{5}{6} = \dfrac{9*5}{16*6} = \dfrac{45}{96} = \dfrac{45 \div 3}{96 \div 3} = \dfrac{15}{32}$, or

$\dfrac{9}{16} * \dfrac{5}{6} = \dfrac{{}^{3}\cancel{9}}{16} * \dfrac{5}{\cancel{6}_{2}} = \dfrac{3*5}{16*2} = \dfrac{15}{32}$

D. $\dfrac{3}{10} * \dfrac{5}{9} = \dfrac{3*5}{10*9} = \dfrac{15}{90} = \dfrac{15 \div 15}{90 \div 15} = \dfrac{1}{6}$, or

$\dfrac{3}{10} * \dfrac{5}{9} = \dfrac{{}^{1}\cancel{3}}{{}_{2}\cancel{10}} * \dfrac{\cancel{5}^{1}}{\cancel{9}_{3}} = \dfrac{1*1}{2*3} = \dfrac{1}{6}$

Dividing Fractions

When you want to divide fractions, change the problem to multiplication by the reciprocal of the divisor. As in all other division, the divisor fraction cannot be 0.

$$\dfrac{1}{3} \div \dfrac{3}{4} = \dfrac{1}{3} * \dfrac{4}{3} = \dfrac{4}{9} \text{ and } \dfrac{3}{4} \div \dfrac{3}{8} = \dfrac{3}{4} * \dfrac{8}{3} = \dfrac{24}{12} = 2$$

Example:

Divide these fractions.

A. $\dfrac{3}{5} \div \dfrac{9}{10}$ B. $\dfrac{5}{6} \div \dfrac{7}{12}$ C. $\dfrac{9}{16} \div \dfrac{3}{8}$ D. $\dfrac{2}{3} \div \dfrac{5}{6}$

Solution:

A. $\dfrac{3}{5} \div \dfrac{9}{10} = \dfrac{3}{5} * \dfrac{10}{9} = \dfrac{3*10}{5*9} = \dfrac{30}{45} = \dfrac{30 \div 15}{45 \div 15} = \dfrac{2}{3}$

B. $\dfrac{5}{6} \div \dfrac{7}{12} = \dfrac{5}{6} * \dfrac{12}{7} = \dfrac{5}{\cancel{6}_{1}} * \dfrac{\cancel{12}^{2}}{7} = \dfrac{5 \times 2}{1 \times 7} = \dfrac{10}{7}$

C. $\dfrac{9}{16} \div \dfrac{3}{8} = \dfrac{9}{16} * \dfrac{8}{3} = \dfrac{\cancel{9}^{3}}{\cancel{16}_{2}} * \dfrac{\cancel{8}^{1}}{\cancel{3}_{1}} = \dfrac{3 \times 1}{2 \times 1} = \dfrac{3}{2}$

D. $\dfrac{2}{3} \div \dfrac{5}{6} = \dfrac{2}{3} * \dfrac{6}{5} = \dfrac{2*6}{3*5} = \dfrac{12}{15} = \dfrac{12 \div 3}{15 \div 3} = \dfrac{4}{5}$

When you multiply or divide fractions and whole numbers, you may wish to write the whole number over 1 before you do the computation, so that, for example, 2 becomes $\dfrac{2}{1}$.

Example:

Do the indicated multiplications and divisions.

A. $\dfrac{3}{4} * 2$ B. $12 * \dfrac{5}{24}$ C. $\dfrac{2}{3} \div 4$ D. $4 \div \dfrac{3}{5}$

Solution:

A. $\dfrac{3}{4} * 2 = \dfrac{3}{4} * \dfrac{2}{1} = \dfrac{3 * 2}{4 * 1} = \dfrac{6}{4} = \dfrac{6 \div 2}{4 \div 2} = \dfrac{3}{2}$

B. $12 * \dfrac{5}{24} = \dfrac{12}{1} * \dfrac{5}{24} = \dfrac{12 * 5}{24 * 1} = \dfrac{60}{24} = \dfrac{60 \div 12}{24 \div 12} = \dfrac{5}{2}$

C. $\dfrac{2}{3} \div 4 = \dfrac{2}{3} \div \dfrac{4}{1} = \dfrac{2}{3} * \dfrac{1}{4} = \dfrac{2 * 1}{3 * 4} = \dfrac{2}{12} = \dfrac{2 \div 2}{12 \div 2} = \dfrac{1}{6}$

D. $4 \div \dfrac{3}{5} = \dfrac{4}{1} \div \dfrac{3}{5} = \dfrac{4}{1} * \dfrac{5}{3} = \dfrac{20}{3}$

When performing multiplications or divisions with mixed numbers, first replace each mixed number with the equivalent improper fraction, and then do the computation.

Example:

Multiply these mixed numbers.

A. $2\dfrac{3}{4} \times \dfrac{9}{11}$ B. $3\dfrac{2}{5} \times 1\dfrac{1}{4}$ C. $4 \times 2\dfrac{7}{8}$ D. $4\dfrac{3}{8} \times 3\dfrac{1}{5}$

Solution:

A. $2\dfrac{3}{4} \times \dfrac{9}{11} = \dfrac{11}{4} \times \dfrac{9}{11} = \dfrac{11 * 9}{4 * 11} = \dfrac{11}{11} * \dfrac{9}{4} = 1 \times \dfrac{9}{4} = \dfrac{9}{4} = 2\dfrac{1}{4}$

B. $3\dfrac{2}{5} \times 1\dfrac{1}{4} = \dfrac{17}{5} \times \dfrac{5}{4} = \dfrac{17 \times 5}{5 \times 4} = \dfrac{5}{5} \times \dfrac{17}{4} = 1 \times \dfrac{17}{4} = \dfrac{17}{4} = 4\dfrac{1}{4}$

C. $4 \times 2\dfrac{7}{8} = \dfrac{4}{1} \times \dfrac{23}{8} = \dfrac{4 \times 23}{1 \times 8} = \dfrac{4}{4} \times \dfrac{23}{2} = 1 \times \dfrac{23}{2} = \dfrac{23}{2} = 11\dfrac{1}{2}$

D. $4\dfrac{3}{8} \times 3\dfrac{1}{5} = \dfrac{35}{8} \times \dfrac{16}{5} = \dfrac{35 \times 16}{8 \times 5} = \dfrac{5}{5} \times \dfrac{8}{8} \times \dfrac{7 \times 2}{1} = 1 \times 1 \times \dfrac{14}{1} = 14$

Example:

Divide these mixed numbers.

A. $15 \div 1\dfrac{7}{8}$ B. $11\dfrac{1}{3} \div 2\dfrac{5}{6}$ C. $3\dfrac{3}{16} \div 2\dfrac{1}{8}$ D. $8\dfrac{2}{3} \div 2$

Solution:

A. $15 \div 1\dfrac{7}{8} = \dfrac{15}{1} \div \dfrac{15}{8} = \dfrac{15}{1} \times \dfrac{8}{15} = \dfrac{15 \times 8}{1 \times 15} = \dfrac{15}{15} \times \dfrac{8}{1} = 1 \times 8 = 8$

B. $11\dfrac{1}{3} \div 2\dfrac{5}{6} = \dfrac{34}{3} \div \dfrac{17}{6} = \dfrac{34}{3} \times \dfrac{6}{17} = \dfrac{34 \times 6}{3 \times 17} = \dfrac{17}{17} \times \dfrac{3}{3} \times \dfrac{2 \times 2}{1 \times 1} = 1 \times 1 \times \dfrac{4}{1} = 4$

C. $3\frac{3}{16} \div 2\frac{1}{8} = \frac{51}{16} \div \frac{17}{8} = \frac{51}{16} \times \frac{8}{17} = \frac{51 \times 8}{16 \times 17} = \frac{17}{17} \times \frac{8}{8} \times \frac{3 \times 1}{1 \times 2} = 1 \times 1 \times \frac{3}{2} =$

$\frac{3}{2} = 1\frac{1}{2}$

D. $8\frac{2}{3} \div 2 = \frac{26}{3} \div \frac{2}{1} = \frac{26}{3} \times \frac{1}{2} = \frac{26 \times 1}{3 \times 2} = \frac{13}{3} \times \frac{2}{2} = \frac{13}{3} \times 1 = \frac{13}{3} = 4\frac{1}{3}$

■■■ 9.10 PRACTICE PROBLEMS

1. Add these fractions.

   A. $\frac{2}{3}$ B. $\frac{3}{4}$ C. $\frac{7}{8}$ D. $\frac{5}{11}$ E. $\frac{7}{15}$

   $+\frac{1}{4}$ $+\frac{2}{5}$ $+\frac{8}{15}$ $+\frac{3}{7}$ $+\frac{3}{5}$

2. Subtract these fraction.

   A. $\frac{5}{6}$ B. $\frac{2}{3}$ C. $\frac{15}{16}$ D. $\frac{3}{4}$ E. $\frac{5}{14}$

   $-\frac{2}{5}$ $-\frac{1}{4}$ $-\frac{1}{6}$ $-\frac{7}{10}$ $-\frac{1}{4}$

3. Multiply these fractions.

   A. $\frac{2}{3} \times \frac{9}{10}$ B. $\frac{2}{5} \times \frac{5}{6}$ C. $\frac{7}{8} \times \frac{5}{6}$

   D. $\frac{1}{6} \times \frac{15}{16}$ E. $\frac{1}{10} \times \frac{5}{32}$

4. Divide these fractions.

   A. $\frac{3}{5} \div \frac{9}{10}$ B. $\frac{1}{4} \div \frac{9}{10}$ C. $\frac{17}{20} \div \frac{4}{5}$ D. $\frac{5}{6} \div \frac{3}{4}$

   E. $\frac{1}{4} \div \frac{2}{3}$

5. Add these mixed numbers.

   A. $7\frac{2}{5} + 9\frac{4}{5}$ B. $4\frac{2}{3} + 6\frac{1}{2}$ C. $12\frac{2}{3} + 5\frac{1}{4}$

   D. $5\frac{2}{3} + 6\frac{5}{6}$ E. $8\frac{1}{12} + 6\frac{1}{6}$

6. Subtract these mixed numbers.

   A. $5\frac{5}{6} - 3\frac{1}{10}$ B. $3\frac{1}{5} - 1\frac{1}{2}$ C. $8\frac{1}{6} - 3\frac{1}{12}$

   D. $8 - 2\frac{4}{7}$ E. $11\frac{3}{4} - 5$

7. Multiply these mixed numbers.

   A. $4\frac{1}{2} \times 2\frac{1}{4}$ B. $6\frac{1}{4} \times \frac{3}{4}$ C. $2 \times 1\frac{9}{16}$

   D. $5\frac{1}{3} \times 1\frac{1}{8}$ E. $2\frac{5}{8} \times 2\frac{2}{5}$

8. Divide these mixed numbers.

   A. $3\frac{1}{7} \div \frac{3}{4}$ B. $14\frac{1}{2} \div 3\frac{5}{8}$ C. $8 \div 1\frac{4}{5}$

   D. $11\frac{7}{8} \div 5$ E. $2\frac{1}{16} \div 1\frac{3}{8}$

9. Compute the value of each of these expressions.

   A. $4\frac{3}{16} + 2\frac{7}{8} + 5\frac{1}{2}$ B. $7\frac{2}{3} + 5\frac{1}{6} + 3\frac{1}{12}$

   C. $\frac{2}{3} + \frac{5}{6} + \frac{5}{8}$ D. $\frac{9}{16} + \frac{3}{8} + \frac{1}{2}$

   E. $2\frac{3}{4} \times 1\frac{1}{8} \times 3\frac{5}{6}$

10. Compute these values using the order of operations.

   A. $\frac{1}{2} + \frac{2}{3} - \frac{1}{4} \times \frac{1}{3} \div \frac{1}{6}$ B. $\frac{2}{3} + \frac{3}{5} - \frac{1}{2} + \frac{2}{3}$

   $-\frac{5}{6}$

   C. $\frac{3}{4} \times \frac{1}{2} + \frac{3}{5} \div \frac{6}{35}$ D. $\frac{3}{4} \div \frac{5}{8} - \frac{4}{5} \times \frac{3}{8}$

   E. $\frac{3}{10} \times \frac{5}{9} \div \frac{3}{4} \times \frac{2}{3}$

▬▬ 9.11   SOLUTIONS

1. A. $\frac{2}{3}+\frac{1}{4}$ LCM(3, 4) = 12

   $\frac{2\times4}{3\times4}=\frac{8}{12}$ $\frac{1\times3}{4\times3}=\frac{3}{12}$

   $\frac{8}{12}+\frac{3}{12}=\frac{11}{12}$

   B. $\frac{3}{4}+\frac{2}{5}$ LCM(4, 5) = 20

   $\frac{3\times5}{4\times5}=\frac{15}{20}$ $\frac{2\times4}{5\times4}=\frac{8}{20}$

   $\frac{15}{20}+\frac{8}{20}=\frac{23}{20}=1\frac{3}{20}$

   C. $\frac{7}{8}+\frac{8}{15}=\frac{105}{120}+\frac{64}{120}=\frac{169}{120}=1\frac{49}{120}$

   D. $\frac{5}{11}+\frac{3}{7}=\frac{35}{77}+\frac{33}{77}=\frac{68}{77}$

   E. $\frac{7}{15}+\frac{3}{5}=\frac{7}{15}+\frac{9}{15}=\frac{16}{15}=1\frac{1}{15}$

2. A. $\frac{5}{6}-\frac{2}{5}$ LCM(6, 5) = 30

   $\frac{5\times5}{6\times5}=\frac{25}{30}$ $\frac{2\times6}{5\times6}=\frac{12}{30}$

   $\frac{25}{30}-\frac{12}{30}=\frac{13}{30}$

   B. $\frac{2}{3}-\frac{1}{4}$ LCM(3, 4) = 12

   $\frac{2\times4}{3\times4}=\frac{8}{12}$ $\frac{1\times3}{4\times3}=\frac{3}{12}$

   $\frac{8}{12}-\frac{3}{12}=\frac{5}{12}$

   C. $\frac{15}{16}-\frac{1}{6}=\frac{45}{48}-\frac{8}{48}=\frac{37}{48}$

   D. $\frac{3}{4}-\frac{7}{10}=\frac{15}{20}-\frac{14}{20}=\frac{1}{20}$

   E. $\frac{5}{14}-\frac{1}{4}=\frac{10}{28}-\frac{7}{28}=\frac{3}{28}$

3. A. $\frac{2}{3}\times\frac{9}{10}=\frac{2\times9}{3\times10}=\frac{2}{2}\times\frac{3}{3}\times\frac{3}{5}=$

   $1\times1\times\frac{3}{5}=\frac{3}{5}$

   B. $\frac{2}{5}\times\frac{5}{6}=\frac{2\times5}{5\times6}=\frac{2}{2}\times\frac{5}{5}\times\frac{1}{3}=1\times1\times\frac{1}{3}=\frac{1}{3}$

   C. $\frac{7}{8}\times\frac{5}{6}=\frac{7\times5}{8\times6}=\frac{35}{48}$

   D. $\frac{1}{6}\times\frac{15}{16}=\frac{1\times15}{6\times16}=\frac{3}{3}\times\frac{1\times5}{2\times16}=1\times\frac{5}{32}=\frac{5}{32}$

   E. $\frac{1}{10}\times\frac{5}{32}=\frac{1\times5}{10\times32}=\frac{5}{5}\times\frac{1}{2\times32}=1\times\frac{1}{64}$

   $=\frac{1}{64}$

4. A. $\frac{3}{5}\div\frac{9}{10}=\frac{3}{5}\times\frac{10}{9}=\frac{3\times10}{5\times9}=\frac{3}{3}\times\frac{5}{5}\times\frac{2}{3}=$

   $1\times1\times\frac{2}{3}=\frac{2}{3}$

   B. $\frac{1}{4}\div\frac{9}{10}=\frac{1}{4}\times\frac{10}{9}=\frac{1\times10}{4\times9}=\frac{2}{2}\times\frac{1\times5}{2\times9}=$

   $1\times\frac{5}{18}=\frac{5}{18}$

   C. $\frac{17}{20}\div\frac{4}{5}=\frac{17}{20}\times\frac{5}{4}=\frac{17\times5}{20\times4}=\frac{5}{5}\times\frac{17\times1}{4\times4}=$

   $1\times\frac{17}{16}=\frac{17}{16}=1\frac{1}{16}$

   D. $\frac{5}{6}\div\frac{3}{4}=\frac{5}{6}\times\frac{4}{3}=\frac{5\times4}{6\times3}=\frac{2}{2}\times\frac{5\times2}{3\times3}=$

   $1\times\frac{10}{9}=\frac{10}{9}=1\frac{1}{9}$

   E. $\frac{1}{4}\div\frac{2}{3}=\frac{1}{4}\times\frac{3}{2}=\frac{1\times3}{4\times2}=\frac{3}{8}$

5. A. $7\frac{2}{5}+9\frac{4}{5}=(7+9)+\left(\frac{2}{5}+\frac{4}{5}\right)=16+\frac{6}{5}=$

   $16+1\frac{1}{5}=17\frac{1}{5}$

   B. $4\frac{2}{3}+6\frac{1}{2}=(4+6)+\left(\frac{2}{3}+\frac{1}{2}\right)=10+$

   $\left(\frac{4}{6}+\frac{3}{6}\right)=10+\frac{7}{6}=10+1\frac{1}{6}=11\frac{1}{6}$

   C. $12\frac{2}{3}+5\frac{1}{4}=(12+5)+\left(\frac{2}{3}+\frac{1}{4}\right)=17+$

   $\left(\frac{8}{12}+\frac{3}{12}\right)=17+\frac{11}{12}=17\frac{11}{12}$

D. $5\frac{2}{3} + 6\frac{5}{6} = 5 + 6 + \frac{4}{6} + \frac{5}{6} = 11 + \frac{9}{6} = 11 + 1\frac{3}{6} = 12\frac{1}{2}$

E. $8\frac{1}{12} + 6\frac{1}{6} = 8 + 6 + \frac{1}{12} + \frac{2}{12} = 14 + \frac{3}{12} = 14 + \frac{1}{4} = 14\frac{1}{4}$

6. A. $5\frac{5}{6} - 3\frac{1}{10} = (5 - 3) + \left(\frac{5}{6} - \frac{1}{10}\right) = 2 + \left(\frac{25}{30} - \frac{3}{30}\right) = 2 + \frac{22}{30} = 2 + \frac{11}{15} = 2\frac{11}{15}$

B. $3\frac{1}{5} - 1\frac{1}{2} = (3 - 1) + \left(\frac{1}{5} - \frac{1}{2}\right) = 2 + \left(\frac{2}{10} - \frac{5}{10}\right) = 2 + \left(-\frac{3}{10}\right) = 1\frac{7}{10}$

C. $8\frac{1}{6} - 3\frac{1}{12} = (8 - 3) + \left(\frac{1}{6} - \frac{1}{12}\right) = 5 + \left(\frac{2}{12} - \frac{1}{12}\right) = 5 + \frac{1}{12} = 5\frac{1}{12}$

D. $8 - 2\frac{4}{7} = (8 - 2) + \left(0 - \frac{4}{7}\right) = 6 + \left(-\frac{4}{7}\right) = 5\frac{3}{7}$

E. $11\frac{3}{4} - 5 = (11 - 5) + \left(\frac{3}{4} - 0\right) = 6 + \frac{3}{4} = 6\frac{3}{4}$

7. A. $4\frac{1}{2} \times 2\frac{1}{4} = \frac{9}{2} \times \frac{9}{4} = \frac{9 \times 9}{2 \times 4} = \frac{81}{8} = 10\frac{1}{8}$

B. $6\frac{1}{4} \times \frac{3}{4} = \frac{25}{4} \times \frac{3}{4} = \frac{25 \times 3}{4 \times 4} = \frac{75}{16} = 4\frac{11}{16}$

C. $2 \times 1\frac{9}{16} = \frac{2}{1} \times \frac{25}{16} = \frac{2 \times 25}{1 \times 16} = \frac{2}{2} \times \frac{25}{8} = 1 \times \frac{25}{8} = \frac{25}{8} = 3\frac{1}{8}$

D. $5\frac{1}{3} \times 1\frac{1}{8} = \frac{16}{3} \times \frac{9}{8} = \frac{16 \times 9}{3 \times 8} = \frac{8}{8} \times \frac{3}{3} \times \frac{2 \times 3}{1} = 1 \times 1 \times \frac{6}{1} = 6$

E. $2\frac{5}{8} \times 2\frac{2}{5} = \frac{21}{8} \times \frac{12}{5} = \frac{21 \times 12}{8 \times 5} = \frac{4}{4} \times \frac{21 \times 3}{2 \times 5} = 1 \times \frac{63}{10} = \frac{63}{10} = 6\frac{3}{10}$

8. A. $3\frac{1}{7} \div \frac{3}{4} = \frac{22}{7} \div \frac{3}{4} = \frac{22}{7} \times \frac{4}{3} = \frac{22 \times 4}{7 \times 3} = \frac{88}{21} = 4\frac{4}{21}$

B. $14\frac{1}{2} \div 3\frac{5}{8} = \frac{29}{2} \div \frac{29}{8} = \frac{29}{2} \times \frac{8}{29} = \frac{29 \times 8}{2 \times 29} = \frac{29}{29} \times \frac{2}{2} \times \frac{4}{1} = 1 \times 1 \times 4 = 4$

C. $8 \div 1\frac{4}{5} = \frac{8}{1} \div \frac{9}{5} = \frac{8}{1} \times \frac{5}{9} = \frac{8 \times 5}{1 \times 9} = \frac{40}{9} = 4\frac{4}{9}$

D. $11\frac{7}{8} \div 5 = \frac{95}{8} \div \frac{5}{1} = \frac{95}{8} \times \frac{1}{5} = \frac{95 \times 1}{8 \times 5} = \frac{5}{5} \times \frac{19}{8} = 1 \times \frac{19}{8} = 1 \times \frac{19}{8} = \frac{19}{8} = 2\frac{3}{8}$

E. $2\frac{1}{16} \div 1\frac{3}{8} = \frac{33}{16} \div \frac{11}{8} = \frac{33}{16} \times \frac{8}{11} = \frac{33 \times 8}{16 \times 11} = \frac{11}{11} \times \frac{8}{8} \times \frac{3}{2} = 1 \times 1 \times \frac{3}{2} = 1\frac{1}{2}$

9. A. $4\frac{3}{16} + 2\frac{7}{8} + 5\frac{1}{2} = (4 + 2 + 5) + \left(\frac{3}{16} + \frac{14}{16} + \frac{8}{16}\right) = 11 + \frac{25}{16} = 11 + 1\frac{9}{16} = 12\frac{9}{16}$

B. $7\frac{2}{3} + 5\frac{1}{6} + 3\frac{1}{12} = (7 + 5 + 3) + \left(\frac{8}{12} + \frac{2}{12} + \frac{1}{12}\right) = 15 + \left(\frac{11}{12}\right) = 15\frac{11}{12}$

C. $\frac{2}{3} + \frac{5}{6} + \frac{5}{8} = \frac{16}{24} + \frac{20}{24} + \frac{15}{24} = \frac{51}{24} = \frac{17}{8} = 2\frac{1}{8}$

D. $\frac{9}{16} + \frac{3}{8} + \frac{1}{2} = \frac{9}{16} + \frac{6}{16} + \frac{8}{16} = \frac{23}{16} = 1\frac{7}{16}$

E. $2\frac{3}{4} \times 1\frac{1}{8} \times 3\frac{5}{6} = \frac{11}{4} \times \frac{9}{8} \times \frac{23}{6} = \frac{11 \times 9 \times 23}{4 \times 8 \times 6} = \frac{3}{3} \times \frac{11 \times 3 \times 23}{4 \times 8 \times 2} = 1 \times \frac{759}{64} = \frac{759}{64} = 11\frac{55}{64}$

10. A. $\frac{1}{2} + \frac{2}{3} - \frac{1}{4} \times \frac{1}{3} \div \frac{1}{6} = \frac{1}{2} + \frac{2}{3} - \frac{1}{12} \times \frac{6}{1} = \frac{1}{2} + \frac{2}{3} - \frac{1}{2} = \frac{2}{3}$

B. $\frac{2}{3} + \frac{3}{5} - \frac{1}{2} + \frac{2}{3} - \frac{5}{6} = \frac{20}{30} + \frac{18}{30} - \frac{15}{30} + \frac{20}{30} - \frac{25}{30} = \frac{58}{30} - \frac{40}{30} = \frac{18}{30} = \frac{3}{5}$

C. $\dfrac{3}{4} \times \dfrac{1}{2} + \dfrac{3}{5} \div \dfrac{6}{35} = \dfrac{3}{8} + \dfrac{3}{5} \times \dfrac{35}{6} =$

$\dfrac{3}{8} + \dfrac{3 \times 35}{5 \times 6} = \dfrac{3}{8} + \dfrac{7}{2} = \dfrac{3}{8} + \dfrac{28}{8} = \dfrac{31}{8} = 3\dfrac{7}{8}$

D. $\dfrac{3}{4} \div \dfrac{5}{8} - \dfrac{4}{5} \times \dfrac{3}{8} = \dfrac{3}{4} \times \dfrac{8}{5} - \dfrac{4 \times 3}{5 \times 8} = \dfrac{3 \times 8}{4 \times 5} - \dfrac{3}{10} =$

$\dfrac{12}{10} - \dfrac{3}{10} = \dfrac{9}{10}$

E. $\dfrac{3}{10} \times \dfrac{5}{9} \div \dfrac{3}{4} \times \dfrac{2}{3} = \dfrac{3 \times 5}{10 \times 9} \div \dfrac{3}{4} \times \dfrac{2}{3} = \dfrac{1}{6} \times \dfrac{4}{3} \times \dfrac{2}{3} =$

$\dfrac{4}{18} \times \dfrac{2}{3} = \dfrac{8}{54} = \dfrac{4}{27}$

9.12 DECIMALS

A decimal is made up of a **whole number** part, which can be 0, a **decimal point**, and a **decimal fraction**. For example, 2.1345, 720.864913, 0.4823, 0.222 . . . , and 0.9090090009 . . . are all examples of decimals.

In 0.631579824, the 6 is in the tenths place, 3 is in the hundredths place, 1 is in the thousandths place, 5 is in the ten-thousandths place, 7 is in the hundred-thousandths place, 9 is in the millionths place, 8 is in the ten-millionths place, 2 is in the hundred-millionths place, and 4 is in the billionths place. You can continue naming decimal places indefinitely, but the ones already named go far beyond what you will need for the GRE.

When comparing decimal fractions, you start at the decimal and compare the digits one by one until you find a difference in the digits. The one with the greater number in this place is the greater number.

Example:

Which decimal is greater?

A. 0.7 or 0.074   B. 0.9086 or 0.908   C. 0.608 or 0.64

Solution:

A. 0.7 has a 7 in the tenths place, while 0.074 has a 0 in the tenths place. 0.7 > 0.074
B. 0.9086 has a 9 in the tenths place, 0.908 has a 9 in the tenths place. 0.9086 has a 0 in the hundredths place; 0.908 has a 0 in the hundredths place. 0.9086 has an 8 in the thousandths place; 0.908 has an 8 in the thousandths place. 0.9086 has an additional digit in the ten-thousandths place, so 0.9086 > 0.908.
C. Both numbers have a 6 in the tenths place. 0.608 has a 0 in the hundredths place, while 0.64 has a 4 in the hundredths place. 0.64 > 0.608.

When converting decimals to fractions, there are two different situations: one in which the decimal is finite, and one in which the decimal is infinite and repeating.

Example:

Write each decimal as a fraction.

A. 0.265   B. 0.41   C. 0.25   D. 0.65   E. 0.104

Solution:

A. 0.265 is 265 thousandths, so $0.265 = \dfrac{265}{1,000} = \dfrac{53}{200}$.

B. 0.41 is 41 hundredths, so $0.41 = \dfrac{41}{100}$.

C. 0.25 is 25 hundredths, so $0.25 = \dfrac{25}{100} = \dfrac{1}{4}$.

D. 0.65 is 65 hundredths, so $0.65 = \dfrac{65}{100} = \dfrac{13}{20}$.

E. 0.104 is thousandths, so $0.104 = \dfrac{104}{1,000} = \dfrac{13}{125}$.

Note that you write the digits in the decimal over a one followed by zeros. The number of zeros is equal to the number of digits in the decimal.

When a decimal has all of its digits in repeating groups, you can write it as a fraction by writing one repeating group over the same number of nines as there are digits in the repeating group.

Example:

Change these repeating decimals to fractions.

A. 0.333 . . .   B. 0.252525 . . .   C. 0.090909 . . .   D. 0.123123123 . . .

Solution:

A. 0.333 . . . has one digit that repeats, 3. Write this repeating value over a 9 since only one digit repeats. $0.333\ldots = \dfrac{3}{9} = \dfrac{1}{3}$.

B. 0.252525 . . . has two digits that repeat, 25. Write the repeating value over two 9s, since two digits repeat. $0.252525\ldots = \dfrac{25}{99}$.

C. 0.090909 . . . has two digits that repeat, 09. $0.090909\ldots = \dfrac{09}{99} = \dfrac{1}{11}$.

D. 0.123123123 . . . has three digits that repeat, 123. $0.123123123\ldots = \dfrac{123}{999} = \dfrac{41}{333}$.

If a decimal has some digits between the decimal point and the start of the first repeating group, you have to modify the procedure above. In this case, you write the nonrepeating digits and one repeating group minus the nonrepeating digits over a group of nines followed by a group of zeros. The number of nines is the same as the number of digits in the repeating group, and the number of zeros is equal to the number of digits that do not repeat. Thus $0.12453453453\ldots = \dfrac{12,453 - 12}{99,900} = \dfrac{12,441}{99,900} = \dfrac{4,147}{33,300}$.

Example:

Change these decimals to fractions.

A. 0.4353535 …   B. 0.142333 …   C. 0.12373737 …

Solution:

A. $0.4353535\ldots = \dfrac{435 - 4}{990} = \dfrac{431}{990}$.

B. $0.142333\ldots = \dfrac{1{,}423 - 142}{9{,}000} = \dfrac{1{,}281}{9{,}000} = \dfrac{427}{3{,}000}$.

C. $0.12373737\ldots = \dfrac{1{,}237 - 12}{9{,}900} = \dfrac{1{,}225}{9{,}900} = \dfrac{49}{396}$.

You can also change fractions to decimals. The easiest fractions to change to decimals are ones whose denominators are powers of 10 such as 10, 100, or 1,000.
$\dfrac{17}{100} = 0.17$, $\dfrac{263}{1{,}000} = 0.263$, $\dfrac{9}{100} = 0.09$.

In each case, the numerator is written as a whole number, so the decimal point would be at the right, and then you count to the left of that decimal point the same number of places as there are zeros in the power of ten.

For all other fractions, do the indicated division until the division terminates with a remainder of 0, or it repeats a remainder. If the division terminates, it is a finite decimal with the decimal in the quotient as the answer. If you get a repeated remainder, it is a repeating decimal. The repeating group is the part of the quotient found from the first time the remainder occurred to the second time it occurred.

It is important to note that a fraction in lowest terms will terminate only when 2 and 5 are the only prime divisors of the denominator. If the denominator has any prime divisor other than 2 or 5, that fraction will yield a repeating decimal.

Table of Equivalents

Fraction	Decimal	Fraction	Decimal
1/20	0.05	1/2	0.5
1/16	0.0625	3/5	0.6
1/12	0.08333…	5/8	0.625
1/10	0.1	2/3	0.666…
1/8	0.125	7/10	0.7
1/6	0.1666…	3/4	0.75
1/5	0.2	4/5	0.8
1/4	0.25	5/6	0.8333…
3/10	0.3	7/8	0.875
1/3	0.333…	9/10	0.9
3/8	0.375	15/16	0.9375
2/5	0.4	19/20	0.95

■■ 9.13 PRACTICE PROBLEMS

1. Arrange these decimals in order with the largest first.

   A. 0.01, 0.001, 0.1, 0.0001

   B. 2.25, 0.253, 0.2485, 2.249

   C. 0.38, 1.5, 0.475, 0.0506

   D. 0.006, 5.02, 0.503, 0.1987

   E. 0.92, 0.89, 0.103, 0.098

2. Change each decimal to a fraction.

   A. 0.75   B. 0.08   C. 0.12   D. 0.35   E. 0.165

3. Change each decimal to a fraction.

   A. 0.888...   B. 0.666...
   C. 0.272727...   D. 0.353535...
   E. 0.145145145...

4. Change each decimal to a fraction.

   A. 0.8666...   B. 0.1555...
   C. 0.14252525...   D. 0.5313131...
   E. 0.2666...

■■ 9.14 SOLUTIONS

1. A. 0.1, 0.01, 0.001, 0.0001

   B. 2.25, 2.249, 0.253, 0.2485

   C. 1.5, 0.475, 0.38, 0.0506

   D. 5.02, 0.503, 0.1987, 0.006

   E. 0.92, 0.89, 0.103, 0.098

2. A. $0.75 = \dfrac{75}{100} = \dfrac{3}{4}$

   B. $0.08 = \dfrac{8}{100} = \dfrac{2}{25}$

   C. $0.12 = \dfrac{12}{100} = \dfrac{3}{25}$

   D. $0.35 = \dfrac{35}{100} = \dfrac{7}{20}$

   E. $0.165 = \dfrac{165}{1,000} = \dfrac{33}{200}$

3. A. $0.888... = \dfrac{8}{9}$

   B. $0.666... = \dfrac{6}{9} = \dfrac{2}{3}$

   C. $0.272727... = \dfrac{27}{99} = \dfrac{3}{11}$

   D. $0.353535... = \dfrac{35}{99}$

   E. $0.145145145... = \dfrac{145}{999}$

4. A. $0.8666... = \dfrac{86-8}{90} = \dfrac{78}{90} = \dfrac{13}{15}$

   B. $0.1555... = \dfrac{15-1}{90} = \dfrac{14}{90} = \dfrac{7}{45}$

   C. $0.14252525... = \dfrac{1,425-14}{9,900} = \dfrac{1,411}{9,900}$

   D. $0.5313131... = \dfrac{531-5}{990} = \dfrac{526}{990} = \dfrac{263}{495}$

   E. $0.2666... = \dfrac{26-2}{90} = \dfrac{24}{90} = \dfrac{4}{15}$

■■ 9.15 ARITHMETIC COMPUTATION TEST 1

Use the following test to assess how well you have mastered the material in this chapter so far. Mark your answers by blackening the corresponding answer oval in each question. An answer key and solutions are provided at the end of the test.

1. Which is equal to $5 - 3(6 - (-2)) \div 4$?

   Ⓐ −1
   Ⓑ 0
   Ⓒ 1
   Ⓓ 2
   Ⓔ 4

2. Which is the value of $-x(2z - 5y)$ when $x = -2$, $y = 4$, and $z = 3$?

   (A) 28
   (B) 14
   (C) 8
   (D) -8
   (E) -28

3. Which property is illustrated by $(9 + 3) + 0 = 0 + (9 + 3)$?

   (A) Associative property
   (B) Commutative property
   (C) Identity property
   (D) Inverse property
   (E) Distributive property

4. Which is equal to $\dfrac{2/3}{5/8}$?

   (A) $\dfrac{5}{12}$

   (B) $\dfrac{25}{24}$

   (C) $\dfrac{16}{15}$

   (D) $\dfrac{16}{5}$

   (E) $\dfrac{12}{5}$

5. Which fraction is greater than $\dfrac{4}{5}$?

   (A) $\dfrac{9}{11}$

   (B) $\dfrac{16}{20}$

   (C) $\dfrac{5}{7}$

   (D) $\dfrac{5}{10}$

   (E) $\dfrac{5}{14}$

6. Which is the sum of $\frac{3}{8}$ and $\frac{5}{12}$?

   (A) $\frac{8}{96}$

   (B) $\frac{8}{20}$

   (C) $\frac{8}{12}$

   (D) $\frac{19}{24}$

   (E) $\frac{76}{20}$

7. Which is equal to $5\frac{1}{3} + 2\frac{3}{4}$?

   (A) $10\frac{1}{4}$

   (B) $8\frac{1}{12}$

   (C) $7\frac{12}{13}$

   (D) $7\frac{4}{7}$

   (E) $7\frac{1}{12}$

8. Which is equal to $\frac{11}{15} - \frac{5}{12}$?

   (A) 2

   (B) $\frac{69}{60}$

   (C) $\frac{16}{27}$

   (D) $\frac{2}{5}$

   (E) $\frac{19}{60}$

9. Which is equal to $15 - 8\frac{1}{4}$?

(A) $6\frac{1}{4}$

(B) $6\frac{3}{4}$

(C) $7\frac{1}{4}$

(D) $7\frac{3}{4}$

(E) $23\frac{1}{4}$

10. Which is equal to $\frac{2}{3} + \frac{3}{4} - \frac{1}{2} + \frac{5}{12}$?

(A) $\frac{9}{17}$

(B) $\frac{3}{4}$

(C) $\frac{4}{3}$

(D) $\frac{11}{6}$

(E) $\frac{7}{3}$

11. Which is equal to $\frac{5}{6} \times \frac{8}{15}$?

(A) $\frac{4}{9}$

(B) $\frac{13}{21}$

(C) $\frac{13}{90}$

(D) $\frac{41}{30}$

(E) $\frac{25}{16}$

12. Which is equal to $2\frac{3}{4} \times 3\frac{1}{11}$?

 (A) $8\frac{1}{2}$

 (B) $6\frac{3}{44}$

 (C) $5\frac{4}{15}$

 (D) $5\frac{3}{44}$

 (E) $1\frac{1}{44}$

13. Which is equal to $\frac{5}{6} \div \frac{3}{5}$?

 (A) $\frac{7}{30}$

 (B) $\frac{4}{15}$

 (C) $\frac{1}{2}$

 (D) $\frac{25}{18}$

 (E) $\frac{43}{30}$

14. Which is equal to $4 \times \frac{3}{5}$?

 (A) $\frac{3}{20}$

 (B) $\frac{12}{20}$

 (C) $\frac{12}{5}$

 (D) $4\frac{3}{5}$

 (E) $\frac{23}{5}$

15. Which is equal to $6 \div \dfrac{8}{15}$?

    (A) $\dfrac{4}{45}$

    (B) $\dfrac{5}{16}$

    (C) $\dfrac{16}{5}$

    (D) $6\dfrac{8}{15}$

    (E) $\dfrac{45}{4}$

16. Which is equal to $\dfrac{2}{3} - \dfrac{1}{2} + \dfrac{1}{4} \times \dfrac{5}{6} \div \dfrac{1}{5}$?

    (A) $\dfrac{125}{72}$

    (B) $\dfrac{29}{24}$

    (C) $-\dfrac{5}{54}$

    (D) $\dfrac{1}{72}$

    (E) $-\dfrac{25}{54}$

17. Which is the greatest value: 12.59, 1.27, 0.56, 9.47, or 0.098?

    (A) 12.59
    (B) 9.79
    (C) 1.27
    (D) 0.56
    (E) 0.098

18. Which digit is in the hundredths place in 273.8941?

    (A) 2
    (B) 3
    (C) 7
    (D) 8
    (E) 9

19. Which is 0.328 written as a fraction?

(A) $\dfrac{164}{5}$

(B) $3\dfrac{28}{100}$

(C) $\dfrac{82}{25}$

(D) $\dfrac{41}{125}$

(E) $\dfrac{41}{1250}$

20. Which fraction is equal to 0.060606 ... ?

(A) $\dfrac{2}{3}$

(B) $\dfrac{2}{30}$

(C) $\dfrac{2}{33}$

(D) $\dfrac{2}{333}$

(E) $\dfrac{2}{330}$

Answer Key

1.	A	11.	A
2.	E	12.	A
3.	B	13.	D
4.	C	14.	C
5.	A	15.	E
6.	D	16.	B
7.	B	17.	A
8.	E	18.	E
9.	B	19.	D
10.	C	20.	C

9.16 SOLUTIONS

1. **A** -1
$$5 - 3(6 - (-2)) \div 4 = 5 - 3(8) \div 4 = 5 - 24 \div 4 = 5 - 6 = -1$$

2. **E** -28
$$-x(2z - 5y) = -(-2)(2(3) - 5(4)) = 2(6 - 20) = 2(-14) = -28$$

3. **B** Commutative property
$(9 + 3) + 0 = 0 + (9 + 3)$ has the order of $(9 + 3)$ and 0 reversed so it illustrates the commutative property.

4. **C** $\dfrac{16}{15}$

$$\frac{2/3}{5/8} = \frac{2}{3} \div \frac{5}{8} = \frac{2}{3} \times \frac{8}{5} = \frac{16}{15}$$

5. **A** $\dfrac{9}{11}$

$$\frac{4}{5}, \frac{9}{11} \quad 4 \times 11 = 44 < 45 = 5 \times 9 \quad \text{so } \frac{4}{5} < \frac{9}{11}$$

6. **D** $\dfrac{19}{24}$

$$\frac{3}{8} + \frac{5}{12} = \frac{9}{24} + \frac{10}{24} = \frac{19}{24}$$

7. **B** $8\dfrac{1}{12}$

$$5\frac{1}{3} + 2\frac{3}{4} = 5 + 2 + \frac{1}{3} + \frac{3}{4} = 7 + \frac{4}{12} + \frac{9}{12} = 7 + \frac{13}{12} = 7 + 1\frac{1}{12} = 8\frac{1}{12}$$

8. **E** $\dfrac{19}{60}$

$$\frac{11}{15} - \frac{5}{12} = \frac{44}{60} - \frac{25}{60} = \frac{19}{60}$$

9. **B** $6\dfrac{3}{4}$

$$15 - 8\frac{1}{4} = 15 - 8 - \frac{1}{4} = 7 - \frac{1}{4} = 6\frac{3}{4}$$

10. **C** $\dfrac{4}{3}$

$$\frac{2}{3} + \frac{3}{4} - \frac{1}{2} + \frac{5}{12} = \frac{8}{12} + \frac{9}{12} - \frac{6}{12} + \frac{5}{12} = \frac{16}{12} = \frac{4}{3}$$

11. **A** $\dfrac{4}{9}$

$$\frac{5}{6} \times \frac{8}{15} = \frac{40}{90} = \frac{4}{9}$$

12. **A** $8\dfrac{1}{2}$

$$2\frac{3}{4} \times 3\frac{1}{11} = \frac{11}{4} \times \frac{34}{11} = \frac{34}{4} = \frac{17}{2} = 8\frac{1}{2}$$

13. **D** $\dfrac{25}{18}$

$$\frac{5}{6} \div \frac{3}{5} = \frac{5}{6} \times \frac{5}{3} = \frac{25}{18}$$

14. **C** $\dfrac{12}{5}$

$$4 \times \frac{3}{5} = \frac{4}{1} \times \frac{3}{5} = \frac{12}{5}$$

15. **E** $\dfrac{45}{4}$

$$6 \div \frac{8}{15} = \frac{6}{1} \div \frac{8}{15} = \frac{6}{1} \times \frac{15}{8} = \frac{90}{8} = \frac{45}{4}$$

16. **B** $\dfrac{29}{24}$

$$\frac{2}{3} - \frac{1}{2} + \frac{1}{4} \times \frac{5}{6} \div \frac{1}{5} = \frac{2}{3} - \frac{1}{2} + \frac{5}{24} \times \frac{5}{1} = \frac{16}{24} - \frac{12}{24} + \frac{25}{24} = \frac{29}{24}$$

17. **A** 12.59
    12.59 is the greatest.

18. **E** 9

The digit in the hundredths place is 9.

19. **D** $\dfrac{41}{125}$

$$0.328 = \frac{328}{1000} = \frac{41}{125}$$

20. **C** $\dfrac{2}{33}$

$$0.060606\ldots = \frac{06}{99} = \frac{2}{33}$$

9.17 SOLVED GRE PROBLEMS

For each question, select the best answer unless otherwise instructed.

Quantity A	Quantity B

1. $\dfrac{5}{6}$ $\dfrac{20}{24}$

(A) Quantity A is greater.
(B) Quantity B is greater.
(C) The two quantities are equal.
(D) The relationship cannot be determined from the given information.

2. Which shows $\dfrac{60}{144}$ in lowest terms?

(A) $\dfrac{1}{4}$

(B) $\dfrac{30}{72}$

(C) $\dfrac{5}{12}$

(D) $\dfrac{15}{36}$

(E) $\dfrac{7}{12}$

For this question, enter your answer in the box.

3. What is the numerator when $\dfrac{3}{5}$ is written as an equivalent fraction with a denominator of 80?

4. Jim has 24 different party favors. How many guests (excluding himself) could attend in order for Jim to portion out the favors equally? Select all that apply.

   A. 5
   B. 8
   C. 10
   D. 23
   E. 24

Quantity A	Quantity B
0.39	0.082

5.

   Ⓐ Quantity A is greater.
   Ⓑ Quantity B is greater.
   Ⓒ The two quantities are equal.
   Ⓓ The relationship cannot be determined from the given information.

■■■ 9.18 SOLUTIONS

1. **C** The two quantities are equal.
$$\frac{20}{24} = \frac{20 \div 4}{24 \div 4} = \frac{5}{6}$$

2. **C** $\frac{5}{12}$
$$\frac{60}{144} = \frac{60 \div 12}{144 \div 12} = \frac{5}{12}$$

3. $\boxed{48}$

Since $80 \div 5 = 16$, multiply the numerator and denominator by 16. So $\frac{3}{5} =$

$\frac{3 \times 16}{5 \times 16} = \frac{48}{80}$ and the new numerator is 48.

4. **B** and **E** 8 and 24
24 is divisible by 1, 2, 3, 4, 6, 8, 12, and 24.

5. **A** Quantity A is greater.
The tenths digit in 0.39 is 3, and the tenths digit in 0.082 is 0. Since $3 > 0$ and $0.39 > 0.082$, Quantity A is greater.

■■■ 9.19 GRE PRACTICE PROBLEMS

For each question, select the best answer unless otherwise instructed.

1. **Which is an example of the associative property of addition?**

   Ⓐ $3 \times (4 + 5) = 3 \times 4 + 3 \cdot 5$
   Ⓑ $13 + 8 = 8 + 13$
   Ⓒ $(8 + 4) + 2 = (4 + 8) + 2$
   Ⓓ $(5 \times 2) \times 7 = 5 \times (2 \times 7)$
   Ⓔ $3 + (8 + 7) = (3 + 8) + 7$

2. Which mixed number is equal to $\dfrac{17}{5}$?

(A) $\dfrac{5}{17}$

(B) $3\dfrac{1}{5}$

(C) 3, remainder 2

(D) $3\dfrac{2}{5}$

(E) 3.4

Quantity A	Quantity B

3. $\dfrac{7}{8}$ $\dfrac{14}{21}$

(A) Quantity A is greater.
(B) Quantity B is greater.
(C) The two quantities are equal.
(D) The relationship cannot be determined from the given information.

For this question, enter your answer in the box.

4. Which digit is in the thousandths place in the number 78,062.45791?

5. Which is the answer to $(-3) \times (-1) \times (+2) \times (+3) \times (-2)$?

(A) -36
(B) -1
(C) 0
(D) 1
(E) 36

Quantity A	Quantity B

6. 0.1232323 . . . $\dfrac{123}{999}$

(A) Quantity A is greater.
(B) Quantity B is greater.
(C) The two quantities are equal.
(D) The relationship cannot be determined from the given information.

Quantity A	Quantity B

7. The units digits of 311.007   The tenths digit of 352.1897

(A) Quantity A is greater.
(B) Quantity B is greater.
(C) The two quantities are equal.
(D) The relationship cannot be determined from the given information.

8. What is the least common denominator for $\dfrac{7}{12}$ and $\dfrac{11}{20}$?

    (A) 32
    (B) 35
    (C) 60
    (D) 120
    (E) 240

	Quantity A	Quantity B
9.	$6\dfrac{2}{3} + 5\dfrac{1}{2}$	$14\dfrac{3}{4} - 2\dfrac{1}{3}$

    (A) Quantity A is greater.
    (B) Quantity B is greater.
    (C) The two quantities are equal.
    (D) The relationship cannot be determined from the given information.

10. **Two-thirds of Mrs. Webster's algebra class are boys. How many girls are in the class of 48 students?**

    (A) 48
    (B) 32
    (C) 24
    (D) 16
    (E) 12

ANSWER KEY

1.	E	6.	A
2.	D	7.	C
3.	A	8.	C
4.	7	9.	B
5.	A	10.	D

9.20 WORD PROBLEMS

When solving a word problem, the following procedure is quite helpful:

1. **Read** the problem to determine the information that is given and what is to be found.
2. **Plan** how you will solve the problem. This may require the use of a formula or an equation.
3. **Solve** the problem using the plan created.
4. **Answer** the question asked. Check your answer with all the facts given in the problem.

Word problems often use keywords that indicate the operation to be used in the problem. These common keywords are useful guides, and it pays to watch out for them as you read each problem. Here is a table of these keywords.

Operation	Keywords
Addition	Sum, total, all together, and, added to, combined, exceeds, increased by, more than, greater than, plus
Subtraction	Minus, difference, decreased by, less, diminished by, reduced by, subtracted from, deducted
Multiplication	Times, twice, doubled, tripled, product, multiplied by
Division	Quotient, divided by, divide

9.21 PRACTICE PROBLEMS

1. In a class, 6 students received A's, 12 received B's, 10 C's, 8 D's, and 4 F's. What part of the class received A's?

2. In 36 times at bat, Jones got 8 singles, 3 doubles, 1 triple, and 3 home runs. What part of the time did Jones get a hit?

3. The slope of a roof is the ratio of the rise of the roof to the run of the roof. What is the slope of a roof that rises 6 ft in a run of 24 ft?

4. If four-ninths of a class of 36 students is girls, how many girls are there in the class?

5. If a hydrochloric acid solution has two-thirds water, how much water is used to make 24 ounces of this solution?

6. A book that regularly sold for $19.49 is sold after being given a $2.98 price reduction. How much did the book sell for?

7. A sales clerk received $218.40 regular pay, $28.50 overtime pay, $36.14 in commissions, and a $25 bonus on a payday. How much money did the clerk receive on that payday?

8. A finance company made a loan of $100 for six monthly payments of $21.55. How much money was repaid to the loan company?

9. Jane drove 331 miles in 6 hours. To the nearest tenth of a mile, what was the average number of miles Jane drove per hour?

10. Harry bought a suit for $194.95 and a shirt for $18.45. How much more than the shirt did the suit cost?

■ 9.22 SOLUTIONS

1. **Read:** You know how many students received each grade, so add them to find out how many students are in the class: $6+12+10+8+4=40$ students in the class.

   Plan: Write the fraction for the number of students receiving A's, 6, divided by the number of students in the class, 40: 6/40

   Solve: Simplify the fraction: $6/40 = 3/20$

   Answer: Did you find the fraction of students getting A's as a part of the class? Yes, the 6 for students getting A's is compared to the 40 for the number of students in the class.

2. 8 singles, 3 doubles, 1 triple, 3 home runs. So $8+3+1+3 = 15$ hits.

   The part of the time Jones got a hit is $\dfrac{15}{36} = \dfrac{5}{12}$.

3. $\text{slope} = \dfrac{\text{rise}}{\text{run}} = \dfrac{6 \text{ ft}}{24 \text{ ft}} = \dfrac{1}{4}$

4. $\dfrac{4}{9}$ girls out of 36 students. $\dfrac{4}{9} \times \dfrac{36}{1} = \dfrac{4 \times 36}{9 \times 1} = \dfrac{4 \times 4}{1} = 16$ girls

5. $\dfrac{2}{3}$ water out of 24 ounces. $\dfrac{2}{3} \times \dfrac{24}{1} = \dfrac{2 \times 24}{3 \times 1} = \dfrac{2 \times 8}{1} = 16$ ounces

6.
$19.49	regular price
− $2.98	price reduction
$16.51	selling price

7.
$218.40	regular pay
$28.50	overtime pay
$36.14	commissions
+ $25	bonus
$308.04	Total received

8.
$21.55	each payment
× 6	payments made
$129.30	Total repaid

9. 331 miles driven in 6 hours

$$
\begin{array}{r}
55.16 \\
6\overline{)331.00} \\
30 \\
\overline{31} \\
30 \\
\overline{10} \\
6 \\
\overline{40} \\
36 \\
\overline{4}
\end{array}
$$

55.16 is 55.2 rounded to the nearest tenth. Jane averaged about 55.2 miles each hour.

10.
$194.95	suit
−$18.45	shirt
$176.50	difference

■ 9.23 RATIO AND PROPORTION

A **ratio** is a quotient of two quantities. The ratio of a to b is a to b, $\dfrac{a}{b}$, or $a : b$. When ratios are used to compare numbers that have units of measure, the units need to be the same for both numbers. You can find the ratio of one part to another, such as the ratio of boys to girls in a class, or of the parts to the whole, such as ratio of boys or girls to the whole class.

If there are 12 boys and 18 girls in a class, then there are $12 + 18 = 30$ students in the class. The ratio of boys to girls is $12 : 18$ which simplifies to $2 : 3$. The ratio of girls to the class is $18 : 30$ or $3 : 5$.

If the ratio of yes votes to no votes is $175 : 350$, then the ratio of the yes votes to the total votes is $175 : (175 + 350) = 175 : 525 = 1 : 3$, and the ratio of the no votes to total votes is $350 : (175 + 350) = 350 : 525 = 2 : 3$.

When the ratio is known, you do not necessarily know the actual amounts for the quantities being compared. The ratio 2 to $3 = 2 : 3 = \dfrac{2}{3} = \dfrac{4}{6} = \dfrac{6}{9} = \dfrac{2n}{3n} = 2n : 3n$, for all counting numbers n.

Example:

State the ratios requested.

A. The ratio of a nickel to a quarter
B. The ratio of 3 hours to a day
C. The ratio of boys to students in a class of 40 students with 22 girls
D. The ratio of the short to long parts of a 10-foot board when the shorter of the two parts is 4 ft

Solution:

A. nickel = 5 cents   quarter = 25 cents

$$\frac{\text{nickel}}{\text{quarter}} = \frac{5}{25} = \frac{1}{5}$$

B. 3 hours   one day = 24 hours

$$\frac{3 \text{ hours}}{\text{one day}} = \frac{3}{24} = \frac{1}{8}$$

C. girls = 22   class = 40   boys = 40 − 22 = 18

$$\frac{\text{boys}}{\text{class}} = \frac{18}{40} = \frac{9}{20}$$

D. shorter part = 4 ft   total = 10 ft   longer part = 10 ft − 4 ft = 6 ft

$$\frac{\text{shorter part}}{\text{longer part}} = \frac{4}{6} = \frac{2}{3}$$

An extended ratio can have more than two parts such as $a : b : c$ or $a : b : c : d$. If a board is 12 ft long and cut into three parts that have a ratio $1 : 2 : 3$, the lengths are $1n : 2n : 3n$. Since the total length is 12 ft, the lengths are 2 ft, 4 ft, and 6 ft because $2 \text{ ft} + 4 \text{ ft} + 6 \text{ ft} = 12 \text{ ft}$. You can work this out from the ratio $1n : 2n : 3n$ by using the equation $1n + 2n + 3n = 12$. Thus, $6n = 12$ and $n = 2$. So the parts are $1n = 1(2) = 2$ ft, $2n = 2(2) = 4$ ft, and $3n = 3(2) = 6$ ft.

Example:

Find the parts for each ratio.

A. A length of cloth is 40 ft long. It is cut into lengths that have the ratio 2 : 3 : 5. What are the lengths of the parts?
B. A banner is 18 ft long and is cut into three parts with the ratio 1 : 3 : 5. What are the lengths of the parts?
C. A board is 48 inches long and is cut into four parts with the ratio 1 : 1 : 3 : 5. Find the lengths of the parts.

Solution:

A. 2 : 3 : 5 ratio and the total length is 40 ft.
   $2n + 3n + 5n = 40$; $10n = 40$; $n = 4$
   $2n = 2(4) = 8$ ft; $3n = 3(4) = 12$ ft; $5n = 5(4) = 20$ ft
   The parts are 8 ft, 12 ft, and 20 ft.
B. 1 : 3 : 5 ratio and the total length is 18 ft.
   $1n + 3n + 5n = 18$; $9n = 18$; $n = 2$
   $1n = 1(2) = 2$ ft; $3n = 3(2) = 6$ ft; $5n = 5(2) = 10$ ft
   The parts are 2 ft, 5 ft, and 10 ft.
C. 1 : 1 : 3 : 5 ratio and the total length is 48 inches.
   $n + n + 3n + 5n = 48$; $10n = 48$; $n = 4.8$
   $n = 4.8$ inches; $n = 4.8$ inches; $3n = 3(4.8) = 14.4$ inches; $5n = 5(4.8) = 24$ inches
   The parts are 4.8 inches, 4.8 inches, 14.4 inches, and 24 inches.

A **proportion** is a statement that two ratios are equal. A proportion can be stated as "*a* is to *b* as *c* is to *d*" and written as $a : b :: c : d$ or $\frac{a}{b} = \frac{c}{d}$. In the proportion, *a* and *d* are called the **extremes** and *b* and *c* are the **means**. To solve a proportion, cross-multiply to get the product of the means equal to the product of the extremes.

Example:

Write each as a proportion.

A. 26 is to 13 is the same as 6 is to 3.
B. 4 is to 12 as 5 is to 15.
C. 45 is to 80 is the same as 18 is to 32.
D. 150 is to 100 as 54 is to 36.

Solution:

A. $\dfrac{26}{13} = \dfrac{6}{3}$

B. $\dfrac{4}{12} = \dfrac{5}{15}$

C. $\dfrac{45}{80} = \dfrac{18}{32}$

D. $\dfrac{150}{100} = \dfrac{54}{36}$

Example:

Do these proportion problems.

A. If three eggs cost 16 cents, how many eggs can you buy for 80 cents?
B. A recipe calls for 2 cups of flour to 3 tablespoons of shortening. How many cups of flour are needed when you use 15 tablespoons of shortening?
C. If 4 stamps cost 90 cents, how much would 24 stamps cost?

Solution:

A. Let n be the number of eggs you can buy for 80 cents. Then 3 is to 16 as n is to 80.

$$\frac{3}{16} = \frac{n}{80}$$

cross-multiply $16(n) = 3(80)$

$$16n = 240$$

$$n = 240 \div 16$$

$$n = 15$$

Fifteen eggs can be bought for 80 cents.

B. Let n be the number of cups of flour needed. Then 2 is to 3 as n is to 15.

$$\frac{2}{3} = \frac{n}{15}$$

cross-multiply $3n = 30$

$$n = 10$$

Ten cups of flour are needed when 15 tablespoons of shortening are used.

C. Let n be the number of cents that 24 stamps cost. Then 4 is to 90 as 24 is to n.

$$\frac{4}{90} = \frac{24}{n}$$

cross-multiply $4n = 90(24)$

$$4n = 2160$$

$$n = 2160 \div 4$$

$$n = 540$$

Twenty-four stamps would cost 540 cents or $5.40.

Two quantities are directly proportional when a constant k times one of them gives the other. The distance d that a car travels is directly proportional to the time t that it travels. So $d = kt$. The amount of sales tax t that you pay is proportional to the cost c of the items purchased. So $t = kc$.

Example:

Write the proportion for each problem.

A. The number of commercials c in a television program is directly proportional to the length of the program m in minutes.
B. The number of miles m that you run is directly proportional to your rate r of speed.
C. The number of books b that you read is directly proportional to the number of hours h that you spend reading.

Solution:

A. $c = km$
B. $m = kr$
C. $b = kh$

When the product of two quantities is a constant, the quantities are said to be inversely proportional. The number of tickets to a play t that you get for \$100 is inversely proportional to the cost c of a ticket to the play. So $tc = \$100$. The number of workers w needed to finish a job in 6 days is inversely proportional to the rate r of work done per day by a worker. So $wr = 6$.

Example:

Write the proportion for each problem.

A. The number n of cookies in a box with volume 100 in^3 is inversely proportional to the size s of the cookies.
B. The number n of mowers it takes to mow a field in 10 hours is inversely proportional to the area a that each mower cuts per hour.
C. The number of movie tickets t that you can get for \$50.00 is inversely proportional to the price p of a ticket.

Solution:

A. $ns = 100$
B. $na = 10$
C. $tp = 50$

Properties of Proportions

For a proportion $\dfrac{a}{b} = \dfrac{c}{d}$:

1. $ad = bc$

2. $\dfrac{b}{a} = \dfrac{d}{c}$

3. $\dfrac{a}{c} = \dfrac{b}{d}$

4. $\dfrac{a+b}{b} = \dfrac{c+d}{d}$

5. $\dfrac{a-b}{b} = \dfrac{c-d}{d}$

6. $\dfrac{a+b}{a-b} = \dfrac{c+d}{c-d}$

Each of these properties is a transformation of the given proportion.

Even if you do not know the amount of each quantity, sometimes you can still find the ratio of the quantities. If you know that $\frac{1}{3}$ of the students like carrots and $\frac{1}{8}$ of the students like turnips, the ratio of students who like carrots, C, to students who like turnips, T, is $\frac{1}{3}$ to $\frac{1}{8}$.

$$\frac{C}{T} = \frac{1/3}{1/8} = \frac{1}{3} \div \frac{1}{8} = \frac{1}{3} \times \frac{8}{1} = \frac{8}{3}$$

Thus, there are 8 students who like carrots for every 3 who like turnips.

9.24 PRACTICE PROBLEMS

1. Write the ratio requested.
   A. Ratio of 5 days to a week
   B. Ratio of girls to boys in a class of 35 students with 18 boys
   C. Ratio of a dime to a quarter
   D. Ratio of a foot to a yard

2. Find the values requested.
   A. If a foot-long hotdog was cut into two parts that have a ratio of 1 : 2, how long is each part?
   B. If a collection of 54 coins, only nickels and dimes, has the ratio of nickels to dimes of 4 : 5, how many of each coin are there?
   C. If a collection of 1,000 nickels, dimes, and quarters has the ratio of nickels to dimes to quarters being 5 : 3 : 2, how many of each coin are in the collection?

3. Write each as a proportion.
   A. 50 is to 12 as 150 is to 36
   B. 14 to 12 is the same as 42 to 36
   C. 45 is to 90 as 225 is to 450

4. Solve these proportions.
   A. A recipe calls for 1 cup of flour to 2 cups of sugar. If the next batch uses 3 cups of flour, how much sugar is needed?
   B. A chili recipe calls for 4 cups of beans to 2 cups of tomatoes. If you use 5 cups of tomatoes, how many cups of beans are needed?
   C. If 5 stamps cost $1.95, how much would 20 stamps cost?
   D. If you get 3 oranges for 70 cents, how many oranges can you get for $3.50?

5. Write the proportion.
   A. The number of dollars d in your pay is directly proportional to the number of hours h that you work.
   B. The number of meals m that you eat on a trip is directly proportional to the number of days d in the trip.
   C. The number of pages p copied is directly proportional to the number of minutes m that the copier runs.
   D. The number of lunches l that you get for $50 is inversely proportional to the cost c of a meal.
   E. The number of candy bars c that you can get for $20 is inversely proportional to the price p of the candy bar.

6. Find the ratio of a to b.

   A. In a sentence, $\frac{2}{5}$ of the letters are a's and $\frac{1}{20}$ of the letters are b's.

   B. $\dfrac{a - b}{b} = \dfrac{17}{2}$

   C. $\dfrac{b}{a} = \dfrac{3}{10}$

   D. $\frac{3}{8}$ of a class are girls and $\frac{5}{8}$ of the class are boys and $\dfrac{a}{b} = \dfrac{\text{girls}}{\text{boys}}$

▬▬ 9.25 SOLUTIONS

1. A. One week = 7 days
   ratio of 5 days to one week is $5 : 7$
   B. 35 students with 18 boys so $35 - 18 = 17$ girls
   ratio of girls to boys is $17 : 18$
   C. dime = 10 cents   quarter = 25 cents
   ratio of dime to quarter is $10 : 25$ or $2 : 5$
   D. one foot = 1 ft   one yard = 3 ft
   ratio of one foot to one yard is $1 : 3$

2. A. ratio $1 : 2$, total length is 12 inches
   $n + 2n = 12$; $3n = 12$; $n = 4$ inches
   The parts are 4 inches and 8 inches.
   B. ratio $4 : 5$, number of coins is 54
   $4n + 5n = 54$; $9n = 54$; $n = 6$
   $4(6) = 24$ nickels; $5(6) = 30$ dimes
   C. 1,000 coins,   ratio $5 : 3 : 2$ for nickels to dimes
   to quarters
   $5n + 3n + 2n = 1000$; $10n = 1,000$;
   $n = 100$
   $5(100) = 500$ nickels; $3(100) = 300$ dimes;
   $2(100) = 200$ quarters

3. A. $50 : 12 :: 150 : 36$ or $\dfrac{50}{12} = \dfrac{150}{36}$

   B. $14 : 12 :: 42 : 36$ or $\dfrac{14}{12} = \dfrac{42}{36}$

   C. $45 : 90 :: 225 : 450$ or $\dfrac{45}{90} = \dfrac{225}{450}$

4. A. 1 cup flour to 2 cups sugar
   $$\frac{1}{2} = \frac{3}{n}$$
   $$1 \times n = 2 \times 3$$
   $$n = 6$$
   Six cups of sugar are needed.

   B. 4 cups of beans to 2 cups of tomatoes
   $$\frac{4}{2} = \frac{n}{5}$$
   $$2n = 20$$
   $$n = 10$$
   Ten cups of beans are needed.

   C. 5 stamps for $1.95
   $$\frac{5}{1.95} = \frac{20}{n}$$
   $$5n = 39.00$$
   $$n = 7.80$$

   Twenty stamps would cost $7.80.

D. Three oranges for 70¢, $3.50 = 350¢
   $$\frac{3}{70} = \frac{n}{350}$$
   $$70n = 1,050$$
   $$n = 15$$

   Fifteen oranges can be bought for $3.50.

5. A. $d = kh$
   B. $m = kd$
   C. $p = km$
   D. $lc = 50$
   E. $cp = 20$

6. A. $\dfrac{a}{b} = \dfrac{2}{5} \div \dfrac{1}{20}$

   $\dfrac{a}{b} = \dfrac{2}{5} \times \dfrac{20}{1}$

   $\dfrac{a}{b} = \dfrac{40}{5}$

   $\dfrac{a}{b} = \dfrac{8}{1}$

   B. $\dfrac{a - b}{b} = \dfrac{17}{2}$

   Use proportion property (4).

   $$\frac{a - b + b}{b} = \frac{17 + 2}{2}$$

   $$\frac{a}{b} = \frac{19}{2}$$

   C. $\dfrac{b}{a} = \dfrac{3}{10}$

   Use proportion property (2).

   $\dfrac{a}{b} = \dfrac{10}{3}$

   D. $\dfrac{a}{b} = \dfrac{\text{girls}}{\text{boys}}$

   $\dfrac{a}{b} = \dfrac{3}{8} \div \dfrac{5}{8}$

   $\dfrac{a}{b} = \dfrac{3}{8} \times \dfrac{8}{5}$

   $\dfrac{a}{b} = \dfrac{3}{5}$

9.26 MOTION AND WORK PROBLEMS

The motion formula $d = rt$ states that the distance traveled is equal to the rate of travel multiplied by time traveled. The rate is distance per unit of time, and the unit of time must match the time units.

$$\text{distance: } d = rt \quad \text{rate: } r = \frac{d}{t} \quad \text{time: } t = \frac{d}{r}$$

If a bicycle travels at 8 miles per hour for 3 hours, then the bicycle has traveled $d = 8$ miles/hour times 3 hours = 24 miles.

However, if a car traveled 50 miles per hour for 30 minutes, it did NOT travel 50×30 miles since 50 is in miles per hour and the time is in minutes. The correct result is found by changing 30 minutes to 0.5 hour. $d = 50$ miles/hour $\times 0.5$ hour = 25 miles.

Example:

Solve each of these motion problems.

A. A driver drove for 10 hours at the average rate of 53 miles per hour (mph) to get from Memphis to Chicago. How far was it from Memphis to Chicago?
B. A plane traveled the 328 miles from Limestone to Lincoln in two hours. What was the plane's average rate?
C. A boat traveled 120 miles from Wabash to Wilson at an average rate of 60 miles per hour. How many hours did the boat trip take?

Solution:

A. $d = rt$
   $d = 53(10)$
   $d = 530$ miles
   It was 530 miles from Memphis to Chicago.
B. $r = \dfrac{d}{t}$
   $r = 328 \div 2$
   $r = 164$ mph
   The average rate of the plane was 164 mph.
C. $t = \dfrac{d}{r}$
   $t = 120 \div 60$
   $t = 2$ hours
   The boat trip took 2 hours.

On the GRE, you need to be able to convert one unit of measure to another. The units will be in the same measurement system only. Also, only common conversions will be used.

Time	U.S. Customary English System Units	Metric Units
60 minutes = 1 hour	12 inches = 1 foot	10 millimeters = 1 centimeter
60 seconds = 1 minute	3 feet = 1 yard	100 centimeters = 1 meter
24 hours = 1 day	5,280 feet = 1 mile	1,000 meters = 1 kilometer
7 days = 1 week	1,760 yards = 1 mile	1,000 milligrams = 1 gram
52 weeks = 1 year		1,000 grams = 1 kilogram
12 months = 1 year	16 ounces = 1 pound	1,000 milliliters = 1 liter

Work problems involve individuals or machines working together to accomplish a task when you know how long it takes each to do the work individually. If a person can do a job in 5 hours, then that person can do $\frac{1}{5}$ of the work in one hour. Thus, the person can do $\frac{2}{5}$ of the job in 2 hours, $\frac{3}{5}$ in 3 hours, and $\frac{4}{5}$ in 4 hours. There is an assumption that a person always works at a constant rate.

The principle behind work problems is that the time worked t times the rate of work r gives the amount of work w done: $w = tr$. For each person working, calculate the part of the work done and then add all of the parts together to get one unit of work completed.

Example:

Solve these work problems.

A. Sue can paint a room in 4 hours, and Sam can paint the same room in 5 hours. How long would it take them to paint the room together?
B. Abe and Ben, working together, can build a cabinet in 6 days. Abe works twice as fast as Ben. How long would it take each of them to build an identical cabinet on his own?
C. A holding tank can be filled by three pipes (A, B, and C) in 1 hour, 1.5 hours, and 3 hours, respectively. In how many minutes can the tank be filled by the three pipes together?
D. A copy machine makes 40 copies per minute. A second copy machine can make 30 copies per minute. If the two machines work together, how long would it take them to produce 1,400 copies?

Solution:

A. Sue paints the room in 4 hours, so she does $\frac{1}{4}$ per hour.

Sam paints the room in 5 hours, so she does $\frac{1}{5}$ per hour.

They work together for t hours.

$$\frac{t}{4} + \frac{t}{5} = 1$$

$$5t + 4t = 20$$

$$9t = 20$$

$$t = 2\frac{2}{9} \text{ hours}$$

It takes Sue and Sam $2\frac{2}{9}$ hours to paint the room together.

B. $t = $ time for Abe. $2t = $ time for Ben

$$\frac{6}{t} + \frac{6}{2t} = 1$$

$$12 + 6 = 2t$$

$$18 = 2t$$

$$9 = t$$

It would take Abe 9 hours working alone, and it would take Ben 18 hours working alone.

C. $t = $ the time, in minutes, it would take the pipes together to fill the tank

$$\frac{t}{60} + \frac{t}{90} + \frac{t}{180} = 1$$

$$3t + 2t + t = 180$$

$$6t = 180$$

$$t = 30$$

It takes the three pipes 30 minutes to fill the tank together.

D. $t = $ the number of minutes together

$$40t + 30t = 1,400$$

$$70t = 1,400$$

$$t = 20$$

It takes the two copiers 20 minutes to produce the 1,400 copies.

When it takes person A x hours to do a task alone and it takes person B y hours to do the same task alone, then the time it will take them together will be more than half of the faster time but less than half of the slower time. When you have a multiple-choice question that has two people or machines working together, use the fact that one-half of the faster time $<$ together time $<$ one-half of the slower time to eliminate some of the given answer choices.

▆▆ 9.27 PRACTICE PROBLEMS

1. A train left Los Angeles at 8 a.m. headed toward San Francisco. At the same time, another train left San Francisco headed toward Los Angeles. The first train travels at 60 mph, and the second train travels at 50 mph. If the distance between Los Angeles and San Francisco is 440 miles, how long would it take the trains to meet?

2. Two planes left Chicago at the same time, one headed for New York and the other headed in the opposite direction for Denver. The plane headed for New York traveled at 600 mph, while the plane headed for Denver traveled at 150 mph. How long did each plane fly before they were 900 miles apart?

3. Two cars left Sioux Falls headed toward Fargo. The first car left at 7 a.m., traveling at 60 mph. The second car left at 8 a.m., traveling at 70 mph. How long would it take the second car to overtake the first car?

4. Carlos can run a mile in 6 minutes, and Kevin can run a mile in 8 minutes. If Carlos gives Kevin a 1-minute head start, how far will Kevin run before Carlos passes him?

5. Bev can dig a ditch in 4 hours, and Jan can dig a ditch in 3 hours. If Bev and Jan dig a ditch together, how long would it take?

6. David can type a paper in 5 hours, but when he works with Jim it only takes 2 hours. How long would it take Jim to type the paper alone?

7. Kim can paint a barn in 5 days, and Alyssa can paint it in 8 days. Kim and Alyssa start painting a barn together, but after 2 days Alyssa gets sick

and Kim finishes the work alone. How long did it take Kim to finish painting the barn?

8. Jay can prune an orchard in 50 hours, and Ray can prune the orchard in 40 hours. How long will it take them, working together, to prune the orchard?

▬▬ 9.28 SOLUTIONS

1. LA train: 60 mph, t hours, traveled $60t$ miles
   SF train: 50 mph, t hours, traveled $50t$ miles
   total miles traveled: 440 miles

   $$60t + 50t = 440$$
   $$110t = 440$$
   $$t = 4 \text{ hours}$$

2. NY plane: 600 mph, t hours, traveled $600t$ miles
   Denver plane: 150 mph, t hours, traveled $150t$ miles
   total miles traveled: 900 miles

   $$600t + 150t = 900$$
   $$750t = 900$$
   $$t = 1.2 \text{ hours}$$

3. First car: 60 mph, t hours, traveled $60t$ miles
   Second car: 70 mph, $(t-1)$ hours, traveled $70(t-1)$ miles
   Second car overtook first car, so distances are equal.

   $$70(t - 1) = 60t$$
   $$70t - 70 = 60t$$
   $$10t = 70$$
   $$t = 7$$
   $$t - 1 = 6 \text{ hours for second car to overtake first car}$$

4. Carlos: 1 mile in 6 minutes, t minutes, $\dfrac{t}{6}$ miles run
   Kevin: 1 mile in 8 minutes, $t + 1$ minutes, $\dfrac{t + 1}{8}$ miles run

Carlos overtakes Kevin so their distances are equal.

$$\frac{t}{6} = \frac{t + 1}{8}$$
$$8(t) = 6(t + 1)$$
$$8t = 6t + 6$$
$$2t = 6$$
$$t = 3 \text{ minutes for Carlos to overtake Kevin}$$

5. Bev: 4 hours, $\dfrac{1}{4}$ per hour

   Jan: 3 hours, $\dfrac{1}{3}$ per hour

   $t =$ hours worked together

   $$\frac{t}{4} + \frac{t}{3} = 1$$
   $$3t + 4t = 12$$
   $$7t = 12$$
   $$t = 1\frac{5}{7} \text{ hours together}$$

6. David: 5 hours, $\dfrac{1}{5}$ per hour

   Jim: t hours, $\dfrac{1}{t}$ per hour

   together: 2 hours

   $$\frac{2}{5} + \frac{2}{t} = 1$$
   $$2t + 10 = 5t$$
   $$-3t = -10$$
   $$t = 3\frac{1}{3} \text{ hours or 3 hours and 20 minutes}$$

7. Kim: 5 days, $\frac{1}{5}$ per day

Alyssa: 8 days, $\frac{1}{8}$ per day

2 days together, x more days for Kim alone

$$\frac{2}{8} + \frac{2+x}{5} = 1$$

$$10 + 16 + 8x = 40$$

$$8x = 14$$

$$x = 1.75 \text{ days}$$

8. Jay: 50 hours, $\frac{1}{50}$ per hour

Ray: 40 hours, $\frac{1}{40}$ per hour

$t = $ the time working together

$$\frac{t}{50} + \frac{t}{40} = 1$$

$$4t + 5t = 200$$

$$9t = 200$$

$$t = 22\frac{2}{9} \text{ hours}$$

■■ 9.29 PERCENTAGE

A **percent** means parts per hundred. The symbol for percent is %. Percentages are used to indicate such things as tax rates, discount rates, and interest rates.

To convert from a percent to a decimal, drop the percent sign and move the decimal point two places to the left. When necessary, add zeros on the left of the percent to allow for the correct location of the decimal point.

To convert from a decimal to a percent, move the decimal point to the right, then add the percent sign. When necessary, add zeros on the right of the decimal to allow for the correct location of the decimal point.

When a percent is to be converted to a fraction, first convert the percent to a decimal, then convert the decimal to a fraction. Similarly, to convert a fraction to a percent, first change the fraction to a decimal, and then change the decimal to a percent.

Example:

Convert these percentages to decimals.

A. 40.5%   B. 3%   C. 17%   D. 7%
E. 1.25%

Solution:

A. 40.5% = 0.405
B. 3% = 0.03
C. 17% = 0.17
D. 7% = 0.07
E. 1.25% = 0.0125

Example:

Convert these decimals to percentages.

A. 0.28   B. 0.4   C. 1.2   D. 0.354
E. 1.32

Solution:

A. $0.28 = 28\%$
B. $0.4 = 40\%$
C. $1.2 = 120\%$
D. $0.354 = 35.4\%$
E. $1.32 = 132\%$

Example:

Convert these fractions to percentages.

A. $\dfrac{1}{4}$ B. $\dfrac{4}{5}$ C. $\dfrac{2}{3}$ D. $\dfrac{7}{20}$ E. $\dfrac{1}{8}$

Solution:

A. $\dfrac{1}{4} = 0.25 = 25\%$

B. $\dfrac{4}{5} = 0.8 = 80\%$

C. $\dfrac{2}{3} = 0.666\ldots = 66.7\%$

D. $\dfrac{7}{20} = 0.35 = 35\%$

E. $\dfrac{1}{8} = 0.125 = 12.5\%$

Example:

Convert these percentages to fractions.

A. 95%   B. 75%   C. 33.3%   D. 16%   E. 50%

Solution:

A. $95\% = 0.95 = \dfrac{95}{100} = \dfrac{19}{20}$

B. $75\% = 0.75 = \dfrac{75}{100} = \dfrac{3}{4}$

C. $33.3\% = 0.333 = \dfrac{333}{1,000} \left(\text{about} \dfrac{1}{3}\right)$

D. $16\% = 0.16 = \dfrac{16}{100} = \dfrac{4}{25}$

E. $50\% = 0.50 = \dfrac{50}{100} = \dfrac{1}{2}$

▬▬ 9.30 PRACTICE PROBLEMS

1. Change each decimal to a percent.

   A. 0.01   B. 0.28   C. 0.6   D. 1.39   E. 1.2

2. Change each fraction to a percent.

   A. $\dfrac{1}{6}$ B. $\dfrac{7}{100}$ C. $\dfrac{27}{50}$ D. $\dfrac{11}{25}$ E. $\dfrac{67}{100}$

3. Change each percent to a decimal.

   A. 6%   B. 16%   C. 27%   D. 40%   E. 134%

4. Change each percent to a fraction.

   A. 9%   B. 38%   C. 81%   D. 148%   E. 70%

▬▬ 9.31 SOLUTIONS

1. A. $0.01 = 1\%$
   B. $0.28 = 28\%$
   C. $0.6 = 60\%$
   D. $1.39 = 139\%$
   E. $1.2 = 120\%$

2. A. $\dfrac{1}{6} = 0.1666\ldots = 16.7\%$

   B. $\dfrac{7}{100} = 0.07 = 7\%$

   C. $\dfrac{27}{50} = 0.54 = 54\%$

   D. $\dfrac{11}{25} = 0.44 = 44\%$

   E. $\dfrac{67}{100} = 0.67 = 67\%$

3. A. $6\% = 0.06$
   B. $16\% = 0.16$
   C. $27\% = 0.27$
   D. $40\% = 0.40 = 0.4$
   E. $134\% = 1.34$

4. A. $9\% = 0.09 = \dfrac{9}{100}$

   B. $38\% = 0.38 = \dfrac{38}{100} = \dfrac{19}{50}$

   C. $81\% = 0.81 = \dfrac{81}{100}$

   D. $148\% = 1.48 = \dfrac{148}{100} = \dfrac{37}{25}$

   E. $70\% = 0.70 = \dfrac{70}{100} = \dfrac{7}{10}$

▬▬ 9.32 PERCENTAGE WORD PROBLEMS

Basic percent problems relate these values: the number, N; the part of the number, W; and the percentage, P. You usually change the percentage to a decimal.

Type 1: Find a percent of a number. $P \times N = W$

Type 2: Find what percent one number is of another number. $W \div N = P$

Type 3: Find a number when a percent of it is known. $W \div P = N$

Example:

Solve these Type 1 percentage problems.

A. What number is 24% of 52?
B. What number is 63% of 75?
C. What number is 4% of 462?
D. What number is 120% of 45?
E. What number is 2.5% of 840?

Solution:

A. $P \times N = W$ $P = 24\% = 0.24$ $N = 52$
    $0.24 \times 52 = 12.48$ $W = 12.48$
B. $P \times N = W$ $P = 63\% = 0.63$ $N = 75$
    $0.63 \times 75 = 47.25$ $W = 47.25$
C. $P \times N = W$ $P = 4\% = 0.4$ $N = 462$
    $0.04 \times 462 = 18.48$ $W = 18.48$
D. $P \times N = W$ $P = 120\% = 1.2$ $N = 45$
    $1.2 \times 45 = 54.0$ $W = 54$
E. $P \times N = W$ $P = 2.5\% = 0.025$ $N = 840$
    $0.025 \times 840 = 21.000$ $W = 21$

Example:

Solve these Type 2 percentage problems.

A. 3 is what percent of 5?
B. 36 is what percent of 96?
C. What percent of 8 is 5?
D. What percent of 90 is 81?
E. What percent of 24 is 54?

Solution:

A. $W \div N = P$ $W = 3$ $N = 5$
    $3 \div 5 = 0.6$ $P = 0.6 = 60\%$
B. $W \div N = P$ $W = 36$ $N = 96$
    $36 \div 96 = 0.375$ $P = 0.375 = 37.5\%$
C. $W \div N = P$ $W = 5$ $N = 8$
    $5 \div 8 = 0.625$ $P = 0.625 = 62.5\%$
D. $W \div N = P$ $W = 81$ $N = 90$
    $81 \div 90 = 0.9$ $P = 0.9 = 90\%$
E. $W \div N = P$ $W = 54$ $N = 24$
    $54 \div 24 = 2.25$ $P = 2.25 = 225\%$

Example:

Solve these Type 3 percentage problems.

A. 45% of what number is 90?
B. 31% of what number is 279?
C. 18 is 60% of what number?
D. 175% of what number is 42?
E. 78 is 156% of what number?

Solution:

A. $W \div P = N$ $W = 90$ $P = 45\% = 0.45$
    $90 \div 0.45 = 200$ $N = 200$
B. $W \div P = N$ $W = 279$ $P = 31\% = 0.31$
    $279 \div 0.31 = 900$ $N = 900$
C. $W \div P = N$ $W = 18$ $P = 60\% = 0.6$
    $18 \div 0.6 = 30$ $N = 30$
D. $W \div P = N$ $W = 42$ $P = 175\% = 1.75$
    $42 \div 1.75 = 24$ $N = 24$

E. $W \div P = N$ $W = 78$ $P = 156\% = 1.56$
   $78 \div 1.56 = 50$ $N = 50$

Many consumer problems involve an add-on that is a percent of the price of a purchase. An item that costs $100 when there is a 7% sales tax costs $100 + 0.07($100) = $107. For these types of problems, use the relationship: Price + Add-on = Total Cost. An add-on can be sales tax, a tip, a commission paid, shipping cost, or luxury tax.

Example:

Solve these percent problems.

A. A store owner paid $60 for a hat and plans to mark up the price by 12%. What would be the selling price for the hat?
B. Jane earns $10.50 per hour. Her boss gave her a 5% raise. What is Jane's new hourly rate of pay?
C. Ralph bought a coat for $39.95 and paid a sales tax of 6% of his purchase. What was the total price of the coat?

Solution:

A. $60 + 0.12($60) = S
   $60 + $7.20 = S
   $67.20 = S
   The selling price of the hat is $67.20.

B. $10.50 + 0.05($10.50) = S
   $10.50 + $0.525 = S
   $11.025 = S
   $11.03 = S   rounded to the nearest cent
   Jane's new hourly wage is $11.03.

C. $39.95 + 0.06($39.95) = C
   $39.95 + $2.397 = C
   $42.347 = C
   $42.35 = C   rounded to the nearest cent
   The total cost of the coat is $42.35.

Some consumer problems involve a reduction based on the selling price of an item. These problems are often discount problems where the selling price is based on the original price reduced by a percentage of itself.
   Original Price − Discount = Selling Price

Example:

Solve these percent problems.

A. A car has a sticker price of $26,410. The customer bargained for 5.5% off. How much did the car sell for?
B. A lamp was usually sold for $60. It was discounted 20%. What was the sale price of the lamp?
C. A table was sold at a 40% discount sale for $150. What was the original price of the table?

Solution:

A. Original price − discount = sale price
$$\$26{,}410 - 0.055(\$26{,}410) = S$$
$$\$26{,}410 - \$1452.55 = S$$
$$\$24{,}957.45 = S$$
The selling price of the car was $24,957.45.

B. $\$60 - 0.2(\$60) = S$
$$\$60 - \$12 = S$$
$$\$48 = S$$
The sale price of the lamp is $48.

C. Original price = 100% of P Discount = 40% of P
Sale price = $1.00P - 0.40P$
$$0.6P = \$150$$
$$P = \$150 \div 0.6$$
$$P = \$250$$
The original price of the table is $250.

In some problems more than one percentage is used. These can be multiple discounts, sales tax and a tip, and a mark-up and then a discount, among many others.

Example:

Solve these percent problems.

A. A vase was purchased at a wholesale warehouse for $120. The vase was marked up 40% in a retail store. The store owner sold the vase at a 25% discount sale. What was the selling price of the vase?

B. In a restaurant, a meal costs $32.75. There is a 6% tax to be added on before you get the bill. If you leave the wait staff a 20% tip, what was the total cost of a meal?

C. The workers at a company took a 15% cut in pay in 1999 and received a 15% pay raise in 2000. If a worker made $19.40 per hour before the 1999 pay cut, how much would this worker be earning after the 2000 pay raise?

Solution:

A. Cost = $120
Retail price = $120 + 0.4($120) = $120 + $48 = $168
Sale price = $168 − 0.25($168) = $168 − $42 = $126
The selling price of the vase is $126.

B. Cost of meal = $32.75
Meal with tax = $32.75 + 0.06($32.75) = $32.75 + $1.965 = $34.715 = $34.72
Meal with tax and tip: = $34.72 + 0.2($34.72) = $34.72 + $6.944 = $41.664 = $41.66.
The total cost of the meal is $41.66.

C. Original pay rate = $19.40
Pay after 1999 cut = $19.40 − 0.15($19.40) = $19.40 − $2.91 = $16.49
Pay after 2000 raise = $16.49 + 0.15($16.49) = $16.49 + $2.4735 = $18.9635 = $18.96
The worker's pay after the raise in 2000 would be $18.96.

■■■ 9.33 PRACTICE PROBLEMS

1. Solve these percentage problems.
   A. 16 is 40% of what number?
   B. What percent of 1,032 is 645?
   C. What number is 120% of 216?
   D. 0.4% of what number is 2?
   E. 28 is what percent of 25?

2. A. What would Sharon's salary be after a 4.5% pay raise, if she makes $25,000 per year now?
   B. Jack weighed 65 kg in January and by July he had lost 13 kg. What percent of his body weight did he lose?
   C. Three hundred and ninety-six students in the freshman class at City College are from Iowa. If 18% of the freshman class is from Iowa, how large is the freshman class at City College?
   D. A real estate agent receives a 6% commission on the sale price of the property. How much would the real estate agent earn on the sale of a lot for $38,000?
   E. Thrifty Mart marks up the price of greeting cards by 42%. If a card costs Thrifty Mart $2, what would the selling price of the card be?

3. A. Deuce Hardware plans a Fourth of July sale in which all garden tools will be marked down 22%. What would be the sale price of a wheelbarrow that regularly sells for $50?
   B. A paring knife has a wholesale price of $1.20. If it is marked up 30%, what would be the retail price for the knife?
   C. A suit sold for $150 at a 25% off sale. What was the original price of the suit?
   D. You select items with a total price of $38.90. If the sales tax is 5%, what is the total cost of these items?
   E. A table has a list price of $1,200. The store discounted the table 30%, but it did not sell. At a clearance sale, they took 20% off the discounted price. What was the price of the table at the clearance sale?
   F. You select a meal with a price of $35. There is a tax of 6% on the meal. If you leave a 20% tip, what is the total cost of the meal?
   G. Jerry Smith was hired at an annual salary of $26,500. He receives a raise of 11% in January, then another 6% raise in June. What is his salary after the raise in June?

■■■ 9.34 SOLUTIONS

1. A. $40\% = 0.4$, $16 \div 0.4 = 40$ $N = 40$
   B. $645 \div 1,032 = 0.625$, $0.625 = 62.5\%$
      $P = 62.5\%$
   C. $1.20(216) = 259.2$ $W = 259.2$
   D. $2 \div 0.004 = 500$ $N = 500$
   E. $28 \div 25 = 1.12$ $1.12 = 112\%$ $P = 112\%$

2. A. $\$25,000 + 0.045(\$25,000) = S$
      $\$25,000 + \$1,125 = S$
      $\$26,125 = S$
      Sharon's salary after the raise is $26,125.
   B. $13 \div 65 = 20\%$
      Jack lost 20% of his body weight.
   C. $396 \div 0.18 = 2,200$
      There are 2,200 freshmen at City College.
   D. $0.06(\$38,000) = \$2,280$
      The real estate agent earned a commission of $2,280.
   E. $\$2 + 0.42(\$2) = S$
      $\$2 + \$0.84 = S$
      $\$2.84 = S$
      Thrifty Mart would sell the card for $2.84.

3. A. $\$50 - 0.22(\$50) = S$
      $\$50 - \$11 = S$
      $\$39 = S$
      The sale price of the wheelbarrow is $39.
   B. $\$1.20 + 0.3(\$1.20) = S$
      $\$1.20 + \$0.36 = S$
      $\$1.56 = S$
      The paring knife would sell for $1.56.
   C. $100\% - 25\% = 75\%$. The sale price is 75% of the original price.
      $\$150 \div 0.75 = \$200$
      The original price of the suit was $200.
   D. $\$38.90 + 0.05(\$38.90) = C$
      $\$38.90 + \$1.945 = C$
      $\$40.845 = C$
      $\$40.85 = C$
      The total cost of the items is $40.85.
   E. List price: $1,200
      Discount price: $\$1,200 - 0.3(\$1,200) = \$1,200 - \$360 = \$840$
      Clearance price: $\$840 - 0.2(\$840) = \$840 - \$168 = \$672$
      At the clearance sale, the table has a price of $672.

F. Meal = $35
Meal plus tax = $35 + 0.06($35) = $35 + $2.10
= $37.10
Meal plus tax and tip: $37.10 +
0.2($37.10) = $37.10 + $7.42 = $44.52
The total cost of the meal was $44.52.

G. Starting salary: $26,500
Salary after first raise: $26,500 +
0.11($26,500) = $26,500 + $2,915 =
$29,415
Salary after second raise: $29,415 +
0.06($29,415) = $29,415 + $1,764.90 =
$31,179.90
His salary after the raise in June was
$31,179.90.

9.35 TYPES OF AVERAGES

The **average**, also called the *arithmetic mean*, is the sum of a set of values divided
by the number of values. The average (arithmetic mean) of the values 8, 16, 4,
12, and 10 is found by finding the SUM = 8 + 16 + 4 + 12 + 10 = 50 and dividing
by the number of values, $N = 5$.
    AVE = SUM $\div$ N. So AVE = 50 $\div$ 5 = 10.

Example:

Find the average for each set of data.

A. 104, 114, 124, 134
B. 27, 22, 17, 37, 22
C. 4, 2, 24, 14, 34, 8, 10, 20
D. 5, 1, 7, 3, 1
E. 25, 2, 5, 6, 5, 23, 22, 7, 10, 15, 21, 23

Solution:

A. SUM = 104 + 114 + 124 + 134 = 476 $N = 4$
    AVE = SUM $\div$ N = 476 $\div$ 4 = 119
B. SUM = 27 + 22 + 17 + 37 + 22 = 125 $N = 5$
    AVE = SUM $\div$ N = 125 $\div$ 5 = 25
C. SUM = 4 + 2 + 24 + 14 + 34 + 8 + 10 + 20 = 116 $N = 8$
    AVE = SUM $\div$ N = 116 $\div$ 8 = 14.5
D. SUM = 5 + 1 + 7 + 3 + 1 = 17 $N = 5$
    AVE = SUM $\div$ N = 17 $\div$ 5 = 3.4
E. SUM = 25 + 2 + 5 + 6 + 5 + 23 + 22 + 7 + 10 + 15 + 21 + 23 = 164 $N = 12$
    AVE = SUM $\div$ N = 164 $\div$ 12 = 13.666 . . . = 13.7

If all values but one are known and the average is also known, then the last
value can be determined precisely. Suppose that 8, 16, 4, and 10 are four of
five values that have an average of 10. If five values have a mean of 10, then
the sum of the values must be 5(10) = 50. The sum of the four known values is
8 + 16 + 4 + 10 = 38. The missing value is 50 − 38 = 12.

Example:

Find the missing value.

A. If the average of six values is 28 and five of the six values are 29, 19, 23, 20,
    and 43, what is the sixth value?
B. If the average of five numbers is 41 and the numbers include 46, 35, 38, and
    41, what is the missing number?

C. If Sara scored 77, 89, 98, 97, 99, and 91 on the first six tests, what does she need to score on the seventh test to get an average of 93 for the seven tests?
D. If Joe scored 77, 80, and 81 on three quizzes, what does he need to score on the next quiz to have an 82 average?

Solution:

A. $6(28) = 168$ $29 + 19 + 23 + 20 + 43 = 134$
   $168 - 134 = 34$
   The sixth number is 34.
B. $5(41) = 205$ $46 + 35 + 38 + 41 = 160$
   $205 - 160 = 45$
   The missing number is 45.
C. $7(93) = 651$ $77 + 89 + 98 + 97 + 99 + 91 = 551$
   $651 - 551 = 100$
   She needs to score 100 on the seventh test.
D. $4(82) = 328$ $77 + 80 + 81 = 238$
   $328 - 238 = 90$
   He needs to score 90 on the fourth quiz.

The **mode** of a set of values is the value that occurs most often. If all values occur the same number of times, there is no mode. If two or more values occur with the greatest frequency, then each of the values is a mode.

Example:

Find the mode for each set of values.

A. 2, 8, 7, 6, 4, 3, 8, 5, 1
B. 3, 7, 2, 6, 4, 1, 6, 5, 7, 9
C. 8, 7, 2, 5, 9. 4, 1, 3
D. 8, 7, 6, 9, 4, 18, 6, 3, 2, 6, 7

Solution:

A. Because 8 occurs twice and all other values occur once, 8 is the mode.
B. Both 6 and 7 occur twice, so 6 and 7 are both modes.
C. All the values occur once, so there is no mode.
D. Because 6 occurs three times and no other value occurs more than twice, 6 is the mode.

The **median** is the middle value, or the average of the two middle values, when the values are arranged in order from least to greatest. The median of 1, 3, 4, 7, 10 is 4 since it is the middle value in the ordered values. The median of 1, 3, 4, 7, 10, 20 is the average of the two middle values 4 and 7, $(4 + 7) \div 2 = 5.5$.

Example:

Find the median for each set of values.

A. 24, 6, 7, 23, 13, 12, 18
B. 17, 15, 9, 13, 21, 32, 41, 7, 12
C. 147, 159, 132, 181, 174, 253

D. 74, 81, 39, 74, 82, 74, 80, 100, 74, 42
E. 11, 38, 73, 91, 16, 51, 39

Solution:

A. The ordered values are 6, 7, 12, 13, 18, 23, 24.
   The middle value of the seven values is 13.
   The median is 13.
B. The ordered values are 7, 9, 12, 13, 15, 17, 21, 32, 41.
   The middle value of the nine values is 15.
   The median is 15.
C. The ordered values are 132, 147, 159, 174, 181, 253.
   The middle two values of the six values are 159 and 174.
   The average of the two middle values is $(159 + 174) \div 2 = 166.5$.
   The median is 166.5.
D. The ordered values are 39, 42, 74, 74, 74, 74, 80, 81, 82, 100.
   The middle two values of the 10 values are 74 and 74.
   The median is 74 since the average of 74 and 74 is 74.
E. The ordered values are 11, 16, 38, 39, 51, 73, 91.
   The middle value of the seven values is 39.
   The median is 39.

When you want the average of two or more averages, you must weight each average with the number of values in the average and then divide the sum of the **weighted averages** by the sum of the numbers for each of the averages. If A is the average of h values, B is the average of i values, and C is the average of j values, then the combined average is the sum of the weighted averages divided by the sum of the weights. SUM $= hA + iB + jC$ and $N = h + i + j$. AVE $= (hA + iB + jC) \div (h + i + j)$.

Example:

Find the weighted averages.

A. The average score on a test for 400 students in Crawford County is 650. The average score on the same test for 600 students in Edwards County is 680. What would be the combined average for the students in Crawford and Edwards Counties?
B. Miss Jackson has 30 students in first period who averaged 85 on the State Algebra Test, 24 students in second period who averaged 90 on the Algebra Test, and 22 students in fifth period who averaged 82 on the Algebra Test. What was the average for these three classes?
C. The average traffic fine was $90 on Monday for the 18 people ticketed, the average fine was $76 on Tuesday for the 20 people ticketed, and the average fine was $80 on Wednesday for the 24 people ticketed. What was the average fine for the 3 days?

Solution:

A. SUM $= 400(650) + 600(680) = 260,000 + 408,000 = 668,000$
   $N = 400 + 600 = 1,000$
   AVE $= 668,000 \div 1,000 = 668$
B. SUM $= 30(85) + 24(90) + 22(82) = 2,550 + 2,160 + 1,804 = 6,514$
   $N = 30 + 24 + 22 = 76$
   AVE $= 6,514 \div 76 = 85.7105 = 85.7$

C. SUM $= 18(\$90) + 20(\$76) + 24(\$80) = \$1{,}620 + \$1{,}520 + \$1{,}920 = \$5{,}060$
$N = 18 + 20 + 24 = 62$
AVE $= \$5{,}060 \div 62 = \$81.6129 = \$81.61$

When consecutive integers are averaged, the average will always equal the average of the first and last numbers in the sequence. The average of 5, 6, 7, 8, and 9 is $35 \div 5 = 7$ in the traditional way and $(5 + 9) \div 2 = 7$ in the way for consecutive integers.

Example:

Find the average of these sequences of consecutive integers.

A. 1, 2, 3, 4, 5, 6, 7, 8
B. 7, 8, 9, 10, 11, 12
C. 25, 26, 27, 28, 29

Solution:

A. AVE $= (1 + 8) \div 2 = 9 \div 2 = 4.5$
B. AVE $= (7 + 12) \div 2 = 19 \div 2 = 9.5$
C. AVE $= (25 + 29) \div 2 = 54 \div 2 = 27$

The **range** is the easiest way to describe how a set of data spreads out. To compute the range, R, subtract the smallest value, Min, from the greatest value, Max. Thus, $R = Max - Min$.

Example:

Find the range for each set of values.

A. 2, 8, 7, 6, 4, 3, 8, 5, 1
B. 3, 7, 2, 6, 4, 7, 6, 5, 7, 9
C. 8, 7, 6, 5, 12, 4, 5, 3
D. 8, 7, 6, 9, 4, 18, 6, 3, 2, 6, 7

Solutions:

A. Max $= 8$ and Min $= 1$, so $R = Max - Min = 8 - 1 = 7$.
B. Max $= 9$ and Min $= 2$, so $R = Max - Min = 9 - 2 = 7$.
C. Max $= 12$ and Min $= 3$, so $R = Max - Min = 12 - 3 = 9$.
D. Max $= 18$ and Min $= 2$, so $R = Max - Min = 18 - 2 = 16$.

9.36 PRACTICE PROBLEMS

1. Find the average for each set of values. Round approximate answers to the nearest tenth.
    A. 12, 18, 16, 10, 6, 14, 17, 18, 20, 13
    B. 11, 14, 28, 36, 10, 16, 20
    C. 40, 32, 17, 38, 16, 40, 53, 70
    D. 22, 29, 30, 21, 20, 24, 25, 20, 22
    E. 120, 130, 70, 90, 50

2. Find the weighted average.
    A. Class A had 40 students and an average of 78. Class B had 25 students with an average of 72. What is the average for these two classes together?
    B. In January, 32 people bought car tags at an average cost of $97.40, and in February, 48 people bought car tags at an average cost of $87.20.

What was the average cost of car tags for these 2 months?

C. Last month Paul's five grocery bills averaged $52.70. This month Paul's four grocery bills averaged $67.50. What was Paul's average grocery bill for the last 2 months?

3. Find the median for each set of values.
   A. 7, 18, 26, 43, 76, 10, 26, 40
   B. 9, 36, 24, 85, 72, 500
   C. 29, 37, 14, 65, 71, 13, 24
   D. 48, 10, 17, 46, 97, 3, 5, 81, 140

4. Find the missing value.
   A. John is trying to average 300 miles each day on his trip. If he traveled 306 miles the first day, 284 miles the second day, and 292 miles the third day, how many miles must he travel the next day to reach his average for the 4 days?
   B. Larry has to have an average bowling score of 240 to maintain his rank on the bowling team. If he has scores of 260, 210, 250, and 230, what must he score on his next game to be able to keep his rank?

C. Susan has scored 90, 78, 96, 94, 88, and 100 on her first six tests of the semester. If she wants to earn a 90 average for the seven tests in the semester, what does she need to score on the last test?

5. Find the average of these sequences of consecutive integers.
   A. 1, 2, 3, 4, 5, 6, 7, 8, 9
   B. 501, 502, 503, 504, 505, 506
   C. 298, 299, 300, 301, 302, 303, 304, 305, 306
   D. 37, 38, 39, 40, 41, 42, 43, 44, 45, 46
   E. 632, 633, 634, 635, 636, 637, 638

6. Find the mode for each set of data.
   A. 2, 8, 6, 3, 9, 4, 6, 5, 9
   B. 4, 3, 7, 2, 8, 9, 4, 6, 4
   C. 3, 9, 4, 11, 17, 41, 11
   D. 2, 8, 7, 13, 15, 19, 4, 21

7. Find the range for each set of data.
   A. 3, 8, 9, 16, 8, 12, 4, 11, 17, 9
   B. 16, 18, 12, 14, 16, 13, 21, 91, 21, 72

■ 9.37 SOLUTIONS

1. A. SUM = 144 $N = 10$
      AVE = $144 \div 10 = 14.4$
   B. SUM = 135 $N = 7$
      AVE = $135 \div 7 = 19.2857 = 19.3$
   C. SUM = 306 $N = 8$
      AVE = $306 \div 8 = 38.25 = 38.3$
   D. SUM = 213 $N = 9$
      AVE = $213 \div 9 = 23.666 = 23.7$
   E. SUM = 460 $N = 5$
      AVE = $460 \div 5 = 92$

2. A. Class A: 40,   AVE = 78   Class B: 25, AVE = 72
      SUM $= 40(78) + 25(72) = 3,120 + 1,800 = 4,920$
      $N = 40 + 25 = 65$
      AVE $= 4,920 \div 65 = 75.6923 = 75.7$
   B. January: 32,   AVE = $97.40   February: 48, AVE = $87.20
      SUM $= 32(\$97.40) + 48(\$87.20) = \$3,116.80 + \$4,185.60 = \$7,302.40$
      $N = 32 + 48 = 80$
      AVE $= \$7,302.40 \div 80 = \$91.28$
   C. Last month: 5, AVE = $52.70   This month: 4, AVE = $67.50
      SUM $= 5(\$52.70) + 4(\$67.50) = \$263.50 + \$270.00 = \$533.50$

$N = 5 + 4 = 9$
AVE $= \$533.50 \div 9 = \$59.27777 = \$59.28$

3. A. Ordered values: 7, 10, 18, 26, 26, 40, 43, 76
      The middle two values of the 8 values are 26 and 26.
      The median is 26.
   B. Ordered values: 9, 24, 36, 72, 85, 500
      The middle two values are 36 and 72.
      $(36 + 72) \div 2 = 54$
      The median is 54.
   C. Ordered values: 13, 14, 24, 29, 37, 65, 71
      The middle value of the 7 values is 29.
      The median is 29.
   D. Ordered values: 3, 5, 10, 17, 46, 48, 81, 97, 140
      The middle value of the 9 values is 46.
      The median is 46.

4. A. AVE = 300 $N = 4$
      SUM $= 300(4) = 1,200$. Given values: $306 + 284 + 292 = 882$
      missing value $= 1,200 - 882 = 318$
      John needs to travel 318 miles on day 4.
   B. AVE = 240 $N = 5$
      SUM $= 240(5) = 1,200$. Given values: $260 + 210 + 250 + 230 = 950$

missing value $= 1,200 - 950 = 250$
Larry needs to bowl a 250 on his next game.

C. AVE $= 90$ $N = 7$
SUM $= 90(7) = 630$. Given values: $90 + 78 +$
$96 + 94 + 88 + 100 = 546$
missing value $= 630 - 546 = 84$
She needs to score an 84 on the last test.

5. A. AVE $= (1 + 9) \div 2 = 5$
   B. AVE $= (501 + 506) \div 2 = 503.5$
   C. AVE $= (298 + 306) \div 2 = 302$
   D. AVE $= (37 + 46) \div 2 = 41.5$
   E. AVE $= (632 + 638) \div 2 = 635$

6. A. Because 6 occurs twice and 9 occurs twice, the modes are 6 and 9.
   B. Because 4 occurs three times, the mode is 4.
   C. Because 11 occurs twice, the mode is 11.
   D. Because every value occurs the same number of times, there is no mode.

7. A. Max $= 17$ and Min $= 3$, so R $=$ Max $-$ Min $=$ $17 - 3 = 14$.
   B. Max $= 91$ and Min $= 12$, so R $=$ Max $-$ Min $=$ $91 - 12 = 79$.

9.38 POWERS AND ROOTS

A term such as $5x^2$ has three parts: the 5 is the **coefficient** of x^2, the x is the **base**, and the 2 is the **exponent**. The coefficient 5 tells you that five x^2's have been added together. The 2 tells you that two factors of x were multiplied to get the x^2.

$$x^3 = x \cdot x \cdot x, \quad 5^4 = 5 \cdot 5 \cdot 5 \cdot 5, \quad 5x^2 = 5 \cdot x \cdot x, \quad x^3 y^2 = x \cdot x \cdot x \cdot y \cdot y$$

When the exponent is 1 it is not necessary to write it, so $3x^1 = 3x$. Also, when the coefficient is 1 you do not write it, so $1x^5 = x^5$. Read x^2 as "x squared," x^3 as "x cubed," x^4 as "x to the fourth," x^5 as "x to the fifth," etc. When you write $nx = x + x + x + \cdots + x$, there are n terms of x added together. When you write $x^n = x \cdot x \cdot x \cdot \ldots \cdot x$, there are n factors of x multiplied together.

Laws of Exponents

If x, a, and b are real numbers, then the following laws hold.

1. $x^a \cdot x^b = x^{a+b}$
2. $(x^a)^b = x^{ab}$
3. $\dfrac{x^a}{x^b} = x^{a-b}$ $x \neq 0$
4. $(xy)^a = x^a \times y^a$
5. $\left(\dfrac{x}{y}\right)^a = \dfrac{x^a}{y^a}$ $y \neq 0$
6. $x^{-a} = \dfrac{1}{x^a}$ $a > 0, x \neq 0$
7. $x^0 = 1$ $x \neq 0$

Example:

Use the laws of exponents to simplify each expression.

A. $(3x)^2$ B. $(-4y)^2$ C. 2^4 D. 5^0 E. $\left(\dfrac{2}{3}\right)^3$

Solution:

A. $(3x)^2 = 3x \cdot 3x = 3 \cdot 3 \cdot x \cdot x = 9x^2$ or $(3x)^2 = 3^2 x^2 = 9x^2$
B. $(-4y)^2 = (-4y)(-4y) = (-4)(-4)y \cdot y = 16y^2$
C. $2^4 = 2 \cdot 2 \cdot 2 \cdot 2 = 16$

D. $5^0 = 1$

E. $\left(\dfrac{2}{3}\right)^3 = \dfrac{2^3}{3^3} = \dfrac{8}{27}$

Example:

A. $x^{-2} * x^5$ B. $x^2 * x^5$ C. $3^4 * 3^9$ D. 2^{-2} E. $(x^2)^3$

Solution:

A. $x^{-2} \cdot x^5 = x^{-2+5} = x^3$
B. $x^2 \cdot x^5 = x^{2+5} = x^7$
C. $3^4 \cdot 3^9 = 3^{4+9} = 3^{13}$

D. $(2)^{-2} = \dfrac{1}{2^2} = \dfrac{1}{4}$

E. $(x^2)^3 = x^{2 \cdot 3} = x^6$ or $(x^2)^3 = x^2 \cdot x^2 \cdot x^2 = x^6$

Scientific notation is a way to write very large or very small numbers by rewriting them as numbers between 1 and 10 times a power of 10. 265,000 is written as 2.65×10^5 and 0.000148 is written as 1.48×10^{-4}.

Example:

Write each number in scientific notation.

A. 2,541,000   B. 360,000   C. 0.00045   D. 0.0000157

Solution:

A. $2,541,000 = 2_\wedge 541000 = 2.541 \times 10^6$
B. $360,000 = 3_\wedge 60000 = 3.6 \times 10^5$
C. $0.00045 = 0.0004_\wedge 5 = 4.5 \times 10^{-4}$
D. $0.0000157 = 0.00001_\wedge 57 = 1.57 \times 10^{-5}$

Use exponents to write numbers in expanded notation. $265 = 2 \times 10^2 + 6 \times 10^1 + 5 \times 10^0$, $0.45 = 4 \times 10^{-1} + 5 \times 10^{-2}$, $24.3 = 2 \times 10^1 + 4 \times 10^0 + 3 \times 10^{-1}$.

Example:

Write each number in expanded notation.

A. 3,472   B. 0.742   C. 21.45   D. 3.568

Solution:

A. $3,472 = 3 \times 10^3 + 4 \times 10^2 + 7 \times 10^1 + 2 \times 10^0$
B. $0.742 = 7 \times 10^{-1} + 4 \times 10^{-2} + 2 \times 10^{-3}$
C. $21.45 = 2 \times 10^1 + 1 \times 10^0 + 4 \times 10^{-1} + 5 \times 10^{-2}$
D. $3.568 = 3 \times 10^0 + 5 \times 10^{-1} + 6 \times 10^{-2} + 8 \times 10^{-3}$

If n is a positive integer, and if a and b are such that $a^n = b$, then a is the nth root of b. $\sqrt[n]{b} = a$. $\sqrt{a}$ is the square root of a and $\sqrt[3]{a}$ is the cube root of a. Fractional exponents can indicate roots. $\sqrt{a} = a^{1/2}$, $\sqrt[3]{a} = a^{1/3}$, $\sqrt[n]{a} = a^{1/n}$.

Example:

Simplify these radicals.

A. $\sqrt{x^2}$ B. $\sqrt{16}$ C. $\sqrt{16x^2}$ D. $\sqrt[3]{x^6}$ E. $\sqrt[3]{125}$

Solution:

A. $\sqrt{x^2} = x$
B. $\sqrt{16} = 4$
C. $\sqrt{16x^2} = \sqrt{4^2 x^2} = \sqrt{(4x)^2} = 4x$
D. $\sqrt[3]{x^6} = \sqrt[3]{(x^2)^3} = x^2$
E. $\sqrt[3]{125} = \sqrt[3]{5 \cdot 5 \cdot 5} = \sqrt[3]{5^3} = 5$

Table of Squares and Cubes

Number	Square	Cube
1	1	1
2	4	8
3	9	27
4	16	64
5	25	125
6	36	216
7	49	343
8	64	512
9	81	729
10	100	1,000
11	121	
12	144	
13	169	
14	196	
15	225	
16	256	
17	289	
18	324	
19	361	
20	400	

In $\sqrt[n]{a}$, n is the **index**, a is the **radicand**, and $\sqrt{}$ is the **radical sign**. When radical expressions are to be added, they must have the same index and the same radicand. Radical expressions with the same index and the same radicand are said to be like expressions and can be added or subtracted by adding or subtracting their coefficients.

Example:

Add or subtract as indicated.

A. $3\sqrt{5} + 7\sqrt{5}$ B. $\sqrt{x} + 7\sqrt{x}$ C. $8\sqrt{3x} + 5\sqrt{3x}$
D. $28\sqrt{x} - 2\sqrt{x}$ E. $11\sqrt{5x} - 15\sqrt{5x}$

Solution:

A. $3\sqrt{5} + 7\sqrt{5} = (3+7)\sqrt{5} = 10\sqrt{5}$

B. $\sqrt{x} + 7\sqrt{x} = 1\sqrt{x} + 7\sqrt{x} = (1+7)\sqrt{x} = 8\sqrt{x}$

C. $8\sqrt{3x} + 5\sqrt{3x} = (8+5)\sqrt{3x} = 13\sqrt{3x}$

D. $28\sqrt{x} - 2\sqrt{x} = (28-2)\sqrt{x} = 26\sqrt{x}$

E. $11\sqrt{5x} - 15\sqrt{5x} = (11-15)\sqrt{5x} = -4\sqrt{5x}$

You can multiply radicals if they have the same index. $\sqrt[n]{a} \cdot \sqrt[n]{b} = \sqrt[n]{a \cdot b}$ when a and b are positive. $\sqrt{2} \times \sqrt{5} = \sqrt{10}$; $\sqrt{a} \times \sqrt{a} = a, a > 0$; $\sqrt{2} \times \sqrt{2} = 2$.

$$\sqrt[n]{a} \div \sqrt[n]{b} = \frac{\sqrt[n]{a}}{\sqrt[n]{b}} = \sqrt[n]{\frac{a}{b}}, a > 0, b > 0; \quad \sqrt{15} \div \sqrt{3} = \frac{\sqrt{15}}{\sqrt{3}} = \sqrt{\frac{15}{3}} = \sqrt{5}$$

Example:

Multiply or divide these radicals.

A. $\sqrt{5x} \cdot \sqrt{7y}$ B. $\sqrt{3xy} \cdot \sqrt{10z}$ C. $\sqrt{7a} \cdot \sqrt{15b}$

D. $\sqrt{21} \div \sqrt{7}$ E. $\sqrt{10x} \div \sqrt{2x}$

Solution:

A. $\sqrt{5x} \cdot \sqrt{7y} = \sqrt{5x \cdot 7y} = \sqrt{35xy}$

B. $\sqrt{3xy} \cdot \sqrt{10z} = \sqrt{3xy \cdot 10z} = \sqrt{30xyz}$

C. $\sqrt{7a} \cdot \sqrt{15b} = \sqrt{7a \cdot 15b} = \sqrt{105ab}$

D. $\sqrt{21} \div \sqrt{7} = \frac{\sqrt{21}}{\sqrt{7}} = \sqrt{\frac{21}{7}} = \sqrt{3}$

E. $\sqrt{10x} \div \sqrt{2x} = \frac{\sqrt{10x}}{\sqrt{2x}} = \sqrt{\frac{10x}{2x}} = \sqrt{5}$

Radical expressions can be simplified by finding the roots of factors.

$$\sqrt{12} = \sqrt{4 \cdot 3} = \sqrt{4} \cdot \sqrt{3} = 2\sqrt{3}$$

$$\sqrt{x^5 y^4 z^3} = \sqrt{x^4 x y^4 z^2 z} = \sqrt{x^4 y^4 z^2} \sqrt{xz} = x^2 y^2 z \sqrt{xz}$$

In square roots, look for the greatest factor that is a perfect square, and then rewrite the radicand as the product of the perfect square factor and the non-perfect square factor.

Example:

Simplify these radicals.

A. $\sqrt{80}$ B. $\sqrt{x^3 y}$ C. $\sqrt{9b^4 c^3}$ D. $\sqrt{169a^5}$ E. $\sqrt{125a^2 b^2}$

Solution:

A. $\sqrt{80} = \sqrt{16 \cdot 5} = \sqrt{16} \cdot \sqrt{5} = 4\sqrt{5}$

B. $\sqrt{x^3 y} = \sqrt{x^2 \cdot xy} = \sqrt{x^2} \cdot \sqrt{xy} = x\sqrt{xy}$

C. $\sqrt{9b^4c^3} = \sqrt{9b^4c^2 \cdot c} = \sqrt{9b^4c^2} \cdot \sqrt{c} = 3b^2c\sqrt{c}$

D. $\sqrt{169a^5} = \sqrt{13^2a^4 \cdot a} = \sqrt{13^2a^4} \cdot \sqrt{a} = 13a^2\sqrt{a}$

E. $\sqrt{125a^2b^2} = \sqrt{5^2a^2b^2 \cdot 5} = \sqrt{5^2a^2b^2} \cdot \sqrt{5} = 5ab\sqrt{5}$

A radical is simplified if it does not contain any perfect square factors or any factors that are fractions or decimals. To eliminate a fraction under the radical sign, make the denominator a perfect square. $\sqrt{\dfrac{1}{2}} = \sqrt{\dfrac{1}{2} \cdot \dfrac{2}{2}} = \sqrt{\dfrac{2}{4}} = \dfrac{\sqrt{2}}{\sqrt{4}} = \dfrac{\sqrt{2}}{2}.$

Example:

Simplify each radical expression.

A. $\sqrt{\dfrac{9}{16}}$　　　B. $\sqrt{\dfrac{2}{3}}$　　　C. $\sqrt{\dfrac{5}{6}}$　　　D. $\sqrt{\dfrac{3}{8}}$　　　E. $\sqrt{\dfrac{6}{50}}$

Solution:

A. $\sqrt{\dfrac{9}{16}} = \dfrac{\sqrt{9}}{\sqrt{16}} = \dfrac{3}{4}$

B. $\sqrt{\dfrac{2}{3}} = \sqrt{\dfrac{2}{3} \cdot \dfrac{3}{3}} = \sqrt{\dfrac{6}{9}} = \dfrac{\sqrt{6}}{\sqrt{9}} = \dfrac{\sqrt{6}}{3}$

C. $\sqrt{\dfrac{5}{6}} = \sqrt{\dfrac{5}{6} \cdot \dfrac{6}{6}} = \sqrt{\dfrac{30}{36}} = \dfrac{\sqrt{30}}{\sqrt{36}} = \dfrac{\sqrt{30}}{6}$

D. $\sqrt{\dfrac{3}{8}} = \sqrt{\dfrac{3}{8} \cdot \dfrac{2}{2}} = \sqrt{\dfrac{6}{16}} = \dfrac{\sqrt{6}}{\sqrt{16}} = \dfrac{\sqrt{6}}{4}$

E. $\sqrt{\dfrac{6}{50}} = \sqrt{\dfrac{3}{25}} = \dfrac{\sqrt{3}}{\sqrt{25}} = \dfrac{\sqrt{3}}{5}$

Example:

Simplify and combine like terms.

A. $\sqrt{27} + \sqrt{48} - \sqrt{12}$　　　B. $5\sqrt{8} - 3\sqrt{18}$　　　C. $2\sqrt{150} - 4\sqrt{54} + 6\sqrt{48}$

Solution:

A. $\sqrt{27} + \sqrt{48} - \sqrt{12} = \sqrt{9}\sqrt{3} + \sqrt{16}\sqrt{3} - \sqrt{4}\sqrt{3} = 3\sqrt{3} + 4\sqrt{3} - 2\sqrt{3} = 5\sqrt{3}$

B. $5\sqrt{8} - 3\sqrt{18} = 5\sqrt{4}\sqrt{2} - 3\sqrt{9}\sqrt{2} = 5 \cdot 2 \cdot \sqrt{2} - 3 \cdot 3 \cdot \sqrt{2} = 10\sqrt{2} - 9\sqrt{2} = \sqrt{2}$

C. $2\sqrt{150} - 4\sqrt{54} + 6\sqrt{48} = 2\sqrt{25}\sqrt{6} - 4\sqrt{9}\sqrt{6} + 6\sqrt{16}\sqrt{3} = 2 \cdot 5 \cdot \sqrt{6} - 4 \cdot 3 \cdot \sqrt{6} + 6 \cdot 4 \cdot \sqrt{3} = 10\sqrt{6} - 12\sqrt{6} + 24\sqrt{3} = -2\sqrt{6} + 24\sqrt{3}$

9.39   STANDARD DEVIATION

The **standard deviation** is another way to look at how values spread out. It focuses on how much the values differ from the mean. To compute the standard deviation, SD, find the difference between each score and the mean (AVE), and square the difference. Then compute the sum of all these squared differences (Sum Sq), divide that by the number (NUM) of values, and, finally, find the square root of that quotient.

$$SD = \sqrt{\frac{Sum\ Sq}{NUM}} = \sqrt{\frac{sum\ (X - AVE)^2}{NUM}}.$$

Example:

Find the standard deviation for each set of values.

A. 0, 5, 5, 10   B. 1, 6, 13, 20   C. 12, 12, 12, 12

Solutions:

A. $AVE = \dfrac{0 + 5 + 5 + 10}{4} = \dfrac{20}{4} = 5$

$SD = \sqrt{\dfrac{(0 - 5)^2 + (5 - 5)^2 + (5 - 5)^2 + (10 - 5)^2}{4}} = \sqrt{\dfrac{25 + 0 + 0 + 25}{4}}$

$= \sqrt{\dfrac{50}{4}} = \dfrac{\sqrt{50}}{\sqrt{4}} = \dfrac{\sqrt{50}}{2} \approx \dfrac{7}{2} = 3.5$

Note: $7^2 = 49 \approx 50,$ *so estimate* $\sqrt{50} \approx 7$

B. $AVE = \dfrac{1 + 6 + 13 + 20}{4} = \dfrac{40}{4} = 10$

$SD = \sqrt{\dfrac{(1 - 10)^2 + (6 - 10)^2 + (13 - 10)^2 + (20 - 10)^2}{4}}$

$= \sqrt{\dfrac{81 + 16 + 9 + 100}{4}} = \dfrac{\sqrt{206}}{\sqrt{4}} \approx \dfrac{14}{2} = 7$

Note: $14^2 = 196,$ *so estimate* $\sqrt{206} \approx 14$

C. $AVE = \dfrac{12 + 12 + 12 + 12}{4} = \dfrac{48}{4} = 12$

$SD = \sqrt{\dfrac{(12 - 12)^2 + (12 - 12)^2 + (12 - 12)^2 + (12 - 12)^2}{4}}$

$= \sqrt{\dfrac{0}{4}} = \sqrt{0} = 0$

Note: Only when all the values are the same will the standard deviation be 0.

▬▬ 9.40   PRACTICE PROBLEMS

1. Simplify these exponential expressions.

    A. $x^5 \cdot x^7$ B. 11^0
    C. $x^{-5} \cdot x^{11}$ D. $(2ab)(5a^2b^3)$
    E. $-3(x)^2$

2. Write each number in scientific notation.

    A. 26,150,000   B. 1,768   C. 0.00247
    D. 0.0286   E. 12.14

3. Write each number in expanded notation.

    A. 384   B. 12.4   C. 1,050   D. 0.014   E. 0.15

4. Simplify these radicals.

    A. $\sqrt{x^6}$ B. $\sqrt{64}$ C. $\sqrt{25x^2}$
    D. $\sqrt[4]{16}$ E. $\sqrt[3]{x^{12}}$

5. Add or subtract as indicated.

    A. $\sqrt{45} + \sqrt{80}$ B. $\sqrt{4x} + \sqrt{49x}$
    C. $\sqrt{192x} - \sqrt{48x}$ D. $\sqrt{605x} - \sqrt{45x}$
    E. $\sqrt{8ab} + \sqrt{50ab}$

6. Multiply or divide as indicated.

    A. $\sqrt{10x} \cdot \sqrt{6x}$ B. $\sqrt{5xy} \cdot \sqrt{2xy}$
    C. $\sqrt{7a} \cdot \sqrt{14ab}$ D. $\sqrt{48} \div \sqrt{27}$
    E. $\sqrt{125x^3} \div \sqrt{5x}$

7. Simplify these radicals.

    A. $\sqrt{\dfrac{5}{9}}$ B. $\sqrt{\dfrac{16}{3}}$ C. $\sqrt{\dfrac{25}{20}}$

    D. $\sqrt{\dfrac{5}{8}}$ E. $\sqrt{\dfrac{3}{5}}$

8. Find the standard deviation for each set of values.

    A. 8, 10, 12, 6, 4, 7, 9, 4, 3   B. 14, 10, 18, 22

▬▬ 9.41   SOLUTIONS

1. A. $x^5 \cdot x^7 = x^{5+7} = x^{12}$
    B. $11^0 = 1$
    C. $x^{-5} \cdot x^{11} = x^{-5+11} = x^6$
    D. $(2ab)(5a^2b^3) = 2 \cdot 5 \cdot a \cdot a^2 \cdot b \cdot b^3 = 10a^3b^4$
    E. $-3(x)^2 = -3x^2$

2. A. $26{,}150{,}000 = 2{\scriptstyle\wedge}6150000 = 2.615 \times 10^7$
    B. $1{,}768 = 1{\scriptstyle\wedge}768 = 1.768 \times 10^3$
    C. $0.00247 = 0.002{\scriptstyle\wedge}47 = 2.47 \times 10^{-3}$
    D. $0.0286 = 0.02{\scriptstyle\wedge}86 = 2.86 \times 10^{-2}$
    E. $12.14 = 1{\scriptstyle\wedge}2.14 = 1.214 \times 10^1$

3. A. $384 = 3 \times 10^2 + 8 \times 10^1 + 4 \times 10^0$
    B. $12.4 = 1 \times 10^1 + 2 \times 10^0 + 4 \times 10^{-1}$
    C. $1050 = 1 \times 10^3 + 5 \times 10^1$
    D. $0.014 = 1 \times 10^{-2} + 4 \times 10^{-3}$
    E. $0.15 = 1 \times 10^{-1} + 5 \times 10^{-2}$

4. A. $\sqrt{x^6} = \sqrt{(x^3)^2} = x^3$

    B. $\sqrt{64} = \sqrt{(8)^2} = 8$

    C. $\sqrt{25x^2} = \sqrt{(5x)^2} = 5x$

    D. $\sqrt[4]{16} = \sqrt[4]{2^4} = 2$

    E. $\sqrt[3]{x^{12}} = \sqrt[3]{(x^4)^3} = x^4$

5. A. $\sqrt{45} + \sqrt{80} = \sqrt{9}\sqrt{5} + \sqrt{16}\sqrt{5} = 3\sqrt{5} + 4\sqrt{5} = 7\sqrt{5}$

    B. $\sqrt{4x} + \sqrt{49x} = \sqrt{4}\sqrt{x} + \sqrt{49}\sqrt{x} = 2\sqrt{x} + 7\sqrt{x} = 9\sqrt{x}$

    C. $\sqrt{192x} - \sqrt{48x} = \sqrt{64}\sqrt{3x} - \sqrt{16}\sqrt{3x} = 8\sqrt{3x} - 4\sqrt{3x} = 4\sqrt{3x}$

    D. $\sqrt{605x} - \sqrt{45x} = \sqrt{121}\sqrt{5x} - \sqrt{9}\sqrt{5x} = 11\sqrt{5x} - 3\sqrt{5x} = 8\sqrt{5x}$

    E. $\sqrt{8ab} + \sqrt{50ab} = \sqrt{4}\sqrt{2ab} + \sqrt{25}\sqrt{2ab} = 2\sqrt{2ab} + 5\sqrt{2ab} = 7\sqrt{2ab}$

6. A. $\sqrt{10x} \cdot \sqrt{6x} = \sqrt{10x \cdot 6x} = \sqrt{60x^2} = \sqrt{4x^2}\sqrt{15} = 2x\sqrt{15}$

    B. $\sqrt{5xy} \cdot \sqrt{2xy} = \sqrt{10x^2y^2} = \sqrt{x^2y^2}\sqrt{10} = xy\sqrt{10}$

    C. $\sqrt{7a} \cdot \sqrt{14ab} = \sqrt{98a^2b} = \sqrt{49a^2}\sqrt{2b} = 7a\sqrt{2b}$

    D. $\sqrt{48} \div \sqrt{27} = \dfrac{\sqrt{48}}{\sqrt{27}} = \sqrt{\dfrac{48}{27}} = \sqrt{\dfrac{16}{9}} = \dfrac{\sqrt{16}}{\sqrt{9}} = \dfrac{4}{3}$

E. $\sqrt{125x^3} \div \sqrt{5x} = \dfrac{\sqrt{125x^3}}{\sqrt{5x}} = \sqrt{\dfrac{125x^3}{5x}} =$

$\sqrt{25x^2} = 5x$

7. A. $\sqrt{\dfrac{5}{9}} = \dfrac{\sqrt{5}}{\sqrt{9}} = \dfrac{\sqrt{5}}{3}$

B. $\sqrt{\dfrac{16}{3}} = \sqrt{\dfrac{16}{3} \cdot \dfrac{3}{3}} = \sqrt{\dfrac{48}{9}} = \dfrac{\sqrt{48}}{\sqrt{9}} =$

$\dfrac{\sqrt{16}\sqrt{3}}{3} = \dfrac{4\sqrt{3}}{3}$

C. $\sqrt{\dfrac{25}{20}} = \sqrt{\dfrac{5}{4}} = \dfrac{\sqrt{5}}{\sqrt{4}} = \dfrac{\sqrt{5}}{2}$

D. $\sqrt{\dfrac{5}{8}} = \sqrt{\dfrac{5}{8} \cdot \dfrac{2}{2}} = \sqrt{\dfrac{10}{16}} = \dfrac{\sqrt{10}}{\sqrt{16}} = \dfrac{\sqrt{10}}{4}$

E. $\sqrt{\dfrac{3}{5}} = \sqrt{\dfrac{3}{5} \cdot \dfrac{5}{5}} = \sqrt{\dfrac{15}{25}} = \dfrac{\sqrt{15}}{\sqrt{25}} = \dfrac{\sqrt{15}}{5}$

8. A. $AVE = \dfrac{8+10+12+6+4+7+9+4+3}{9} =$

$\dfrac{63}{9} = 7$

$SD = \sqrt{\dfrac{(8-7)^2 + (10-7)^2 + (12-7)^2}{9}}$

$+ \dfrac{(6-7)^2 + (4-7)^2 + (7-7)^2 + (9-7)^2}{9}$

$+ \dfrac{(4-7)^2 + (3-7)^2}{9}$

$= \sqrt{\dfrac{1+9+25+1+9+0+4+9+16}{9}}$

$= \sqrt{\dfrac{74}{9}} = \dfrac{\sqrt{74}}{\sqrt{9}} = \dfrac{\sqrt{74}}{3} \approx \dfrac{9}{3} = 3$

B. $AVE = \dfrac{14+10+18+22}{4} = \dfrac{64}{4} = 16$

$SD = \sqrt{\dfrac{(14-16)^2 + (10-16)^2}{4}}$

$+ \dfrac{(18-16)^2 + (22-16)^2}{4}$

$= \sqrt{\dfrac{4+36+4+36}{4}} = \sqrt{\dfrac{80}{4}} = \dfrac{\sqrt{80}}{\sqrt{4}}$

$= \dfrac{\sqrt{80}}{2} \approx \dfrac{9}{2} = 4.5$

9.42 SIMPLE PROBABILITY

Probability is the likelihood of an event happening by chance alone. The probability of an event is the ratio of the number of times in which the event occurs (favorable outcomes) to the number of possible outcomes (total outcomes). The probability of an event, E, is the ratio of the number of favorable outcomes, f, to the total number of outcomes, t, so $P(E) = \dfrac{f}{t}$.

Examples:

A. What is the probability of rolling a 4 when a six-sided die is rolled?
B. What is the probability of getting heads when a coin is tossed?
C. What is the probability of guessing the correct answer on a multiple-choice question with five answer choices?

Solutions:

A. Because a die is a number cube with the numbers (or dots for the numbers) 1, 2, 3, 4, 5, and 6 on its six faces, there is one favorable outcome, 4, $f = 1$, there are 6 total outcomes, and $t = 6$.

$$P(4) = \dfrac{1}{6}$$

B. Because a coin has two sides, $P\,(heads) = \dfrac{1}{2}$.

C. Because there is only one correct answer among five answer choices, $P\,(correct) = \dfrac{1}{5}$.

Because the least number of times something can happen is 0, the smallest probability is 0. The maximum number of times something can occur is every time, so it would be the same as the total outcomes, and the greatest probability is 1. Thus, the probability of an event is between 0 and 1 inclusive, so $0 \le P(E) \le 1$.

Examples:

A. What is the probability of getting a number less than 7 when rolling a six-sided die?
B. What is the probability of getting a 6 when tossing a coin?

Solutions:

A. Because all of the outcomes for rolling a die are less than 7, $P(less\,than\,7) = \dfrac{6}{6} = 1$.

B. Because the only outcomes for tossing a coin are heads and tails, getting a 6 is impossible, $P\,(6) = \dfrac{0}{2} = 0$.

To determine the number of outcomes for an event, a counting technique is often used that says if the first part of an event can be done in a ways, the second part of the event can be done in b ways, and the third part of the event can be done in c ways, the event can be done in a times b times c ways. Thus, the number of ways the event can happen is $a \cdot b \cdot c$ ways. This counting technique can be used with events with any number of parts.

Examples:

A. A dessert is made up of a scoop of ice cream, a topping, and one kind of nut. The restaurant has five flavors of ice cream, three toppings, and two kinds of nuts available. How many different desserts can the restaurant make?
B. In how many ways can the coins come up when five coins are tossed?
C. How many outcomes are possible when a pair of six-sided dice is rolled?

Solutions:

A. A dessert is ice cream, topping and nuts, so $5 \times 3 \times 2 = 30$ ways to make a dessert. There are 30 desserts possible.
B. The event is that five coins are tossed, so there are $2 \times 2 \times 2 \times 2 \times 2 = 32$ ways. Thus, there are 32 possible outcomes when five coins are tossed.
C. The event is that a pair of dice is rolled, so there are $6 \times 6 = 36$ ways to get a result. Thus, there are 36 possible outcomes when a pair of dice is rolled.

Finding the probability of an event, E, is a three-part activity. First, determine the number of favorable outcomes for the event. Second, determine the total number of possible outcomes for the event. Third, compute the ratio of the number of favorable outcomes to the total number of outcomes. $P(E) = \dfrac{f}{t}$.

Examples:

A. What is the probability of getting exactly one heads when three coins are tossed?
B. What is the probability of getting a sum of 4 when a pair of six-sided dice is rolled?
C. A bag contains 7 red marbles and 5 green marbles. What is the probability that a marble selected at random from the bag will be green?
D. A bag contains 8 blue marbles and 10 yellow marbles. What is the probability that two marbles selected at random from the bag will both be yellow?
E. A bag contains 5 orange marbles and 8 white marbles. A marble is selected at random, its color is noted, and it is then returned to the bag before a second marble is selected at random from the bag. What is the probability that both marbles selected will be orange?

Solutions:

A. The one heads could be on the first coin or on the second coin or on the third coin, so there are three favorable outcomes. Each coin can come up in two ways, so there are $2 \times 2 \times 2 = 8$ ways the three coins can come up. $P(exactly\ one\ heads) = P(1\ H) = \dfrac{3}{8}$.
B. The outcomes for a sum of 4 on a pair of dice are 1 and 3, 2 and 2, and 3 and 1. The total number of outcomes is $6 \times 6 = 36$. $P(sum\ of\ 4) = P(4) = \dfrac{3}{36} = \dfrac{1}{12}$.
C. There are 5 green marbles in the bag. There are $7 + 5 = 12$ marbles in the bag. $P(green) = \dfrac{5}{12}$.
D. There are 10 yellow marbles in the bag, so there are 10 ways to get the first yellow marble. Since the first yellow marble is no longer available, there are only 9 choices for the second yellow marble. Thus, there are $10 \times 9 = 90$ ways to get 2 yellow marbles. There are $8 + 10 = 18$ marbles that can be selected first, and $18 - 1 = 17$ marbles available for the second marble. Thus, there are $18 \times 17 = 306$ ways to select 2 marbles. $P(2\ yellow) = \dfrac{90}{306} = \dfrac{5}{17}$.
E. There are 5 orange marbles for the first selection, and since the first marble is returned to the bag, there are 5 orange marbles available for the second selection. Thus, there are $5 \times 5 = 25$ ways to select 2 orange marbles. There are $5 + 8 = 13$ choices for the first marble and still 13 choices for the second. So there are $13 \times 13 = 169$ choices for selecting two marbles from the bag. $P(2\ orange) = \dfrac{25}{169}$.

▄▄▄ 9.43 PRACTICE PROBLEMS

1. If a six-sided die is rolled, what is the probability of getting a multiple of 3?
2. Two coins are tossed. What is the probability of getting at least one heads?
3. If four coins are tossed, what is the probability of getting exactly one tails?
4. A club consists of 10 males and 15 females. One of the club members is to be selected at random. What is the probability that the person selected will be a female?
5. A club consists of 15 males and 6 females. If two members of the club are selected at random, what is the probability that they both will be females?
6. A pair of six-sided dice is rolled. What is the probability that the result will be a double (both dice have the same number)?
7. A pair of six-sided dice is rolled. What is the probability of getting a sum of 5 or 10?
8. A bag contains 12 red marbles and 18 blue marbles. Two marbles are selected at random from the bag. What is the probability that neither marble will be blue?

▄▄▄ 9.44 SOLUTIONS

1. Because 3 and 6 are both multiples of 3 and there are 6 total outcomes on the die, P (*multiple of* 3) $= \dfrac{2}{6} = \dfrac{1}{3}$.

2. The outcomes for tossing two coins are HH, HT, TH, and TT. P (*at least one heads*) $= \dfrac{3}{4}$.

3. Four coins are tossed, so there are $2 \times 2 \times 2 \times 2 = 16$ possible outcomes. There are exactly four outcomes with exactly one tails: THHH, HTHH, HHTH, and HHHT. P (*exactly one T*) $= \dfrac{4}{16} = \dfrac{1}{4}$.

4. There are 15 females and $10 + 15 = 25$ members of the club. P (*female*) $= \dfrac{15}{25} = \dfrac{3}{5}$.

5. There are $15 + 6 = 21$ members of the club. There are $6 \times 5 = 30$ ways to select 2 of the 6 females from the club. There are $21 \times 20 = 420$ ways to select 2 members of the 30 members of the club. P $(FF) = \dfrac{30}{420} = \dfrac{1}{14}$.

6. The doubles are 1 and 1, 2 and 2, 3 and 3, 4 and 4, 5 and 5, 6 and 6, so there are 6 favorable outcomes. There are $6 \times 6 = 36$ possible outcomes. P (*double*) $= \dfrac{6}{36} = \dfrac{1}{6}$.

7. There are four ways to get a sum of 5: 1 and 4, 2 and 3, 3 and 2, 4 and 1. There are three ways to get a 10: 4 and 6, 5 and 5, 6 and 4. There are $4 + 3 = 7$ ways to get a 5 or a 10. There are $6 \times 6 = 36$ total outcomes. P (*5 or 10*) $= \dfrac{7}{36}$.

8. There are $12 + 18 = 30$ marbles in the bag. There are $12 \times 11 = 132$ ways to get 2 red (not blue) marbles. There are $30 \times 29 = 870$ ways to get 2 marbles from the bag. P(*not blue*) $= \dfrac{132}{870} = \dfrac{22}{145}$.

■■■ 9.45 ARITHMETIC COMPUTATION TEST 2

Use the following test to assess how well you have mastered the material in this chapter. Mark your answers by blackening the corresponding answer oval in each question. An answer key and solutions are provided at the end of the test.

1. A book that regularly sold for $36.29 was given a $3.55 reduction. How much did the book sell for after the price reduction?

    (A) $3.55
    (B) $10.22
    (C) $32.74
    (D) $39.84
    (E) $139.48

2. The angles of a triangle have a sum of 180°. If the angles have a ratio of 1 : 2 : 6, what is the measure of the largest angle?

    (A) 20°
    (B) 40°
    (C) 80°
    (D) 90°
    (E) 120°

3. Which ratio is the same as 45 to 80? Select all that apply.

    [A] 9 to 16
    [B] 63 to 112
    [C] 99 to 176
    [D] 90 to 180
    [E] 108 to 192

4. If three oranges sell for 80 cents, what would 12 oranges cost?

    (A) $9.60
    (B) $4.80
    (C) $3.20
    (D) $2.40
    (E) $1.60

5. A car traveled 60 mph from Belnap to Lincoln and 50 mph from Lincoln to Belnap. If the whole trip was 660 miles, how long did the round trip take?

    (A) 6 hours
    (B) 11 hours
    (C) 12.1 hours
    (D) 13.2 hours
    (E) 24.2 hours

6. One copy machine can make 20 copies a minute and a second copy machine makes 15 copies a minute. If the two copiers work together, how long would it take them to make 2,100 copies?

   (A) 60 minutes
   (B) 105 minutes
   (C) 122.5 minutes
   (D) 140 minutes
   (E) 225 minutes

7. What number is 60% of 45?

   (A) 2.7
   (B) 7.5
   (C) 18
   (D) 27
   (E) 75

8. James bought a sweater for $29.95 and paid 7% tax on it. What was the total cost of the sweater?

   (A) $2.10
   (B) $20.97
   (C) $27.85
   (D) $32.05
   (E) $50.92

9. What is the average of 8, 6, 12, 9, 16, 14, 20, and 3?

   (A) 8
   (B) 9
   (C) 9.5
   (D) 11
   (E) 12

10. What is the median of 6, 9, 4, 21, 27, 8, 3, and 2?

    (A) 6
    (B) 7
    (C) 8
    (D) 21
    (E) 24

11. Which shows 2,540,000 written in scientific notation?

    (A) 0.254×10^7
    (B) 2.54×10^6
    (C) 25.4×10^5
    (D) 25.4×10^{-5}
    (E) 2.54×10^{-6}

12. Which is equal to $(-4x^3)^2$?

   (A) $-16x^6$
   (B) $-4x^6$
   (C) $-8x^5$
   (D) $4x^6$
   (E) $16x^6$

13. Which shows $\sqrt{96a^3b^4c^9}$ simplified completely?

   (A) $4ab^2c^4\sqrt{6ac}$
   (B) $2ab^2c^4\sqrt{6ac}$
   (C) $4ab^2c^3\sqrt{6a}$
   (D) $2ab^2c^4\sqrt{24ac}$
   (E) $4b^2\sqrt{6a^3c^9}$

14. Which shows $\sqrt{\dfrac{7}{12}}$ simplified completely?

   (A) $\dfrac{\sqrt{7}}{12}$

   (B) $\dfrac{\sqrt{21}}{12}$

   (C) $\dfrac{\sqrt{21}}{6}$

   (D) $\dfrac{\sqrt{14}}{12}$

   (E) $\dfrac{\sqrt{11}}{4}$

15. Which shows $\sqrt{45} + \sqrt{245} - \sqrt{320}$ simplified completely?

   (A) $-12\sqrt{2}$
   (B) $-12\sqrt{5}$
   (C) $2\sqrt{5}$
   (D) $5\sqrt{2}$
   (E) $18\sqrt{5}$

Quantity A	Quantity B

16. The percent increase from 2 to 3   The percent decrease from 3 to 2

   (A) Quantity A is greater.
   (B) Quantity B is greater.
   (C) The two quantities are equal.
   (D) The relationship cannot be determined from the given information.

17. What is the mode of the values 11, 38, 16, 81, 38, 16, 11, and 38?

    (A) 11
    (B) 16
    (C) 21
    (D) 30
    (E) 38

18. What is the standard deviation for 2, 7, 14, and 21?

    (A) 7.2
    (B) 11
    (C) 12
    (D) 51.5
    (E) 206

19. If 30% of the five-year-olds surveyed liked butterscotch candy, and 98 of the five-year-olds surveyed did NOT like butterscotch candy, how many five-year-olds were surveyed?

    (A) 30
    (B) 42
    (C) 70
    (D) 98
    (E) 140

20. What is the range of 10, 5, 4, 3, 1, 2, 3, and 11?

    (A) 3.5
    (B) 5
    (C) 6
    (D) 7
    (E) 10

21. Wendy's Warehouse purchased a table for $200 for resale. If Wendy's marks up the table price by 60%, what is the selling price of the table?

    (A) $80
    (B) $120
    (C) $260
    (D) $280
    (E) $320

22. A recipe calls for 3 ounces of butter for each 4 cups of flour used. If 12 ounces of butter are used, how many cups of flour are needed?

    (A) 16
    (B) 13
    (C) 9
    (D) 4
    (E) 1

23. Randy tosses five coins. What is the probability that only one coin will come up heads?

   (A) $\dfrac{1}{2}$

   (B) $\dfrac{5}{16}$

   (C) $\dfrac{1}{5}$

   (D) $\dfrac{5}{32}$

   (E) $\dfrac{5}{64}$

24. The Central High glee club has 5 girls and 4 boys as members. If a member is selected at random, what is the probability that a girl will be selected?

   (A) $\dfrac{1}{5}$

   (B) $\dfrac{1}{4}$

   (C) $\dfrac{4}{9}$

   (D) $\dfrac{5}{9}$

   (E) $\dfrac{4}{5}$

25. Cassie rolls a six-sided die three times. What is the probability that she will roll exactly two fives?

   (A) $\dfrac{2}{3}$

   (B) $\dfrac{1}{2}$

   (C) $\dfrac{1}{3}$

   (D) $\dfrac{1}{6}$

   (E) $\dfrac{5}{72}$

Answer Key

1. C	6. A	11. B	16. A	21. E
2. E	7. D	12. E	17. E	22. A
3. A,B,C, and E	8. D	13. A	18. A	23. D
4. C	9. D	14. C	19. E	24. D
5. C	10. B	15. C	20. E	25. E

9.46 SOLUTIONS

1. **C** $32.74
   $36.29 $3.55 = $32.74

2. **E** 120°
   1 : 2 : 6 and sum 180° $n + 2n + 6n = 180°$ $9n = 180°$ $n = 20°$
   The angles are 20°, 40°, and 120°, so the largest is 120°.

3. **A, B, C, and E** 90 to 180 is the only one not equal.
   $$\frac{45}{80} = \frac{63}{112} = \frac{99}{176} = \frac{108}{192} = \frac{9}{16} \neq \frac{90}{180} = \frac{1}{2}$$

4. **C** $3.20
   3 for 80 cents 12 = 4 × 3   12 for 4 × 80 = 320 cents = $3.20

5. **C** 12.1 hours
   330 miles at each rate $330 \div 50 = 6.6$ hours $330 \div 60 = 5.5$ hours
   5.5 hours + 6.6 hours = 12.1 hours

6. **A** 60 minutes
   $20t + 15t = 2100$, $35t = 2100$, $t = 60$ minutes

7. **D** 27
   60% of 45 = 0.6 × 45 = 27.0 = 27

8. **D** $32.05
   $29.95 + 0.07($29.95) = $29.95 + $2.0965 = $32.0465 = $32.05 to the
   nearest cent.

9. **D** 11
   SUM = 8 + 6 + 12 + 9 + 16 + 14 + 20 + 3 = 88, $N = 8$
   AVE = SUM ÷ N = 88 ÷ 8 = 11

10. **B** 7
    Ordered values are: 2, 3, 4, 6, 8, 9, 21, 27.
    The middle values of the 8 values are 6 and 8.   (6 + 8) ÷ 2 = 14 ÷ 2 = 7
    The median is 7.

11. **B** 2.54×10^6
    $2,540,000 = 2.54 \times 10^6$

12. **E** $16x^6$

$$(-4x^3)^2 = (-4)^2 (x^3)^2 = 16x^6$$

13. **A** $4ab^2c^4\sqrt{6ac}$

$$\sqrt{96a^3b^4c^9} = \sqrt{16 \cdot 6 \cdot a^2 \cdot a \cdot b^4 \cdot c^8 \cdot c} = \sqrt{16a^2b^4c^8}\sqrt{6ac} = 4ab^2c^4\sqrt{6ac}$$

14. **C** $\dfrac{\sqrt{21}}{6}$

$$\sqrt{\frac{7}{12}} = \sqrt{\frac{7}{12} \cdot \frac{3}{3}} = \sqrt{\frac{21}{36}} = \frac{\sqrt{21}}{\sqrt{36}} = \frac{\sqrt{21}}{6}$$

15. **C** $2\sqrt{5}$

$$\sqrt{45} + \sqrt{245} - \sqrt{320} = \sqrt{9}\sqrt{5} + \sqrt{49}\sqrt{5} - \sqrt{64}\sqrt{5} = 3\sqrt{5} + 7\sqrt{5} - 8\sqrt{5} = 2\sqrt{5}$$

16. **A** Quantity A is greater.
    In Quantity A, the difference between 2 and 3 is 1, so the percent increase is 1 divided by the original number, 2. The percent increase is 50%. In Quantity B, the difference between 3 and 2 is also 1, but the original number is 3, so the percent decrease is 33.3%.

17. **E** 38
    Because 38 occurs three times, the mode is 38.

18. **A** 7.2

$$AVE = \frac{2 + 7 + 14 + 21}{4} = \frac{44}{4} = 11$$

$$SD = \sqrt{\frac{(2-11)^2 + (7-11)^2 + (14-11)^2 + (21-11)^2}{4}}$$

$$= \sqrt{\frac{206}{4}} = \sqrt{51.5} \approx 7$$

Since $7^2 = 49$, $\sqrt{51.5}$ is approximately equal to 7.

19. **E** 140
    If 30% liked the candy, then 70% did not. 70% of what number is equal to 98?
    $0.70 \cdot x = 98.$ $x = 140.$

20. **E** 10
    Range = Maximum − Minimum = 11 − 1 = 10.

21. **E** 320

    The amount of mark-up is 60% of $200, so the mark-up is $0.60 \times 200 = 120$. The selling price is cost plus mark-up, so the selling price is $200 + \$120 = \$320$.

22. **A** 16

    The ratio is 3 ounces of butter to 4 cups of flour. Because 12 ounce of butter was used, the recipe has been multiplied by 4. Thus, 4 times the flour is needed and $4(4) = 16$ cups of flour.

23. **D** $\dfrac{5}{32}$

    There are $2 \times 2 \times 2 \times 2 \times 2 = 32$ ways for five coins to turn up. The one heads could be on any of the five tosses. $P(1H) = \dfrac{5}{32}$.

24. **D** $\dfrac{5}{9}$

    There are 5 females in the club, and $4 + 5 = 9$ members in the club. $P(female) = \dfrac{5}{9}$.

25. **E** $\dfrac{5}{72}$

    There are $6 \times 6 \times 6 = 216$ outcomes for rolling a die three times. There are five "non-5" outcomes, and the non-5 can be in any one of the three rolls, so there are 15 favorable outcomes. $\dfrac{15}{216} = \dfrac{5}{72}$.

9.47 SOLVED GRE PROBLEMS

For each question, select the best answer unless otherwise instructed.

1. **Noah has 8 A's, 2 B's, 6 C's, 1 D, and 3 F's on homework in a physics class. What percentage of his homework grades are B's?**

   (A) 10%
   (B) 12%
   (C) 13%
   (D) 20%
   (E) 40%

   Data Set: 12, 4, 6, 3, 16, 10, 17, 9, 4

Quantity A	Quantity B

2. The mean for the data set   The median for the data set

   (A) Quantity A is greater.
   (B) Quantity B is greater.
   (C) The two quantities are equal.
   (D) The relationship cannot be determined from the given information.

A class has 12 boys and 18 girls.

	Quantity A	**Quantity B**
3.	The ratio of boys to girls	The ratio of girls to the class

(A) Quantity A is greater.
(B) Quantity B is greater.
(C) The two quantities are equal.
(D) The relationship cannot be determined from the given information.

For this question, enter your answer in the box.

4. **Ben can paint a room in 6 hours and Carol can paint the same room in 4 hours. How many hours would it take them together to paint the room?**

5. Which shows $\sqrt{\dfrac{12}{10}}$ simplified completely?

(A) $\sqrt{12}$

(B) $\sqrt{\dfrac{6}{5}}$

(C) $\dfrac{2\sqrt{3}}{\sqrt{10}}$

(D) $\dfrac{\sqrt{30}}{5}$

(E) $\dfrac{\sqrt{120}}{10}$

■ 9.48 SOLUTIONS

1. **A** 10%
   There are two grades of B and $8 + 2 + 6 + 1 + 3 = 20$ grades. The percentage of Bs is $\dfrac{2}{20} = \dfrac{1}{10} = 0.1 = 10\%$.

2. **C** The two quantities are equal.
   The mean is $AVE = \dfrac{SUM}{NUM} = \dfrac{81}{9} = 9$. The data values arranged in order are: 3, 3, 4, 6, 9, 10, 12, 16, 17. The median is 9.

3. **A** Quantity A is greater.
   The ratio of boys to girls is $\dfrac{12}{18} = \dfrac{2}{3}$. The ratio of girls to the class is $\dfrac{18}{30} = \dfrac{3}{5}$.

4. $\boxed{2.4}$

Let H be the number of hours it takes them to paint the room together. Ben can paint the room in 6 hours, so he paints $\frac{1}{6}$ of the room per hour and $\frac{H}{6}$ in H hours. Carol can paint the room in 4 hours, so she does $\frac{1}{4}$ of the room per hour and $\frac{H}{4}$ in H hours. Together they paint 100% of the room and 100% = 1. So, $\frac{H}{6} + \frac{H}{4} = 1$. Thus, $2H + 3H = 12$ and $H = 2.4$ hours. Type 2.4 in the box.

5. **D** $\dfrac{\sqrt{30}}{5}$

$$\sqrt{\frac{12}{10}} = \sqrt{\frac{6}{5}} = \sqrt{\frac{6}{5} \times \frac{5}{5}} = \sqrt{\frac{30}{25}} = \frac{\sqrt{30}}{\sqrt{25}} = \frac{\sqrt{30}}{5}.$$

9.49 GRE PRACTICE PROBLEMS

For each question, select the best answer unless otherwise instructed.

1. **On a rent-to-own plan, Jack bought a television that had a cash price of $245. He paid $10.50 a week for 36 weeks to buy the television using the plan. How much would Jack have saved if he had paid cash for the television?**

   (A) $3.69
   (B) $6.81
   (C) $133
   (D) $252
   (E) $378

   Apples are 6 for $1.95 and oranges are 8 for $2.05.

Quantity A	Quantity B
The cost of 4 apples	The cost of 5 oranges

2.

   (A) Quantity A is greater.
   (B) Quantity B is greater.
   (C) The two quantities are equal.
   (D) The relationship cannot be determined from the given information.

3. **Maria drove 572 miles at an average speed of 48 miles per hour. To the nearest minute, how long did the trip take?**

   (A) 11 hours
   (B) 11 hours, 9 minutes
   (C) 11 hours, 55 minutes
   (D) 12 hours
   (E) 12 hours, 9 minutes

4. **For which number is 175% of it equal to 357?**

   (A) 89.25
   (B) 204
   (C) 267.75
   (D) 476
   (E) 624.75

   Data Set: 8, 2, 4, 9, 1, 8, 7, 4, 3, 7, 9, 8, 4, 5, 2, 9, 5

Quantity A	**Quantity B**
Mode of the data set	Range of the data set

5.

   (A) Quantity A is greater.
   (B) Quantity B is greater.
   (C) The two quantities are equal.
   (D) The relationship cannot be determined from the given information.

For this question, enter your answer in the box.

6. **What is the median for the data set 8, 4, 5, 8, 2, 7, 6, 4, 8, 1, 3, and 6?**

   ☐

7. **Which expression is equal to $(5x^2 y^3)^4$?**

   (A) $5x^6 y^7$
   (B) $5x^8 y^{12}$
   (C) $625x^6 y^7$
   (D) $625x^8 y^{12}$
   (E) $625x^{16} y^{81}$

8. **Which number shows 0.00008724 written in scientific notation?**

   (A) 8.724×10^5
   (B) 8.724×10^{-5}
   (C) 8.724×10^{-6}
   (D) 87.24×10^7
   (E) 87.24×10^{-7}

Quantity A	**Quantity B**
$\sqrt[3]{\dfrac{27}{125}}$	$\sqrt{\dfrac{49}{121}}$

9.

   (A) Quantity A is greater.
   (B) Quantity B is greater.
   (C) The two quantities are equal.
   (D) The relationship cannot be determined from the given information.

10. Which expression shows $\sqrt{98} - 2\sqrt{72} + \sqrt{128}$ simplified completely?

    (A) $-672\sqrt{2}$

    (B) $-336\sqrt{2}$

    (C) $3\sqrt{2}$

    (D) $9\sqrt{2}$

    (E) $-5\sqrt{2} + 4\sqrt{32}$

ANSWER KEY

1. C
2. A
3. C
4. B
5. D
6. 5.5
7. D
8. B
9. B
10. C

CHAPTER 10
ALGEBRA

10.1 ALGEBRAIC EXPRESSIONS

An **algebraic expression** is a combination of letters and numbers that are used to represent numbers; thus,

$$x^3 - 5xy + 2x - 3y^4, \quad 2ab^2c^5 \quad \text{and} \quad \frac{5xy + 2ab}{5x^2 - 2a^3} \quad \text{are algebraic expressions.}$$

A **term** consists of products and quotients of letters and numbers. Thus, $2xy$, $5x^2/3y^3$, and $-4a^5$ are terms. The algebraic expression $3x^2y - 4xy$ consists of two terms.

A **monomial** is a one-term algebraic expression. Example of monomials are $3x$, 7, x^2/y, $7y$.

A **binomial** is a two-term algebraic expression. Examples of binomials are $x + 2$, $5x + 3y$, $7x^2y^2 + 3ab$.

A **trinomial** is a three-term algebraic expression. Examples of trinomials are $3x + 2y + 3$, $x^2 + 5xy - y^2$, $x^3 - ab + xy$.

The term **polynomial** is used when talking about an algebraic expression without stating a specific number of terms.

In the term $7x^2y^3$, the 7 is the **coefficient** of x^2y^3. **Like terms** are terms that have the same variable parts. For example, $7x^2y^3$ and $-2x^2y^3$ are like terms, but $7x^2y^3$ and $3xy^3$ are unlike terms. You combine like terms by adding their coefficients. Unlike terms cannot be combined.

Example:

Simplify each expression by combining like terms.

A. $7x + 3y^3 - 3x + xy + 5y^3$
B. $3x - 2y^3 + 7xy + 2xy - 5x - 6y^3$

Solution:

A. $7x + 3y^3 - 3x + xy + 5y^3 = 7x - 3x + 3y^3 + 5y^3 + xy = 4x + 8y^3 + xy$
B. $3x - 2y^3 + 7xy + 2xy - 5x - 6y^3 = 3x - 5x - 2y^3 - 6y^3 + 7xy + 2xy = -2x - 8y^3 + 9xy$

10.2 EXPONENTS REVISITED

If n is a positive integer, then a^n represents the product of n factors of a.

Thus, $a^5 = a \cdot a \cdot a \cdot a \cdot a$ and $b^3 = b \cdot b \cdot b$. In a^n, a is the **base**, and n is the **exponent**.

If $n = 2$, a^2 is read "a squared," and if $n = 3$, a^3 is read "a cubed." a^n is read "a to the nth power."

Example:

Show the meaning of each expression.

A. x^3　　　　　　　　　　B. 3^4　　　　　　　　　　C. $(-2)^5$

Solution:

A. $x^3 = x \cdot x \cdot x$
B. $3^4 = 3 \cdot 3 \cdot 3 \cdot 3$
C. $(-2)^5 = (-2) \cdot (-2) \cdot (-2) \cdot (-2) \cdot (-2) = -32$

If n is a positive integer, then $a^{-n} = \dfrac{1}{a^n}, a \neq 0$.

Example:

Show the meaning of each expression.

A. 3^{-4}　　　　　B. 2^{-3}　　　　　C. $-5y^{-3}$　　　　　D. $(x + y)^{-1}$

Solution:

A. $3^{-4} = \dfrac{1}{3^4} = \dfrac{1}{81}$　　　　　　　　　　B. $2^{-3} = \dfrac{1}{2^3} = \dfrac{1}{8}$

C. $-5y^{-3} = \dfrac{-5}{y^3}$　　　　　　　　　　D. $(x + y)^{-1} = \dfrac{1}{x + y}$

10.3 ROOTS REVISITED

If n is a positive integer and if a and b are such that $a^n = b$, then n is called the **nth root** of b. If b is positive, there is only one positive number a such that $a^n = b$. You write this positive number as $\sqrt[n]{b}$ and call it the **principal nth root** of b. If b is negative and n is even, there is no positive nth root of b. However, if n is odd, there is a negative nth root of b.

Example:

Simplify each of these roots.

A. $\sqrt[3]{27}$　　　　B. $\sqrt{25}$　　　　C. $\sqrt[4]{16}$　　　　D. $\sqrt[3]{-8}$　　　　E. $\sqrt[4]{-16}$

Solution:

A. $\sqrt[3]{27} = \sqrt[3]{3^3} = 3$
B. $\sqrt{25} = \sqrt{5^2} = 5$
C. $\sqrt[4]{16} = \sqrt[4]{2^4} = 2$
D. $\sqrt[3]{-8} = \sqrt[3]{(-2)^3} = -2$
E. $\sqrt[4]{-16}$ is not a real number since n is even and b is negative.

10.4 GENERAL LAWS OF EXPONENTS

If m and n are real numbers, then the following laws hold.

1. $a^m \cdot a^n = a^{m+n}$
2. $(a^m)^n = a^{mn}$
3. $\dfrac{a^m}{a^n} = a^{m-n}$, if $a \neq 0$
4. $(ab)^m = a^m b^m$
5. $\left(\dfrac{a}{b}\right)^m = \dfrac{a^m}{b^m}$, if $b \neq 0$
6. $a^{\frac{m}{n}} = \sqrt[n]{a^m}$, if $a \geq 0$
7. $a^0 = 1$, if $a \neq 0$

Example:

Simplify these expressions by using the laws of exponents.

A. $x^6 \div x^4$
B. $(x^2 \cdot y^5)^3$
C. $(x^{1/3})^3$
D. 5^0
E. $(x^{-3})^{-4}$

Solution:

A. $x^6 \div x^4 = x^{6-4} = x^2$
B. $(x^2 \cdot y^5)^3 = (x^2)^3(y^5)^3 = x^6 y^{15}$
C. $(x^{1/3})^3 = x^{3/3} = x^1 = x$
D. $5^0 = 1$
E. $(x^{-3})^{-4} = x^{-3(-4)} = x^{12}$

Example:

Simplify these expressions using the laws of exponents, for $a \neq 0$, $x \neq 0$.

A. $\dfrac{x^5}{x^8}$
B. $8^{-2/3}$
C. $x^{-5/2}$
D. $\dfrac{\sqrt{x}}{\sqrt[3]{x}}$
E. $4^{3/2}$

Solution:

A. $\dfrac{x^5}{x^8} = x^{5-8} = x^{-3} = \dfrac{1}{x^3}$

B. $8^{-2/3} = (8^{2/3})^{-1} = \left(\sqrt[3]{8^2}\right)^{-1} = \left(\sqrt[3]{64}\right)^{-1} = 4^{-1} = \dfrac{1}{4}$

C. $x^{-5/2} = \dfrac{1}{x^{5/2}}$

D. $\dfrac{\sqrt{x}}{\sqrt[3]{x}} = \dfrac{x^{1/2}}{x^{1/3}} = x^{1/2-1/3} = x^{3/6-2/6} = x^{1/6} = \sqrt[6]{x}$

E. $4^{3/2} = (4^{1/2})^3 = (2)^3 = 8$

▬▬ 10.5 PRACTICE PROBLEMS

1. Simplify these algebraic expressions.

   A. $17x - 18y + 2xy - 15x + 11y - 8xy$
   B. $5x^2 - 3x + 6 - 2x^2 + 18x - 1$
   C. $15x^2 - 3y^2 + 4z^2 - 8z^2 + 5y^2 + 6x^2$

2. Simplify these exponential expressions.

   A. x^{-3} B. $(-2)^3$ C. $(3x)^3$ D. $(-2x)^2$

3. Simplify each of the roots.

   A. $\sqrt[3]{8}$ B. $\sqrt[3]{-125}$ C. $\sqrt[4]{625}$
   D. $\sqrt{121}$ E. $\sqrt{81}$

4. Simplify these expressions.

   A. $x^3 \cdot x^5$ B. $(x^{1/4})^8$
   C. $(x^3 \cdot y^5)^3$ D. $7^{-3} \cdot 7^8 \cdot 7^{-2}$
   E. $x^{10} \div x^2$

5. Simplify these expressions.

   A. $\dfrac{x^7}{x^9}$ B. $x^{-2/3}$ C. $\dfrac{x^{1/6}}{x^{1/2}}$ D. $\sqrt[4]{4}$

▬▬ 10.6 SOLUTIONS

1. A. $17x - 18y + 2xy - 15x + 11y - 8xy = 17x - 15x - 18y + 11y + 2xy - 8xy = 2x - 7y - 6xy$
   B. $5x^2 - 3x + 6 - 2x^2 + 18x - 1 = 5x^2 - 2x^2 - 3x + 18x + 6 - 1 = 3x^2 + 15x + 5$
   C. $15x^2 - 3y^2 + 4z^2 - 8z^2 + 5y^2 + 6x^2 = 15x^2 + 6x^2 - 3y^2 + 5y^2 + 4z^2 - 8z^2 = 21x^2 + 2y^2 - 4z^2$

2. A. $x^{-3} = \dfrac{1}{x^3}$
   B. $(-2)^3 = (-2)(-2)(-2) = -8$
   C. $(3x)^3 = (3x)(3x)(3x) = 27x^3$
   D. $(-2x)^2 = (-2x)(-2x) = 4x^2$

3. A. $\sqrt[3]{8} = \sqrt[3]{2^3} = 2$
   B. $\sqrt[3]{-125} = \sqrt[3]{(-5)^3} = -5$
   C. $\sqrt[4]{625} = \sqrt[4]{5^4} = 5$
   D. $\sqrt{121} = \sqrt{11^2} = 11$
   E. $\sqrt{81} = \sqrt{9^2} = 9$

4. A. $x^3 \cdot x^5 = x^{3+5} = x^8$
   B. $(x^{1/4})^8 = x^{8/4} = x^2$
   C. $(x^3 \cdot y^5)^3 = (x^3)^3(y^5)^3 = x^{3 \cdot 3}y^{5 \cdot 3} = x^9 y^{15}$
   D. $7^{-3} \cdot 7^8 \cdot 7^{-2} = 7^{-3+8-2} = 7^3 = 7 \cdot 7 \cdot 7 = 343$
   E. $x^{10} \div x^2 = x^{10-2} = x^8$

5. A. $\dfrac{x^7}{x^9} = x^{7-9} = x^{-2} = \dfrac{1}{x^2}$
   B. $x^{-2/3} = \dfrac{1}{x^{2/3}} = \dfrac{1}{\sqrt[3]{x^2}}$
   C. $\dfrac{x^{1/6}}{x^{1/2}} = x^{1/6-1/2} = x^{1/6-3/6} = x^{-2/6} = x^{-1/3} = \dfrac{1}{x^{1/3}} = \dfrac{1}{\sqrt[3]{x}}$
   D. $\sqrt[4]{4} = \sqrt[4]{(2)^2} = (2^2)^{1/4} = 2^{2/4} = 2^{1/2} = \sqrt{2}$

▬▬ 10.7 TABLES OF POWERS AND ROOTS

Number	Square	Cube	Square Root	Cube Root
1	1	1	1	1
2	4	8	1.414	1.260
3	9	27	1.732	1.442
4	16	64	2	1.587
5	25	125	2.236	1.710
6	36	216	2.449	1.817
7	49	343	2.646	1.913

Continued

Number	Square	Cube	Square Root	Cube Root
8	64	512	2.828	2
9	81	729	3	2.080
10	100	1,000	3.16	2.154
11	121	1,331	3.317	2.224

Additional Squares

Number	Square	Number	Square
12	144	19	361
13	169	20	400
14	196	21	441
15	225	22	484
16	256	23	529
17	289	24	576
18	324	25	625

■ 10.8 RADICAL EXPRESSIONS

A radical $\sqrt[n]{a}$ has three parts: the **index** n, the **radical sign** $\sqrt{}$, and the **radicand** a. To simplify a radical expression, make the index as small as possible and have no factors that are nth powers, no fractions in the radicand, and no radical expressions in the denominator.

Example:

Which radical expression needs to be simplified?

A. $\sqrt{9x}$ 　　　　B. $\sqrt{3x^5}$ 　　　　C. $\sqrt{\dfrac{1}{4}}$ 　　　　D. $\dfrac{3}{\sqrt{x}}$

E. $\sqrt{\dfrac{5}{8}}$

Solution:

A. $\sqrt{9x}$ 　Has a perfect square factor of 9

B. $\sqrt{3x^5}$ 　Has a perfect square factor of x^4

C. $\sqrt{\dfrac{1}{4}}$ 　Has a perfect square factor of $\dfrac{1}{4}$

D. $\dfrac{3}{\sqrt{x}}$ 　Has a radical in the denominator of the fraction

E. $\sqrt{\dfrac{5}{8}}$ 　Has a common fraction in the radicand

Reducing the Index

A radical can be written with a fractional exponent, $\sqrt[n]{x^a} = x^{\frac{a}{n}}$, whenever $GCD(a, n) > 1$. To simplify the index, reduce the fractional exponent. If $GCD(a, n) = 1$, the index is as small as it can get.

Example:

Reduce the index wherever possible.

A. $\sqrt[4]{a^2}$　　　B. $\sqrt[6]{a^4b^2}$　　　C. $\sqrt{a^3}$　　　D. $\sqrt{a^4}$　　　E. $\sqrt[5]{a^2b^3}$

Solution:

A. $\sqrt[4]{a^2} = a^{2/4} = a^{1/2} = \sqrt{a}$

B. $\sqrt[6]{a^4b^2} = \sqrt[6]{(a^2b)^2} = (a^2b)^{2/6} = (a^2b)^{1/3} = \sqrt[3]{a^2b}$

C. $\sqrt{a^3} = \sqrt{a^3}$, since $GCD(2, 3) = 1$, so the index cannot be reduced.

D. $\sqrt{a^4} = a^{4/2} = a^2$, so no radical in answer.

E. $\sqrt[5]{a^2b^3} = \sqrt[5]{(a^2b^3)^1}$, since $GCD(1, 5) = 1$, so the index cannot be reduced.

Removing the Perfect Square Factors in the Radicand

If the radicand has a factor with an exponent greater than or equal to the index, then the radicand can be simplified. $\sqrt{a^2b} = \sqrt{a^2}\sqrt{b} = a\sqrt{b}$, and $\sqrt{a^3b} = \sqrt{a^2}\sqrt{ab} = a\sqrt{ab}$.

Example:

Simplify each radicand.

A. $\sqrt{75}$　　　B. $\sqrt{27a}$　　　C. $\sqrt{12xy}$　　　D. $\sqrt{4xy^2}$　　　E. $\sqrt{200x^5}$

Solution:

A. $\sqrt{75} = \sqrt{3 \cdot 5^2} = \sqrt{5^2}\sqrt{3} = 5\sqrt{3}$

B. $\sqrt{27a} = \sqrt{3^3a} = \sqrt{3^2}\sqrt{3a} = 3\sqrt{3a}$

C. $\sqrt{12xy} = \sqrt{2^2 \cdot 3xy} = \sqrt{2^2}\sqrt{3xy} = 2\sqrt{3xy}$

D. $\sqrt{4xy^2} = \sqrt{2^2xy^2} = \sqrt{2^2y^2}\sqrt{x} = 2y\sqrt{x}$

E. $\sqrt{200x^5} = \sqrt{(2 \cdot 10^2 \cdot x^4 \cdot x)} = \sqrt{(10^2 \cdot x^4)}\sqrt{2x} = 10x^2\sqrt{2x}$

Fractions in the Radicand

If there is a decimal in the radicand, write the decimal factor as a common fraction. When there is a common fraction in the radicand, multiply the numerator and denominator by a common factor that makes the denominator a perfect square. For example,

$$\sqrt{\frac{1}{2}} = \sqrt{\frac{1}{2} \cdot \frac{2}{2}} = \sqrt{\frac{2}{4}} = \frac{\sqrt{2}}{\sqrt{4}} = \frac{\sqrt{2}}{2}$$

Example:

Simplify each radicand.

A. $\sqrt{\dfrac{8}{9}}$ B. $\sqrt{\dfrac{7}{3}}$ C. $\sqrt{\dfrac{3}{8}}$ D. $\sqrt{\dfrac{3}{10}}$ E. $\sqrt{0.2}$

Solution:

A. $\sqrt{\dfrac{8}{9}} = \dfrac{\sqrt{8}}{\sqrt{9}} = \dfrac{\sqrt{4}\sqrt{2}}{3} = \dfrac{2\sqrt{2}}{3}$

B. $\sqrt{\dfrac{7}{3}} = \sqrt{\dfrac{7}{3}\cdot\dfrac{3}{3}} = \sqrt{\dfrac{21}{9}} = \dfrac{\sqrt{21}}{\sqrt{9}} = \dfrac{\sqrt{21}}{3}$

C. $\sqrt{\dfrac{3}{8}} = \sqrt{\dfrac{3}{8}\cdot\dfrac{2}{2}} = \sqrt{\dfrac{6}{16}} = \dfrac{\sqrt{6}}{\sqrt{16}} = \dfrac{\sqrt{6}}{4}$

D. $\sqrt{\dfrac{3}{10}} = \sqrt{\dfrac{3}{10}\cdot\dfrac{10}{10}} = \sqrt{\dfrac{30}{100}} = \dfrac{\sqrt{30}}{\sqrt{100}} = \dfrac{\sqrt{30}}{10}$

E. $\sqrt{0.2} = \sqrt{\dfrac{2}{10}} = \sqrt{\dfrac{1}{5}} = \sqrt{\dfrac{1}{5}\cdot\dfrac{5}{5}} = \sqrt{\dfrac{5}{25}} = \dfrac{\sqrt{5}}{\sqrt{25}} = \dfrac{\sqrt{5}}{5}$

Radicals in the Denominator of a Fraction

If a fraction has a radical in the denominator, multiply the numerator and denominator of the fraction by the radical expression that will make the denominator a perfect nth root. For example,

$$\dfrac{5}{\sqrt{2}} = \dfrac{5}{\sqrt{2}}\cdot\dfrac{\sqrt{2}}{\sqrt{2}} = \dfrac{5\sqrt{2}}{\sqrt{4}} = \dfrac{5\sqrt{2}}{2}$$

Example:

Simplify these radical expressions.

A. $\dfrac{4}{\sqrt{7}}$ B. $\dfrac{3}{\sqrt{2}}$ C. $\dfrac{5}{\sqrt{3}}$ D. $\dfrac{\sqrt{2}}{\sqrt{15}}$

Solution:

A. $\dfrac{4}{\sqrt{7}} = \dfrac{4}{\sqrt{7}}\cdot\dfrac{\sqrt{7}}{\sqrt{7}} = \dfrac{4\sqrt{7}}{\sqrt{49}} = \dfrac{4\sqrt{7}}{7}$

B. $\dfrac{3}{\sqrt{2}} = \dfrac{3}{\sqrt{2}}\cdot\dfrac{\sqrt{2}}{\sqrt{2}} = \dfrac{3\sqrt{2}}{\sqrt{4}} = \dfrac{3\sqrt{2}}{2}$

C. $\dfrac{5}{\sqrt{3}} = \dfrac{5}{\sqrt{3}}\cdot\dfrac{\sqrt{3}}{\sqrt{3}} = \dfrac{5\sqrt{3}}{\sqrt{9}} = \dfrac{5\sqrt{3}}{3}$

D. $\dfrac{\sqrt{2}}{\sqrt{15}} = \dfrac{\sqrt{2}}{\sqrt{15}}\cdot\dfrac{\sqrt{15}}{\sqrt{15}} = \dfrac{\sqrt{30}}{15}$

10.9 PRACTICE PROBLEMS

1. Simplify these radical expressions.

   A. $\sqrt{18}$ B. $\sqrt[3]{40}$ C. $\sqrt[3]{81}$ D. $\sqrt{648}$

2. Simplify these radical expressions.

   A. $\sqrt[4]{9a^2}$ B. $\sqrt[6]{125a^3b^3}$ C. $\sqrt{a^3}$ D. $\sqrt[9]{a^3}$

3. Simplify these expressions.

   A. $\sqrt{7a^3y^2}$ B. $\sqrt{8b^4x^3}$ C. $\sqrt{25x^6}$ D. $\sqrt{8x^5y^7}$

4. Simplify these radical expressions.

   A. $\dfrac{10\sqrt{6}}{5\sqrt{2}}$ B. $\dfrac{3}{\sqrt{5}}$ C. $\sqrt{\dfrac{5}{16}}$ D. $\sqrt{\dfrac{7}{18}}$

10.10 SOLUTIONS

1. A. $\sqrt{18} = \sqrt{9}\sqrt{2} = 3\sqrt{2}$

   B. $\sqrt[3]{40} = \sqrt[3]{8} \cdot \sqrt[3]{5} = 2 \cdot \sqrt[3]{5}$

   C. $\sqrt[3]{81} = \sqrt[3]{3^4} = \sqrt[3]{3^3} \cdot \sqrt[3]{3} = 3\sqrt[3]{3}$

   D. $\sqrt{648} = \sqrt{(2^3 \cdot 3^4)} = \sqrt{(2^2 \cdot 3^4)}\sqrt{2} =$
   $2 \cdot 3^2 \sqrt{2} = 18\sqrt{2}$

2. A. $\sqrt[4]{9a^2} = \sqrt[4]{(3a)^2} = (3a)^{2/4} = (3a)^{1/2} = \sqrt{(3a)} = \sqrt{3a}$

   B. $\sqrt[6]{125a^3b^3} = \sqrt[6]{(5ab)^3} = (5ab)^{3/6} = (5ab)^{1/2} = \sqrt{5ab}$

   C. $\sqrt{a^3} = \sqrt{a^2}\sqrt{a} = a\sqrt{a}$

   D. $\sqrt[9]{a^3} = (a)^{3/9} = a^{1/3} = \sqrt[3]{a}$

3. A. $\sqrt{7a^3y^2} = \sqrt{(a^2y^2)}\sqrt{7a} = ay\sqrt{7a}$

   B. $\sqrt{8b^4x^3} = \sqrt{4b^4x^2}\sqrt{2x} = 2b^2x\sqrt{2x}$

   C. $\sqrt{25x^6} = \sqrt{5^2(x^3)^2} = 5x^3$

   D. $\sqrt{8x^5y^7} = \sqrt{4x^4y^6}\sqrt{2xy} = 2x^2y^3\sqrt{2xy}$

4. A. $\dfrac{10\sqrt{6}}{5\sqrt{2}} = \dfrac{10}{5} \cdot \dfrac{\sqrt{6}}{\sqrt{2}} = 2 \cdot \sqrt{\dfrac{6}{2}} = 2\sqrt{3}$

   B. $\dfrac{3}{\sqrt{5}} = \dfrac{3}{\sqrt{5}} \cdot \dfrac{\sqrt{5}}{\sqrt{5}} = \dfrac{3\sqrt{5}}{5}$

   C. $\sqrt{\dfrac{5}{16}} = \dfrac{\sqrt{5}}{\sqrt{16}} = \dfrac{\sqrt{5}}{4}$

   D. $\sqrt{\dfrac{7}{18}} = \sqrt{\dfrac{7}{18} \cdot \dfrac{2}{2}} = \sqrt{\dfrac{14}{36}} = \dfrac{\sqrt{14}}{\sqrt{36}} = \dfrac{\sqrt{14}}{6}$

10.11 OPERATIONS WITH RADICALS

When adding or subtracting radicals, you can combine them if they are **like radicals**: they have the same index and the same radicand. Like radicals are combined by combining their coefficients.

Example:

Simplify and combine these radical expressions.

A. $\sqrt{54} - \sqrt{24} + \sqrt{96}$

B. $5\sqrt{40} - 3\sqrt{90}$

C. $15\sqrt{10} - \sqrt{250} + 8\sqrt{90}$

D. $\sqrt[3]{16a^3} - 3\sqrt[3]{a^3}$

E. $2\sqrt{108y} - \sqrt{27y} + \sqrt{363y}$

Solution:

A. $\sqrt{54} - \sqrt{24} + \sqrt{96} = \sqrt{9}\sqrt{6} - \sqrt{4}\sqrt{6} + \sqrt{16}\sqrt{6} = 3\sqrt{6} - 2\sqrt{6} + 4\sqrt{6} = 5\sqrt{6}$

B. $5\sqrt{40} - 3\sqrt{90} = 5\sqrt{4}\sqrt{10} - 3\sqrt{9}\sqrt{10} = 5(2)\sqrt{10} - 3(3)\sqrt{10} = 10\sqrt{10} - 9\sqrt{10} = \sqrt{10}$

C. $15\sqrt{10} - \sqrt{250} + 8\sqrt{90} = 15\sqrt{10} - \sqrt{25}\sqrt{10} + 8\sqrt{9}\sqrt{10} = 15\sqrt{10} - 5\sqrt{10} + 8(3)\sqrt{10} = 10\sqrt{10} + 24\sqrt{10} = 34\sqrt{10}$

D. $\sqrt[3]{16a^3} - 3\sqrt[3]{a^3} = \sqrt[3]{8a^3} \cdot \sqrt[3]{2} - 3a = 2a\sqrt[3]{2} - 3a$

E. $2\sqrt{108y} - \sqrt{27y} + \sqrt{363y} = 2\sqrt{36}\sqrt{3y} - \sqrt{9}\sqrt{3y} + \sqrt{121}\sqrt{3y} = 2(6)\sqrt{3y} - 3\sqrt{3y} + 11\sqrt{3y} = 12\sqrt{3y} + 8\sqrt{3y} = 20\sqrt{3y}$

When n is a positive integer, and a and b are positive real numbers, $\sqrt[n]{a}\sqrt[n]{b} = \sqrt[n]{ab}$.

For example, $\sqrt{2}\sqrt{3} = \sqrt{2 \cdot 3} = \sqrt{6}$, $\sqrt{2x} \cdot \sqrt{5y} = \sqrt{2x \cdot 5y} = \sqrt{10xy}$

Also, $\dfrac{\sqrt[n]{a}}{\sqrt[n]{b}} = \sqrt[n]{\dfrac{a}{b}}$

Example:

Multiply these radical expressions.

A. $\sqrt{8} \cdot \sqrt{6}$
B. $(5\sqrt{3})(4\sqrt{5})$
C. $\sqrt{5}(3\sqrt{7} - 2\sqrt{5})$
D. $\sqrt{5}(\sqrt{15} + \sqrt{10})$
E. $(3\sqrt{5x})(5\sqrt{10x})$

Solution:

A. $\sqrt{8} \cdot \sqrt{6} = \sqrt{48} = \sqrt{16 \cdot 3} = \sqrt{16}\sqrt{3} = 4\sqrt{3}$

B. $(5\sqrt{3})(4\sqrt{5}) = 5 \cdot 4 \cdot \sqrt{3} \cdot \sqrt{5} = 20\sqrt{15}$

C. $\sqrt{5}(3\sqrt{7} - 2\sqrt{5}) = \sqrt{5}(3\sqrt{7}) + \sqrt{5}(-2\sqrt{5}) = 3\sqrt{35} - 2\sqrt{25} = 3\sqrt{35} - 10$

D. $\sqrt{5}(\sqrt{15} + \sqrt{10}) = \sqrt{5}\sqrt{15} + \sqrt{5}\sqrt{10} = \sqrt{75} + \sqrt{50} = \sqrt{25}\sqrt{3} + \sqrt{25}\sqrt{2} = 5\sqrt{3} + 5\sqrt{2}$

E. $(3\sqrt{5x})(5\sqrt{10x}) = 3 \cdot 5 \cdot \sqrt{5x}\sqrt{10x} = 15\sqrt{50x^2} = 15\sqrt{25x^2}\sqrt{2} = 15(5x)\sqrt{2} = 75x\sqrt{2}$

Example:

Divide the radical expressions.

A. $\dfrac{\sqrt{18}}{\sqrt{2}}$
B. $\dfrac{\sqrt{108y^3}}{\sqrt{3y}}$
C. $\dfrac{\sqrt{3}}{\sqrt{21}}$
D. $\dfrac{\sqrt{5}}{\sqrt{6}}$
E. $\dfrac{8}{3\sqrt{11x}}$

Solution:

A. $\dfrac{\sqrt{18}}{\sqrt{2}} = \sqrt{\dfrac{18}{2}} = \sqrt{9} = 3$

B. $\dfrac{\sqrt{108y^3}}{\sqrt{3y}} = \sqrt{\dfrac{108y^3}{3y}} = \sqrt{36y^2} = 6y$

C. $\dfrac{\sqrt{3}}{\sqrt{21}} = \sqrt{\dfrac{3}{21}} = \sqrt{\dfrac{1}{7}} = \sqrt{\dfrac{1}{7} \cdot \dfrac{7}{7}} = \sqrt{\dfrac{7}{49}} = \dfrac{\sqrt{7}}{\sqrt{49}} = \dfrac{\sqrt{7}}{7}$

D. $\dfrac{\sqrt{5}}{\sqrt{6}} = \sqrt{\dfrac{5}{6}} = \sqrt{\dfrac{5}{6} \cdot \dfrac{6}{6}} = \sqrt{\dfrac{30}{36}} = \dfrac{\sqrt{30}}{\sqrt{36}} = \dfrac{\sqrt{30}}{6}$

E. $\dfrac{8}{3\sqrt{11x}} = \dfrac{8}{3\sqrt{11x}} \cdot \dfrac{\sqrt{11x}}{\sqrt{11x}} = \dfrac{8\sqrt{11x}}{3(11x)} = \dfrac{8\sqrt{11x}}{33x}$

10.12 PRACTICE PROBLEMS

1. Simplify each expression by combining like terms.

   A. $5\sqrt{11}+3\sqrt{11}$ B. $3\sqrt{54}+4\sqrt{6}$

   C. $4\sqrt{72}-5\sqrt{8}+\sqrt{50}$ D. $\sqrt{75y}-\sqrt{3y}-\sqrt{12y}$

   E. $5\sqrt{108x}-\sqrt{27y}$

2. Find each product and simplify.

   A. $\sqrt{50}\cdot\sqrt{6}$ B. $(2\sqrt{x})(-7\sqrt{y})$

   C. $\sqrt[3]{9}\sqrt[3]{18}$ D. $(\sqrt{x+1})(-3\sqrt{x+1})$

   E. $\sqrt{3}(\sqrt{7x}-\sqrt{15y})$

3. Simplify each expression completely.

   A. $\dfrac{3}{\sqrt{6}}$ B. $\dfrac{6}{\sqrt{21}}$ C. $\sqrt{\dfrac{1}{6t}}$ D. $\dfrac{\sqrt{27}}{\sqrt{54}}$ E. $\dfrac{\sqrt{7}}{\sqrt{27}}$

10.13 SOLUTIONS

1. A. $5\sqrt{11}+3\sqrt{11}=8\sqrt{11}$

   B. $3\sqrt{54}+4\sqrt{6}=3\sqrt{9}\sqrt{6}+4\sqrt{6}=3(3)\sqrt{6}+$
   $4\sqrt{6}=9\sqrt{6}+4\sqrt{6}=13\sqrt{6}$

   C. $4\sqrt{72}-5\sqrt{8}+\sqrt{50}=4\sqrt{36}\sqrt{2}-5\sqrt{4}\sqrt{2}+$
   $\sqrt{25}\sqrt{2}=4(6)\sqrt{2}-5(2)\sqrt{2}+5\sqrt{2}=24\sqrt{2}-$
   $10\sqrt{2}+5\sqrt{2}=19\sqrt{2}$

   D. $\sqrt{75y}-\sqrt{3y}-\sqrt{12y}=\sqrt{25}\sqrt{3y}-\sqrt{3y}-$
   $\sqrt{4}\sqrt{3y}=5\sqrt{3y}-\sqrt{3y}-2\sqrt{3y}=2\sqrt{3y}$

   E. $5\sqrt{108x}-\sqrt{27x}=5\sqrt{36}\sqrt{3x}-\sqrt{9}\sqrt{3x}=$
   $5(6)\sqrt{3x}-3\sqrt{3x}=30\sqrt{3x}-3\sqrt{3x}=27\sqrt{3x}$

2. A. $\sqrt{50}\cdot\sqrt{6}=\sqrt{300}=\sqrt{100}\sqrt{3}=10\sqrt{3}$

   B. $(2\sqrt{x})(-7\sqrt{y})=-14\sqrt{xy}$

   C. $\sqrt[3]{9}\sqrt[3]{18}=\sqrt[3]{162}=\sqrt[3]{27}\sqrt[3]{6}=3\sqrt[3]{6}$

   D. $(\sqrt{x+1})(-3\sqrt{x+1})=-3(x+1)=-3x-3$

   E. $\sqrt{3}(\sqrt{7x}-\sqrt{15y})=\sqrt{21x}-\sqrt{45y}=\sqrt{21x}-$
   $3\sqrt{5y}$

3. A. $\dfrac{3}{\sqrt{6}}=\dfrac{3}{\sqrt{6}}\cdot\dfrac{\sqrt{6}}{\sqrt{6}}=\dfrac{3\sqrt{6}}{6}=\dfrac{\sqrt{6}}{2}$

   B. $\dfrac{6}{\sqrt{21}}=\dfrac{6}{\sqrt{21}}\cdot\dfrac{\sqrt{21}}{\sqrt{21}}=\dfrac{6\sqrt{21}}{21}=\dfrac{2\sqrt{21}}{7}$

   C. $\sqrt{\dfrac{1}{6t}}=\sqrt{\dfrac{1}{6t}\cdot\dfrac{6t}{6t}}=\sqrt{\dfrac{6t}{(6t)^2}}=\dfrac{\sqrt{6t}}{6t}$

   D. $\dfrac{\sqrt{27}}{\sqrt{54}}=\sqrt{\dfrac{27}{54}}=\sqrt{\dfrac{1}{2}}=\sqrt{\dfrac{1}{2}\cdot\dfrac{2}{2}}=\sqrt{\dfrac{2}{4}}=\dfrac{\sqrt{2}}{2}$

   E. $\dfrac{\sqrt{7}}{\sqrt{27}}=\dfrac{\sqrt{7}}{\sqrt{27}}\cdot\dfrac{\sqrt{3}}{\sqrt{3}}=\dfrac{\sqrt{21}}{\sqrt{81}}=\dfrac{\sqrt{21}}{9}$

10.14 ALGEBRA TEST 1

Use the following test to assess how well you have mastered the material in this chapter so far. Mark your answers by blackening the corresponding answer oval in each question. An answer key and solutions are provided at the end of the test.

1. Which is equal to $(x^2+2x+3)-(2x^2+x-5)-(-x^2-3x+1)$?

   Ⓐ $4x+7$

   Ⓑ $4x-1$

   Ⓒ $-2x^2+4x+7$

   Ⓓ $2x^2+4x-1$

   Ⓔ $-2x^2+9$

2. Which is equal to $(x^3)^2 + 5^3(5^4)$?

   (A) $x^9 + 5^7$
   (B) $x^9 + 25^7$
   (C) $x^6 + 25^7$
   (D) $x^6 + 25^{12}$
   (E) $x^6 + 5^7$

3. Which is $\sqrt{300x^5y^9}$ completely simplified?

   (A) $10x^2y^4\sqrt{3xy}$
   (B) $10x^2y^2\sqrt{3xy}$
   (C) $10x^2y^3\sqrt{3x}$
   (D) $3x^2y^4\sqrt{10xy}$
   (E) $5xy\sqrt{8x^3y^5}$

4. Which is $2\sqrt{108y} - \sqrt{27y} + \sqrt{363y}$ completely simplified?

   (A) $14\sqrt{3y}$
   (B) $14\sqrt{3}$
   (C) $9\sqrt{3y} + \sqrt{363y}$
   (D) $20\sqrt{3y}$
   (E) $20\sqrt{y}$

5. Which number is equal to -3^4?

   (A) -81
   (B) -12
   (C) $\dfrac{1}{81}$
   (D) $\dfrac{1}{12}$
   (E) 81

6. Which number is equal to 5^{-3}?

   (A) -125
   (B) -15
   (C) $\dfrac{1}{125}$
   (D) $\dfrac{1}{15}$
   (E) 125

7. Which number is equal to $\sqrt[3]{27}$?

   (A) 3
   (B) 5
   (C) 3^3
   (D) $3\sqrt[3]{3}$
   (E) $9\sqrt[3]{3}$

8. Which number is equal to $2^3 \times 3^2$?

   (A) 36
   (B) 72
   (C) 5^5
   (D) 6^5
   (E) 6^6

9. Which expression is equal to $\left(x^{-2}\right)^{-5}$, for $x \neq 0$?

   (A) x^7
   (B) x^{10}
   (C) $\dfrac{1}{x^7}$
   (D) $\dfrac{1}{x^{10}}$
   (E) $\dfrac{1}{x^{32}}$

10. Which expression is equivalent to $x^{\frac{3}{4}}$, $x > 0$?

    (A) $x\sqrt[3]{x}$
    (B) $\sqrt[3]{x^4}$
    (C) $\sqrt[4]{x^3}$
    (D) $\left(x^3\right)^4$
    (E) $\left(x^4\right)^3$

11. Which expression is equivalent to $(5ax)^0$, $a \neq 0$, $x \neq 0$?

    (A) 0
    (B) 1
    (C) 5
    (D) $5\,ax$
    (E) $\dfrac{1}{5ax}$

12. Which number is equal to $\sqrt[4]{25}$?

    (A) $\sqrt[4]{5}$
    (B) $\sqrt{5}$
    (C) $\sqrt{25}$
    (D) 5
    (E) 25

13. Which number is equal to $(-4)^0$?

    (A) -4
    (B) -1
    (C) 0
    (D) 1
    (E) 4

14. Which expression is equivalent to $\sqrt[12]{x^4 y^2}$?

(A) $\sqrt{xy}$

(B) $\sqrt[3]{x^2 y}$

(C) $\sqrt[3]{xy^2}$

(D) $\sqrt[6]{x^4 y}$

(E) $\sqrt[6]{x^2 y}$

15. Which expression is equivalent to $\sqrt{18xy}$?

(A) $2\sqrt{14xy}$

(B) $4\sqrt{14xy}$

(C) $3\sqrt{xy}$

(D) $3\sqrt{2xy}$

(E) $9\sqrt{2xy}$

16. Which shows $\sqrt{\dfrac{7}{3}}$ written in simplest form?

(A) $\dfrac{\sqrt{21}}{9}$

(B) $\dfrac{3\sqrt{21}}{21}$

(C) $\dfrac{7}{\sqrt{21}}$

(D) $\dfrac{\sqrt{21}}{3}$

(E) $\sqrt{21}$

17. Which shows $\sqrt{\dfrac{5}{8}}$ written in simplest form?

(A) $\dfrac{\sqrt{10}}{4}$

(B) $\dfrac{\sqrt{40}}{8}$

(C) $\dfrac{5}{2\sqrt{10}}$

(D) $\dfrac{\sqrt{40}}{8}$

(E) $\dfrac{2\sqrt{10}}{64}$

18. Which shows $\dfrac{\sqrt{2}}{\sqrt{30}}$ written in simplest form?

(A) $\sqrt{\dfrac{2}{30}}$

(B) $\dfrac{1}{\sqrt{15}}$

(C) $\dfrac{\sqrt{15}}{15}$

(D) $\sqrt{\dfrac{6}{90}}$

(E) $\dfrac{\sqrt{6}}{30}$

19. Which shows $\sqrt[3]{648}$ written in simplest form?

(A) $18\sqrt{2}$

(B) $6\sqrt{3}$

(C) $3\sqrt[3]{24}$

(D) $2\sqrt[3]{81}$

(E) $6\sqrt[3]{3}$

20. Which shows $\sqrt{18} \times \sqrt{6}$ in simplest form?

(A) $3\sqrt{6}$

(B) $6\sqrt{3}$

(C) $3\sqrt{12}$

(D) $\sqrt{108}$

(E) $3\sqrt{2} \times \sqrt{6}$

21. Which shows $5\sqrt{13} + 3\sqrt{13}$ in simplest form?

(A) $15\sqrt{169}$

(B) 195

(C) 104

(D) $8\sqrt{26}$

(E) $8\sqrt{13}$

22. Which shows $\sqrt{60} \times \sqrt{6}$ simplified completely?

(A) $\sqrt{66}$

(B) $6\sqrt{10}$

(C) $6\sqrt{11}$

(D) $10\sqrt{6}$

(E) 60

23. Which shows $4\sqrt{50} - 5\sqrt{18} + 3\sqrt{72}$ simplified completely?

    (A) $-60\sqrt{2}$
    (B) $2\sqrt{2}$
    (C) $8\sqrt{2}$
    (D) $10\sqrt{2}$
    (E) $23\sqrt{2}$

24. Which shows $\sqrt{80} \times \sqrt{75}$ simplified completely?

    (A) $12\sqrt{5}$
    (B) $15\sqrt{20}$
    (C) $10\sqrt{60}$
    (D) $20\sqrt{15}$
    (E) $\sqrt{6000}$

25. Which shows $\sqrt{5a^3} + \sqrt{45a^3}$ simplified completely?

    (A) $a(\sqrt{5} + \sqrt{45})$
    (B) $a\sqrt{50}$
    (C) $4a\sqrt{5a}$
    (D) $20a^2$
    (E) $4a\sqrt{10a}$

ALGEBRA TEST 1

Answer Key

1. A	6. C	11. B	16. D	21. E
2. E	7. A	12. B	17. A	22. B
3. A	8. B	13. D	18. C	23. E
4. D	9. B	14. E	19. E	24. D
5. A	10. C	15. D	20. B	25. C

▬ 10.15 SOLUTIONS

1. **A** $4x + 7$
$(x^2+2x+3)-(2x^2+x-5)-(-x^2-3x+1) = x^2+2x+3-2x^2-x+5+x^2+3x-1 = 0x^2 + 4x + 7 = 4x + 7.$

2. **E** $x^6 + 5^7$
$(x^3)^2 + 5^3 \cdot 5^4 = x^{3\cdot2} + 5^{3+4} = x^6 + 5^7.$

3. **A** $10x^2y^4\sqrt{3xy}$
$\sqrt{300x^5y^9} = \sqrt{3 \cdot 100 \cdot x^4 \cdot x \cdot y^8 \cdot y} = \sqrt{100x^4y^8}\sqrt{3xy} = 10x^2y^4\sqrt{3xy}.$

4. **D** $20\sqrt{3y}$
$2\sqrt{108y} - \sqrt{27y} + \sqrt{363y} = 2\sqrt{36}\sqrt{3y} - \sqrt{9}\sqrt{3y} + \sqrt{121}\sqrt{3y} = 2(6)\sqrt{3y} - 3\sqrt{3y} + 11\sqrt{3y} = 12\sqrt{3y} - 3\sqrt{3y} + 11\sqrt{3y} = 20\sqrt{3y}.$

5. **A** -81
$-3^4 = -3 \times -3 \times -3 \times -3 = -81.$

6. **C** $\dfrac{1}{125}$
$5^{-3} = \dfrac{1}{5^3} = \dfrac{1}{125}.$

7. **A** 3
$\sqrt[3]{27} = \sqrt[3]{3^3} = 3^{\frac{3}{3}} = 3^1 = 3.$

8. **B** 72
$2^3 \cdot 3^2 = 8 \cdot 9 = 72.$

9. **B** x^{10}
$\left(x^{-2}\right)^{-5} = x^{-2\times-5} = x^{10}.$

10. **C** $\sqrt[4]{x^3}$
$x^{\frac{3}{4}} = \left(x^3\right)^{\frac{1}{4}} = \sqrt[4]{x^3}.$

11. **B** 1
$(5ax)^0 = 5^0 \cdot a^0 \cdot x^0 = 1 \cdot 1 \cdot 1 = 1.$

12. **B** $\sqrt{5}$

$\sqrt[4]{25} = \sqrt[4]{5^2} = (5^2)^{\frac{1}{4}} = 5^{\frac{2}{4}} = 5^{\frac{1}{2}} = \sqrt{5}.$

13. **D** 1

$(-4)^0 = 1.$

14. **E** $\sqrt[6]{x^2y}$

$\sqrt[12]{x^4y^2} = \sqrt[12]{(x^2y)^2} = (x^2y)^{\frac{2}{12}} = (x^2y)^{\frac{1}{6}} = \sqrt[6]{x^2y}.$

15. **D** $3\sqrt{2xy}$

$\sqrt{18xy} = \sqrt{9} \cdot \sqrt{2xy} = 3\sqrt{2xy}.$

16. **D** $\dfrac{\sqrt{21}}{3}$

$\sqrt{\dfrac{7}{3}} = \sqrt{\dfrac{7}{3} \cdot \dfrac{3}{3}} = \sqrt{\dfrac{21}{9}} = \dfrac{\sqrt{21}}{\sqrt{9}} = \dfrac{\sqrt{21}}{3}.$

17. **A** $\dfrac{\sqrt{10}}{4}$

$\sqrt{\dfrac{5}{8}} = \sqrt{\dfrac{5}{8} \cdot \dfrac{2}{2}} = \sqrt{\dfrac{10}{16}} = \dfrac{\sqrt{10}}{\sqrt{16}} = \dfrac{\sqrt{10}}{4}.$

18. **C** $\dfrac{\sqrt{15}}{15}$

$\dfrac{\sqrt{2}}{\sqrt{30}} = \sqrt{\dfrac{2}{30}} = \sqrt{\dfrac{1}{15}} = \sqrt{\dfrac{1}{15} \cdot \dfrac{15}{15}} = \sqrt{\dfrac{15}{15^2}} = \dfrac{\sqrt{15}}{\sqrt{15^2}} = \dfrac{\sqrt{15}}{15}.$

19. **E** $6\sqrt[3]{3}$

$\sqrt[3]{648} = \sqrt[3]{8 \cdot 81} = \sqrt[3]{8 \cdot 27 \cdot 3} = \sqrt[3]{8 \cdot 27} \cdot \sqrt[3]{3} = 2 \cdot 3 \cdot \sqrt[3]{3} = 6\sqrt[3]{3}.$

20. **B** $6\sqrt{3}$

$\sqrt{18} \cdot \sqrt{6} = \sqrt{18 \cdot 6} = \sqrt{6 \cdot 6 \cdot 3} = \sqrt{6 \cdot 6} \cdot \sqrt{3} = 6\sqrt{3}.$

21. **E** $8\sqrt{13}$

$5\sqrt{13} + 3\sqrt{13} = (5 + 3)\sqrt{13} = 8\sqrt{13}.$

22. **B** $6\sqrt{10}$

$\sqrt{60} \cdot \sqrt{6} - \sqrt{60 \cdot 6} = \sqrt{6 \cdot 6 \cdot 10} - \sqrt{6 \cdot 6}\sqrt{10} = 6\sqrt{10}.$

23. **E** $23\sqrt{2}$

$4\sqrt{50} - 5\sqrt{18} + 3\sqrt{72} = 4\sqrt{25} \times \sqrt{2} - 5\sqrt{9} \times \sqrt{2} + 3\sqrt{36} \times \sqrt{2} =$
$4 \times 5 \times \sqrt{2} - 5 \times 3 \times \sqrt{2} + 3 \times 6 \times \sqrt{2} = 20\sqrt{2} - 15\sqrt{2} + 18\sqrt{2} = 23\sqrt{2}.$

24. **D** $20\sqrt{15}$

$\sqrt{80} \cdot \sqrt{75} = \sqrt{16}\sqrt{5}\sqrt{25}\sqrt{3} = 4\sqrt{5} \cdot 5\sqrt{3} = 4(5)\sqrt{5 \cdot 3} = 20\sqrt{15}.$

25. **C** $4a\sqrt{5a}$

$\sqrt{5a^3} + \sqrt{45a^3} = \sqrt{a^2}\sqrt{5a} + \sqrt{9a^2}\sqrt{5a} = a\sqrt{5a} + 3a\sqrt{5a} = 4a\sqrt{5a}.$

10.16 SOLVED GRE PROBLEMS

For each question, select the best answer unless otherwise instructed.

	Quantity A	Quantity B
1.	$\sqrt[3]{64}$	$\sqrt{36}$

(A) Quantity A is greater.
(B) Quantity B is greater.
(C) The two quantities are equal.
(D) The relationship cannot be determined from the given information.

2. **Which shows $2\sqrt{540}$ written in simplest form?**

(A) $6\sqrt{15}$
(B) $8\sqrt{15}$
(C) $12\sqrt{15}$
(D) $6\sqrt{60}$
(E) $4\sqrt{135}$

For this question, enter your answer in the box.

3. **What is the value of $\left(x^3\right)^2$ when $x = 3$?**

4. **Which is equal to $3^4 \times 4^2$?**

(A) 96
(B) 1,296
(C) 7^6
(D) 7^8
(E) 12^8

	Quantity A	Quantity B
5.	$\sqrt[3]{16x^6}$	$\sqrt[3]{8x^3}$

(A) Quantity A is greater.
(B) Quantity B is greater.
(C) The two quantities are equal.
(D) The relationship cannot be determined from the given information.

10.17 SOLUTIONS

1. **B** Quantity B is greater.
   $\sqrt[3]{64} = \sqrt[3]{4^3} = 4$. $\sqrt{36} = 6$. $6 > 4$, so $\sqrt{36} > \sqrt[3]{64}$.

2. **C** $12\sqrt{15}$
   $2\sqrt{540} = 2\sqrt{36 \times 15} = 2 \times 6 \times \sqrt{15} = 12\sqrt{15}$.

3. 729

    When $x = 3$, $\left(x^3\right)^2 = \left(3^3\right)^2 = 27^2 = 729$.

4. **B** 1,296

    $3^4 \times 4^2 = 81 \times 16 = 1,296$.

5. **D** The relationship cannot be determined from the given information.
    If $x = 0$, $\sqrt[3]{16x^6} = 0$ and $\sqrt[3]{8x^3} = 0$. So $\sqrt[3]{16x^6} = \sqrt[3]{8x^3}$. If x does not equal zero, the quantities will not be equal.

10.18 GRE PRACTICE PROBLEMS

For each question, select the best answer unless otherwise instructed.

1. **What is the value of $(-5)^3$?**

    (A) -125

    (B) -25

    (C) $-\dfrac{1}{125}$

    (D) 25

    (E) 125

2. **What is $\sqrt[3]{243}$?**

    (A) 3

    (B) $3 \times \sqrt[3]{9}$

    (C) $9 \times \sqrt[3]{3}$

    (D) $9\sqrt{3}$

    (E) 27

	Quantity A	**Quantity B**
3.	$5^2 \times 2^3$	$4^2 \times 3^3$

(A) Quantity A is greater.

(B) Quantity B is greater.

(C) The two quantities are equal.

(D) The relationship cannot be determined from the given information.

For this question, enter your answer in the box.

4. **What is the value of $\left(-5 \times 4^2\right)^0$?**

5. **Which number is equal to $\sqrt[6]{x^4y^2}$?**

   (A) xy

   (B) $\sqrt[0]{xy}$

   (C) $\sqrt[3]{x^2y}$

   (D) $\sqrt[3]{x^4y}$

   (E) $\sqrt[3]{x^2y^2}$

Quantity A	**Quantity B**
$(16)^{\frac{3}{4}}$	$\left(-2^{-3}\right)^{-2}$

6.

   (A) Quantity A is greater.

   (B) Quantity B is greater.

   (C) The two quantities are equal.

   (D) The relationship cannot be determined from the given information.

Quantity A	**Quantity B**
$\left(3^{-2}\right)^0$	$\left(4^0\right)^8$

7.

   (A) Quantity A is greater.

   (B) Quantity B is greater.

   (C) The two quantities are equal.

   (D) The relationship cannot be determined from the given information.

8. **Which expression is equal to $\sqrt{18x^9}$?**

   (A) $3x^4\sqrt{2x}$

   (B) $9x^8\sqrt{2x}$

   (C) $x^3\sqrt{18}$

   (D) $3x^3\sqrt{2}$

   (E) $9\sqrt{2x^9}$

Quantity A	**Quantity B**
$\sqrt[3]{4^6 \times 3^3}$	$\sqrt[4]{2^8 \times 5^4}$

9.

   (A) Quantity A is greater.

   (B) Quantity B is greater.

   (C) The two quantities are equal.

   (D) The relationship cannot be determined from the given information.

10. Which shows $\sqrt{\dfrac{3}{5}}$ written in simplest form?

(A) $\dfrac{9}{25}$

(B) $\dfrac{3}{5}$

(C) $\dfrac{\sqrt{60}}{10}$

(D) $\dfrac{\sqrt{15}}{5}$

(E) $\dfrac{3}{\sqrt{15}}$

ANSWER KEY

1. A	6. B
2. B	7. C
3. B	8. A
4. $\boxed{1}$	9. A
5. C	10. D

10.19 TRANSLATING VERBAL EXPRESSIONS INTO ALGEBRAIC EXPRESSIONS

Algebra frequently involves translating English expressions and statements into algebraic expressions and statements.

Keywords for Operations

Operations	Keywords
Addition, +	Sum, total, plus, combined, joined, more than, all together, added to
Subtraction, −	Difference, less than, subtracted from, reduced by, decreased by, minus
Multiplication, ×	Times, product, twice, double, triple, of, multiplied by
Division, ÷	Divided by, ratio of

Example:

Translate each into an algebraic expression.

A. The sum of x and 2 is less than 10.
B. The product of 3 and x is greater than 15.
C. Two less than the product of 10 and x does not equal 100.
D. The product of 5 and y increased by 1 equals 7.
E. Three more than the quotient of w and 4 equals 13.

Solution:

A. $x + 2 < 10$
B. $3x > 15$
C. $10x - 2 \neq 100$
D. $5y + 1 = 7$
E. $\dfrac{w}{4} + 3 = 13$

10.20 EVALUATING ALGEBRAIC EXPRESSIONS

Evaluate an algebraic expression by replacing each variable by its value, and then use the order of operations to find the value of the expression.

Example:

Evaluate each expression.

A. $40 - 9x^2$, for $x = -3$

B. $9y - 16y^2$, for $y = 2$

C. $x^2 - 3xy - y + 4$, for $x = 2$ and $y = 3$

D. $5xy - 2x^2y + 7$, for $x = -1$ and $y = -2$

E. $w^3 - w + 8$, for $w = 4$

Solution:

A. $40 - 9x^2 = 40 - 9(-3)^2 = 40 - 9(9) = 40 - 81 = -41$

B. $9y - 16y^2 = 9(2) - 16(2)^2 = 18 - 16(4) = 18 - 64 = -46$

C. $x^2 - 3xy - y + 4 = (2)^2 - 3(2)(3) - (3) + 4 = 4 - 18 - 3 + 4 = -13$

D. $5xy - 2x^2y + 7 = 5(-1)(-2) - 2(-1)^2(-2) + 7 = 10 + 4 + 7 = 21$

E. $w^3 - w + 8 = 4^3 - 4 + 8 = 64 - 4 + 8 = 68$

10.21 EVALUATING FORMULAS

A formula is a special equation in which the left side tells what you are finding, and the right side tells you how to determine the value of the quantity.

To evaluate $C = \dfrac{5}{9}(F - 32)$ when $F = 95$, replace F by 95 to get:

$$C = \frac{5}{9}(95 - 32) = \frac{5}{9}(63) = 5(7) = 35$$

Example:

Evaluate each formula.

A. $P = 2(l + w)$ when $l = 20$ and $w = 15$.

B. $S = P(1 + rt)$ when $P = \$4,000$, $r = 7\%$, and $t = 5$ years.

C. $A = s^2$, when $s = 15$ ft.

D. $P = 6s$, when $s = 10$ cm.

E. $A = lw$, when $l = 7$ and $w = 5$.

Solution:

A. $P = 2(l + w) = 2(20 + 15) = 2(35) = 70$

B. $S = P(l + rt) = \$4,000(1 + 0.07(5)) = \$4,000(1.35) = \$5,400$

C. $A = s^2 = (15 \text{ ft})^2 = 225 \text{ ft}^2$

D. $P = 6s = 6(10 \text{ cm}) = 60 \text{ cm}$

E. $A = lw = 7(5) = 35$

10.22 PRACTICE PROBLEMS

1. Translate each statement into an algebraic expression.

A. Twice a number x plus the product of 2 more than the number x and 3 is less than 5.

B. A number n less 4 times the number is n plus 5.

C. The difference of x and 8 is less than the product of x and 8.

D. The cube of a number n less the square of n is 8.

E. The product of a number x and 7 decreased by 4 is 35.

2. Evaluate each expression when $x = 3$ and $y = -2$.

  A. $5x(2y + 7)$
  B. $2x + 3y - xy$
  C. $x^2 - y^2$
  D. $5(y + 5) - (3 + 2x)$
  E. $x^3 y^4$

3. Evaluate each formula.

  A. $I = prt$, for $p = \$8,000, r = 4\%$, and $t = 3$ years
  B. $F = 1.4C + 32$, for $C = 30$
  C. $V = lwh$, for $l = 8, w = 5$, and $h = 6$
  D. $P = 3s$, for $s = 6$
  E. $P = 2l + 2w$, for $l = 16$ ft, $w = 10$ ft

▬▬ 10.23 SOLUTIONS

1. A. $2x + (x + 2)3 < 5$
  B. $n - 4n = n + 5$
  C. $x - 8 < 8x$
  D. $n^3 - n^2 = 8$
  E. $7x - 4 = 35$

2. A. $5x(2y + 7) = 5 \cdot 3(2 \cdot (-2) + 7) = 15(-4 + 7) =$
    $15(3) = 45$
  B. $2x + 3y - xy = 2(3) + 3(-2) - 3(-2) = 6 - 6 + 6 =$
    6
  C. $x^2 - y^2 = (3)^2 - (-2)^2 = 9 - 4 = 5$

  D. $5(y + 5) - (3 + 2x) = 5(-2 + 5) - (3 + 2 \cdot 3) =$
    $5(3) - (3 + 6) = 15 - 9 = 6$
  E. $x^3 y^4 = (3)^3(-2)^4 = 27(16) = 432$

3. A. $I = prt = (\$8,000)(0.04)3 = \$960$
  B. $F = 1.4C + 32 = 1.4(30) + 32 = 42 + 32 = 74$
  C. $V = lwh = 8(5)(6) = 240$
  D. $P = 3s = 3(6) = 18$
  E. $P = 2l + 2w = 2(16 \text{ ft}) + 2(10 \text{ ft}) = 32 \text{ ft} + 20 \text{ ft} =$
    52 ft

▬▬ 10.24 ADDITION AND SUBTRACTION OF ALGEBRAIC EXPRESSIONS

To combine two algebraic expressions by addition or subtraction, find the like terms and then add or subtract their coefficients as indicated. You align the problems so that the like terms form a column. When you get the total for a column, write the total and the sign for the term, even if the total has a positive coefficient.

Example:

Add these expressions.

A. $\quad 5x^2 + 3x + 2$
  $+ \underline{3x^2 + 2x + 7}$

B. $\quad 2x^2 - 7x + 8$
  $+ \underline{3x^2 - 8x - 12}$

C. $\quad 3y^3 - 5y^2 + 6y + 4$
  $\quad 2y^3 \qquad + 8y - 7$
  $+ \underline{\qquad 7y^2 - 15y - 5}$

Solution:

A. $\quad 5x^2 + 3x + 2$
  $+ \underline{3x^2 + 2x + 7}$
  $\quad 8x^2 + 5x + 9$

B. $\quad 2x^2 - 7x + 8$
  $+ \underline{3x^2 - 8x - 12}$
  $\quad 5x^2 - 15x - 4$

C. $\quad 3y^3 - 5y^2 + 6y + 4$
  $\quad 2y^3 \qquad + 8y - 7$
  $+ \underline{\qquad 7y^2 - 15y - 5}$
  $\quad 5y^3 + 2y^2 - y - 8$

Example:

Subtract these expressions.

A. $\begin{array}{r} -5a^2b + 3ab^2 + 11 \\ -\ -7a^2b + 8ab^2 - 18 \\ \hline \end{array}$

B. $\begin{array}{r} 3x^2 - 4x + 6 \\ -\ \ x^2 - 5x + 8 \\ \hline \end{array}$

C. $\begin{array}{r} 9y^3 + 5y^2 - 4 \\ -\ 4y^3 - 5y\ \ + 8 \\ \hline \end{array}$

Solution:

You change the signs of the terms in the subtrahend and add. That is, you change from subtraction to the addition of the opposite.

A. $\begin{array}{r} -5a^2b + 3ab^2 + 11 \\ -\ -7a^2b + 8ab^2 - 18 \\ \hline 2a^2b - 5ab^2 + 29 \end{array}$

B. $\begin{array}{r} 3x^2 - 4x + 6 \\ -\ \ x^2 - 5x + 8 \\ \hline 2x^2 + x\ \ - 2 \end{array}$

C. $\begin{array}{r} 9y^3 + 5y^2 \qquad\ \ - 4 \\ -\ 4y^3 \qquad\ \ - 5y\ \ + 8 \\ \hline 5y^3 + 5y^2 +\ 5y - 12 \end{array}$

◼ 10.25 MULTIPLICATION OF ALGEBRAIC EXPRESSIONS

To multiply an algebraic expression by a monomial, multiply each term in the algebraic expression by the monomial. For example, $3y^2(4y^3 + 7y) = 3y^2(4y^3) + 3y^2(7y) = 3 \cdot 4 \cdot y^2 \cdot y^3 + 3 \cdot 7 \cdot y^2 \cdot y = 12y^{2+3} + 21y^{2+1} = 12y^5 + 21y^3$. Thus, $3y^2(4y^3 + 7y) = 12y^5 + 21y^3$.

Example:

Find the product of these expressions.

A. $7x(5x - 4)$ B. $3y^4(y^2 - 5y + 7)$ C. $-3y^2(-5y^3 - 7y^2 + 12)$

Solution:

A. $7x(5x - 4) = 7x(5x) + 7x(-4) = 35x^2 - 28x$
B. $3y^4(y^2 - 5y + 7) = 3y^4(y^2) + 3y^4(-5y) + 3y^4(7) = 3y^{4+2} - 15y^{4+1} + 21y^4 = 3y^6 - 15y^5 + 21y^4$
C. $-3y^2(-5y^3 - 7y^2 + 12) = -3y^2(-5y^3) - 3y^2(-7y^2) - 3y^2(12) = 15y^{2+3} + 21y^{2+2} - 36y^2 = 15y^5 + 21y^4 - 36y^2$

When an algebraic expression is multiplied by a binomial, multiply each term in the algebraic expression by each term of the binomial.

Example:

Find the product of these algebraic expressions.

A. $(3x - 2)(2x + 5)$ B. $(3x - 5)(2x^2 + 7x - 5)$

Solution:

A. $(3x-2)(2x+5) = 3x(2x+5) - 2(2x+5) = 6x^2 + 15x - 4x - 10 = 6x^2 + 11x - 10$

B. $(3x - 5)(2x^2 + 7x - 5) = 3x(2x^2) + 3x(7x) + 3x(-5) + (-5)(2x^2) + (-5)(7x) + (-5)(-5) = 6x^3 + 21x^2 - 15x - 10x^2 - 35x + 25 = 6x^3 + 11x^2 - 50x + 25$

Example:

Multiply these expressions.

A. $3x^3 - 5xy + 5y^3$
    $\times \quad\quad x - y$

B. $2y - 7$
    $\times\ y + 5$

C. $x^2 + 2xy - y^2$
    $\times\ x - xy + y$

Solution:

A. $3x^3 - 5xy + 5y^3$
    $\times \quad x - y$

$$3x^4 \qquad\quad - 5x^2y + 5xy^3$$
$$\quad - 3x^3y \qquad\qquad + 5xy^2 - 5y^4$$
$$3x^4 - 3x^3y - 5x^2y + 5xy^3 + 5xy^2 - 5y^4$$

B. $2y - 7$
    $\times\ y + 5$

$$2y^2 - 7y$$
$$\quad + 10y - 35$$
$$2y^2 + 3y - 35$$

C. $x^2 + 2xy - y^2$
    $\times\ x - xy + y$

$$x^3 + 2x^2y - xy^2$$
$$\qquad - x^3y - 2x^2y^2 + xy^3$$
$$\quad x^2y + 2xy^2 \qquad\qquad - y^3$$
$$x^3 + 3x^2y + xy^2 - x^3y - 2x^2y^2 + xy^3 - y^3$$

When you multiply two binomials, you get four terms. $(a + b)(c + d) = ac + ad + bc + bd$, where ac is the product of the **First** terms of the binomials; ad is the product of the **Outer** terms of the two binomials; bc is the product of the **Inner** terms of the binomials; and bd is the product of the **Last** terms of the binomials. Because the first letters of the words spell **FOIL**, the method of multiplying binomials is called the FOIL method. Frequently, the outer product and inner products are like terms and they are combined into a single term.

Using the FOIL method to multiply $(4x - 3y)(x - y)$, you get $4x^2 - 4xy - 3xy + 3y^2 = 4x^2 - 7xy + 3y^2$.

Example:

Multiply these binomials using the FOIL method.

A. $(2x + y)(x + 3y)$ B. $(3x - 2y)(2x + 3y)$ C. $(10x - 7)(5x - 8)$

D. $(9x - 7y)(5x - 4y)$ E. $(5x - 7y)(5x + 7y)$

Solution:

A. $(2x + y)(x + 3y) = (2x)(x) + (2x)(3y) + y(x) + y(3y) = 2x^2 + 6xy + xy + 3y^2 = 2x^2 + 7xy + 3y^2$

B. $(3x - 2y)(2x + 3y) = (3x)(2x) + (3x)(3y) + (-2y)(2x) + (-2y)(3y) = 6x^2 + 9xy - 4xy - 6y^2 = 6x^2 + 5xy - 6y^2$

C. $(10x - 7)(5x - 8) = 50x^2 - 80x - 35x + 56 = 50x^2 - 115x + 56$

D. $(9x - 7y)(5x - 4y) = 45x^2 - 36xy - 35xy + 28y^2 = 45x^2 - 71xy + 28y^2$

E. $(5x - 7y)(5x + 7y) = 25x^2 + 35xy - 35xy - 49y^2 = 25x^2 - 49y^2$

Some types of binomials are multiplied so often that we have patterns for the way the answer will look. The square of a binomial is one such special product. $(a + b)^2 = a^2 + 2ab + b^2$ and $(a - b)^2 = a^2 - 2ab + b^2$. The product of the sum and difference of two terms is another special product. $(a + b)(a - b) = a^2 - b^2$.

Example:

Multiply these special products.

A. $(2x - 3y)^2$ B. $(5x + 2y)^2$ C. $(2x + 3y)(2x - 3y)$ D. $(9x - y)(9x + y)$

Solution:

A. $(2x - 3y)^2 = (2x)^2 - 2(2x)(3y) + (3y)^2 = 4x^2 - 12xy + 9y^2$

B. $(5x + 2y)^2 = (5x)^2 + 2(5x)(2y) + (2y)^2 = 25x^2 + 20xy + 4y^2$

C. $(2x + 3y)(2x - 3y) = (2x)^2 - (3y)^2 = 4x^2 - 9y^2$

D. $(9x - y)(9x + y) = (9x)^2 - (y)^2 = 81x^2 - y^2$

Sometimes you need to multiply more than two binomials. To do that, use the FOIL method to multiply the first two binomials and then multiply that product by the third binomial. Occasionally, you have the cube of a binomial, which is a special product.

$(a + b)^3 = a^3 + 3a^2b + 3ab^2 + b^3$.
$(a - b)^3 = a^3 - 3a^2b + 3ab^2 - b^3$.

Example:

Multiply these binomials.

A. $(2x - 3y)(x + y)(x - 5y)$ B. $(3x + 2y)(x - y)(2x + y)$
C. $(2x - y)^3$ D. $(3x + 5y)^3$

Solution:

A. $(2x - 3y)(x + y)(x - 5y) = (2x^2 - xy - 3y^2)(x - 5y) = 2x^3 - x^2y - 3xy^2 - 10x^2y + 5xy^2 + 15y^3 = 2x^3 - 11x^2y + 2xy^2 + 15y^3$

B. $(3x + 2y)(x - y)(2x + y) = (3x^2 - xy - 2y^2)(2x + y) = 6x^3 - 2x^2y - 4xy^2 + 3x^2y - xy^2 - 2y^3 = 6x^3 + x^2y - 5xy^2 - 2y^3$

C. $(2x - y)^3 = (2x)^3 - 3(2x)^2y + 3(2x)(y)^2 - (y)^3 = 8x^3 - 12x^2y + 6xy^2 - y^3$

D. $(3x + 5y)^3 = (3x)^3 + 3(3x)^2(5y) + 3(3x)(5y)^2 + (5y)^3 = 27x^3 + 3(9x^2)5y + 9x(25y^2) + 125y^3 = 27x^3 + 135x^2y + 225xy^2 + 125y^3$

■■■ 10.26 DIVISION OF ALGEBRAIC EXPRESSIONS

To divide one monomial by another monomial, use integer divisions and the laws of exponents.

For example, $\dfrac{6x^2y}{2xy} = \dfrac{6}{2} \cdot \dfrac{x^2}{x} \cdot \dfrac{y}{y} = 3x \cdot 1 = 3x$

and $\dfrac{12x^6y^7}{-4x^4y^9} = \dfrac{12}{-4} \cdot \dfrac{x^6}{x^4} \cdot \dfrac{y^7}{y^9} = -3 \cdot \dfrac{x^2}{y^2} = \dfrac{-3x^2}{y^2}$

Example:

Divide these monomials.

A. $14x^3yz \div 7x^5yz^2$ B. $9xy^2 \div -3x^4y$ C. $10y^7 \div 4y^3$ D. $\dfrac{36x^2y^7z^3}{15x^2y^5z^5}$

Solution:

A. $\dfrac{14x^3yz}{7x^5yz^2} = \dfrac{14}{7} \cdot \dfrac{x^3}{x^5} \cdot \dfrac{y}{y} \cdot \dfrac{z}{z^2} = 2 \cdot \dfrac{1}{x^2} \cdot 1 \cdot \dfrac{1}{z} = \dfrac{2}{x^2z}$

B. $\dfrac{9xy^2}{-3x^4y} = \dfrac{9}{-3} \cdot \dfrac{x}{x^4} \cdot \dfrac{y^2}{y} = -3 \cdot \dfrac{1}{x^3} \cdot \dfrac{y}{1} = \dfrac{-3y}{x^3}$

C. $\dfrac{10y^7}{4y^3} = \dfrac{10}{4} \cdot \dfrac{y^7}{y^3} = \dfrac{5}{2} \cdot \dfrac{y^4}{1} = \dfrac{5y^4}{2}$

D. $\dfrac{36x^2y^7z^3}{15x^2y^5z^5} = \dfrac{36}{15} \cdot \dfrac{x^2}{x^2} \cdot \dfrac{y^7}{y^5} \cdot \dfrac{z^3}{z^5} = \dfrac{12}{5} \cdot 1 \cdot \dfrac{y^2}{1} \cdot \dfrac{1}{z^2} = \dfrac{12y^2}{5z^2}$

Example:

Divide the polynomial by the monomial.

A. $(9ab^2 - 6a^2) \div 3a$ B. $(15x^2y^5 - 25x^4y^3) \div (-5xy^2)$

C. $(4x^2y^3 - 16xy^3 + 4xy) \div (2xy)$ D. $[16(a+b)^4 + 12(a+b)^3] \div [2(a+b)]$

Solution:

A. $\dfrac{9ab^2 - 6a^2}{3a} = \dfrac{9ab^2}{3a} + \dfrac{-6a^2}{3a} = 3b^2 - 2a$

B. $\dfrac{15x^2y^5 - 25x^4y^3}{-5xy^2} = \dfrac{15x^2y^5}{-5xy^2} + \dfrac{-25x^4y^3}{-5xy^2} = -3xy^3 + 5x^3y$

C. $\dfrac{4x^2y^3 - 16xy^3 + 4xy}{2xy} = \dfrac{4x^2y^3}{2xy} + \dfrac{-16xy^3}{2xy} + \dfrac{4xy}{2xy} = 2xy^2 - 8y^2 + 2$

D. $\dfrac{16(a+b)^4 + 12(a+b)^3}{2(a+b)} = \dfrac{16(a+b)^4}{2(a+b)} + \dfrac{12(a+b)^3}{2(a+b)} = 8(a+b)^3 + 6(a+b)^2$

10.27 PRACTICE PROBLEMS

1. Add these algebraic expressions.

A. $\quad 2x^2 + \ y^2 - x + y$
$\quad\quad -x^2 + 3y^2 - x$
$\quad + \ \ x^2 - 4y^2 + x - 2y$

B. $\quad x^2 - \ xy + 2yz + 3z^2$
$\quad + x^2 + 2xy - 3yz - 8z^2$

C. $\quad 5x^3 - \ 4x^2 + \ 7x - \ 4$
$\quad + 3x^3 + 11x^2 - 17x + 12$

2. Subtract these algebraic expressions.

A. $\quad\ 3xy - 2yz + 4xz$
$\quad - \ -2xy + \ yz + 3xz$

B. $\quad 3x^4 - 7x^2 + 8x$
$\quad - \ 5x^4 \quad\quad\quad - 7x + 12$

C. $\quad 3x^2 - 5xy - \ y^2$
$\quad - \ \ x^2 - 5xy + 14y^2$

3. Multiply these algebraic expressions.

A. $\quad 5x^2 + 3xy + y^2$
$\quad \times \quad\quad\quad 2x - y$

B. $\quad 5x^3 + 2x - 4$
$\quad \times \quad\quad\quad 3x - 2$

C. $\quad 3x + 7y$
$\quad \times \ 5x - 2y$

4. Multiply these algebraic expressions.

A. $(4x^2y^5)(-2xy)$ 　　B. $(a^2b^3c)(abc)(ab^3c^3)$
C. $(-3x^2y)(3xy)(-4xy)$

5. Multiply these binomials.

A. $(y + 3)(y + 7)$ 　　B. $(2x + 5)(3x - 7)$
C. $(2x - y)(x + 5y)$ 　　D. $(2x + 3y)(2x - 3y)$
E. $(x + 3y)^2$

6. Multiply these binomials.

A. $(x + 2)(3x - 1)(2x + 3)$ 　　B. $(5x - 2y)^3$

7. Divide as indicated.

A. $\dfrac{-12a^4b^7c^3}{3a^2b^4c}$ 　　B. $\dfrac{-16r^{10}s^3t^4}{-4r^7st^3}$

C. $\dfrac{4ab^3 - 6a^2bc + 12a^3b^2c^4}{-2ab}$

10.28 SOLUTIONS

1. A. $\quad\ 2x^2 + \ y^2 - x + \ y$
$\quad\quad -x^2 + 3y^2 - x$
$\quad + \ \ \ x^2 - 4y^2 + x - 2y$
$\quad\ \ \overline{\ 2x^2 \quad\quad\quad - x - y}$

B. $\quad x^2 - \ xy + 2yz + 3z^2$
$\quad + x^2 + 2xy - 3yz - 8z^2$
$\quad \overline{\ 2x^2 + xy - yz - 5z^2}$

C. $\quad 5x^3 - \ 4x^2 + \ 7x - \ 4$
$\quad + 3x^3 + 11x^2 - 17x + 12$
$\quad \overline{\ 8x^3 + \ 7x^2 - 10x + \ 8}$

2. A. $\quad\ 3xy - 2yz + 4xz$
$\quad - \ -2xy + \ yz + 3xz$
$\quad \overline{\ 5xy - 3yz + xz}$

B. $\quad 3x^4 - 7x^2 + \ 8x$
$\quad - \ 5x^4 \quad\quad\quad - 7x + 12$
$\quad \overline{-2x^4 - 7x^2 + 15x - 12}$

C. $\quad 3x^2 - 5xy - \ y^2$
$\quad - \ \ x^2 - 5xy + 14y^2$
$\quad \overline{\ 2x^2 \quad\quad\quad - 15y^2}$

3. A. $\quad 5x^2 + 3xy + y^2$
$\quad \times \quad\quad\quad 2x - y$
$\quad \overline{10x^3 + 6x^2y + 2xy^2}$
$\quad\quad\quad - 5x^2y - 3xy^2 - y^3$
$\quad \overline{10x^3 + \ x^2y - \ xy^2 - y^3}$

B. $\quad 5x^3 + 2x - 4$
$\quad \times \quad\quad\quad 3x - 2$
$\quad \overline{15x^4 \quad\quad\quad + 6x^2 - 12x}$
$\quad\quad - 10x^3 \quad\quad\quad - \ 4x + 8$
$\quad \overline{15x^4 - 10x^3 + 6x^2 - 16x + 8}$

C. $\quad 3x + 7y$
$\quad \times \ 5x - 2y$
$\quad \overline{15x^2 + 35xy}$
$\quad\quad\quad - \ 6xy - 14y^2$
$\quad \overline{15x^2 + 29xy - 14y^2}$

4. A. $(4x^2y^5)(-2xy) = -8x^3y^6$
B. $(a^2b^3c)(abc)(ab^3c^3) = a^4b^7c^5$
C. $(-3x^2y)(3xy)(-4xy) = 36x^4y^3$

5. A. $(y+3)(y+7) = y^2+7y+3y+21 = y^2+10y+21$
B. $(2x + 5)(3x - 7) = 6x^2 - 14x + 15x - 35 =$
$6x^2 + x - 35$

C. $(2x - y)(x + 5y) = 2x^2 + 10xy - xy - 5y^2 = 2x^2 + 9xy - 5y^2$

D. $(2x + 3y)(2x - 3y) = (2x)^2 - (3y)^2 = 4x^2 - 9y^2$

E. $(x + 3y)^2 = x^2 + 2(x)(3y) + (3y)^2 = x^2 + 6xy + 9y^2$

6. A. $(x + 2)(3x - 1)(2x + 3) = [3x^2 + 5x - 2](2x + 3) = 6x^3 + 10x^2 - 4x + 9x^2 + 15x - 6 = 6x^3 + 19x^2 + 11x - 6$

B. $(5x - 2y)^3 = (5x)^3 - 3(5x)^2(2y) + 3(5x)(2y)^2 - (2y)^3 = 125x^3 - 150x^2y + 60xy^2 - 8y^3$

7. A. $\dfrac{-12a^4b^7c^3}{3a^2b^4c} = -\dfrac{12}{3} \cdot \dfrac{a^4}{a^2} \cdot \dfrac{b^7}{b^4} \cdot \dfrac{c^3}{c} = -4a^2b^3c^2$

B. $\dfrac{-16r^{10}s^3t^4}{-4r^7s\,t^3} = \dfrac{-16}{-4} \cdot \dfrac{r^{10}}{r^7} \cdot \dfrac{s^3}{s} \cdot \dfrac{t^4}{t^3} = 4r^3s^2t$

C. $\dfrac{4ab^3 - 6a^2bc + 12a^3b^2c^4}{-2ab} = \dfrac{4ab^3}{-2ab} + \dfrac{-6a^2bc}{-2ab} + \dfrac{12a^3b^2c^4}{-2ab} = -2b^2 + 3ac - 6a^2bc^4$

▬ 10.29 ALGEBRAIC FRACTIONS

One concern with algebraic fractions is that you must be sure that the fraction is defined. Thus, you need to know what values of the variable will result in the denominator becoming zero and then exclude those values.

Example:

For what values of the variable will the fraction not be defined?

A. $\dfrac{3x}{x - 7}$ B. $\dfrac{y + 6}{y + 10}$ C. $\dfrac{3x - 1}{(x + 6)(x - 5)}$ D. $\dfrac{3y - 5}{(y - 1)(2y + 1)}$

Solution:

A. $\dfrac{3x}{x - 7}$ Denominator is zero when $x = 7$

B. $\dfrac{y + 6}{y + 10}$ Denominator is zero when $y = -10$

C. $\dfrac{3x - 1}{(x + 6)(x - 5)}$ Denominator is zero when either factor is zero, so when $x = -6$ or $x = 5$

D. $\dfrac{3y - 5}{(y - 1)(2y + 1)}$ Denominator is zero when $y = 1$ or $y = -\dfrac{1}{2}$

To simplify fractions to lowest terms, use the distributive property to find the common factor between numerator and denominator.

Example:

Reduce each fraction to lowest terms.

A. $\dfrac{15x}{3x^2}$ B. $\dfrac{25x^2y^5}{45x^3y^2}$ C. $\dfrac{m^2 + m}{m^2 - m}$ D. $\dfrac{mn}{m^2 - 5m}$

Solution:

A. $\dfrac{15x}{3x^2} = \dfrac{3 \cdot 5x}{3 \cdot x \cdot x} = \dfrac{3x}{3x} \cdot \dfrac{5}{x} = 1 \cdot \dfrac{5}{x} = \dfrac{5}{x}$

B. $\dfrac{25x^2y^5}{45x^3y^2} = \dfrac{5x^2y^2 \cdot 5y^3}{5x^2y^2 \cdot 9x} = \dfrac{5x^2y^2}{5x^2y^2} \cdot \dfrac{5y^3}{9x} = 1 \cdot \dfrac{5y^3}{9x} = \dfrac{5y^3}{9x}$

C. $\dfrac{m^2 + m}{m^2 - m} = \dfrac{m \cdot m + m \cdot 1}{m \cdot m - m \cdot 1} = \dfrac{m(m + 1)}{m(m - 1)} = \dfrac{m + 1}{m - 1}$

D. $\dfrac{mn}{m^2 - 5m} = \dfrac{mn}{m(m - 5)} = \dfrac{n}{m - 5}$

10.30 FACTORING ALGEBRAIC EXPRESSIONS

When factoring polynomials, first look for the common monomial factor in each term. For example, $15x + 35 = 5(3x) + 5(7) = 5(3x + 7)$.

Example:

Factor each polynomial by factoring out the common monomial factor.

A. $2y^3 + 4y^2$ B. $4y^5 + 8y^3$ C. $3x^3y^4 - 9x^2y^3 - 6xy^2$ D. $-2x^2y - 8xy^2$

Solution:

A. $2y^3 + 4y^2 = 2y^2(y) + 2y^2(2) = 2y^2(y + 2)$
B. $4y^5 + 8y^3 = 4y^3(y^2) + 4y^3(2) = 4y^3(y^2 + 2)$
C. $3x^3y^4 - 9x^2y^3 - 6xy^2 = 3xy^2(x^2y^2 - 3xy - 2)$
D. $-2x^2y - 8xy^2 = -2xy(x + 4y)$

When there are four-term expressions, you can often factor the expression by grouping the terms into two groups of two terms and then factoring out a common monomial factor to get a common quantity. Finally, the quantity is factored out of each term. For example,

$$ax + 4x + ay + 4y = (ax + 4x) + (ay + 4y) = x(a + 4) + y(a + 4) = (a + 4)(x + y)$$

Example:

Factor each polynomial using grouping.

A. $5xy + 5xz + 2y + 2z$ B. $ac - ad - 2c + 2d$
C. $3xm + 3ym - 2x - 2y$ D. $3a^2 + 3ab + a + b$

Solution:

A. $5xy + 5xz + 2y + 2z = (5xy + 5xz) + (2y + 2z) = 5x(y + z) + 2(y + z) = (y + z)(5x + 2)$
B. $ac - ad - 2c + 2d = (ac - ad) + (-2c + 2d) = a(c - d) - 2(c - d) = (c - d)(a - 2)$
C. $3xm + 3ym - 2x - 2y = (3xm + 3ym) + (-2x - 2y) = 3m(x + y) - 2(x + y) = (x + y)(3m - 2)$
D. $3a^2 + 3ab + a + b = (3a^2 + 3ab) + (a + b) = 3a(a + b) + 1(a + b) = (a + b)(3a + 1)$

Special products can be used to factor algebraic expressions. Since $(a + b)(a - b) = a^2 - b^2$, you can factor the difference of two squares into the sum and the difference of the square roots of the terms. A perfect-square trinomial has two perfect-square terms, and the third term, usually in the middle, is plus or minus two times the square roots of the other terms. Since $(a + b)^2 = a^2 + 2ab + b^2$ and $(a - b)^2 = a^2 - 2ab + b^2$, you have the squares of two terms and twice the products of those terms.

Example:

Factor these algebraic expressions using special products.

A. $25a^2 - 9$ B. $x^2 + 4x + 4$ C. $4 - x^2$ D. $4x^2 - 20xy + 25y^2$

Solution:

A. $25a^2 - 9 = (5a)^2 - 3^2 = (5a + 3)(5a - 3)$
B. $x^2 + 4x + 4 = (x)^2 + 2(x)(2) + 2^2 = (x + 2)^2$
C. $4 - x^2 = 2^2 - x^2 = (2 + x)(2 - x)$
D. $4x^2 - 20xy + 25y^2 = (2x)^2 - 2(2x)(5y) + (5y)^2 = (2x - 5y)^2$

Other expressions can be factored using the FOIL method, especially when the first term is just the variable squared. For example, $x^2 + 6x + 5$ can be factored using the fact that you want factors of 5 that add up to 6. $(x+1)(x+5) = x^2 + 6x + 5$.

Example:

Factor each trinomial.

A. $x^2 - 9x + 20$ B. $x^2 + 7x + 12$ C. $x^2 - 29x - 30$

Solution:

A. $x^2 - 9x + 20$ Since the first term is x^2, each binomial will have a first term of x.

$(x\ \ )(x\ \ )$
$(x -)(x -)$ Since the constant term is positive and the middle term is negative, the signs for the binomials are " $-$." You need two factors of 20 that add together to total 9. $1(20) = 20, 1 + 20 = 21; 2(10) = 20, 2 + 10 = 12;$

$(x - 5)(x - 4)$ $5(4) = 20, 5 + 4 = 9.$
You can check the answer by FOIL.

B. $x^2 + 7x + 12$ Since the first term is x^2, each binomial will have a first term of x.

$(x\ \ )(x\ \ )$
$(x +)(x +)$ Since the constant term is positive and the middle
$(x + 3)(x + 4)$ term is positive, the signs for the binomials are " +." Two factors of 12 that have a sum of 7 are 3 and 4.

C. $x^2 - 29x - 30$ Since the first term is x^2, each binomial will have a first term of x.

$(x\ \ )(x\ \ )$
$(x +)(x -)$ Since the constant term is negative, one binomial has a " +" sign, and the other has a " $-$" sign. You need two factors of 30 that have a difference of -29.

$(x + 1)(x - 30)$ $1(30) = 30; 1 - 30 = -29.$ Therefore, the 1 is placed with the binomial that has the " +" sign, and the 30 is placed with the other binomial.

When the trinomial is in the form of $ax^2 + bx + c$, use the "ac" method to find two numbers that add together to yield b. For example, $6m^2 - 11m - 10$, $a = 6, b = -11, c = -10$, and $ac = -60$. You need two factors of -60, one positive and one negative, that will add together to yield -11. Now $4(15) = 60$ and $4 - 15 = -11$, so you want 4 and -15 as your factors of -60. $6m^2 - 11m - 10 =$

$6m^2 + (4 - 15)m - 10 = 6m^2 + 4m - 15m - 10$. Then use factoring by grouping, $6m^2 + 4m - 15m - 10 = (6m^2 + 4m) + (-15m - 10) = 2m(3m + 2) - 5(3m + 2) = (3m + 2)(2m - 5)$.

Example:

Factor the trinomials.

A. $12a^2 + 5ab - 3b^2$ B. $6x^2 + 13x + 6$ C. $4x^2 - 35xy - 9y^2$

Solution:

A. $12a^2 + 5ab - 3b^2$ $a = 12, b = 5, c = -3, ac = 12(-3) = -36$
    $12a^2 + (-4 + 9)ab - 3b^2$ You need two factors of -36 that sum to 5.

    $12a^2 - 4ab + 9ab - 3b^2$ $-4(9) = -36$ and $-4 + 9 = 5$.
    $(12u^2 - 4ub) + (9ab - 3b^2)$ Group the terms.
    $4a(3a - b) + 3b(3a - b)$ Factor each binomial.
    $(3a - b)(4a + 3b)$ Factor the binomial out of each term.
    Thus, $12a^2 + 5ab - 3b^2 = (3a - b)(4a + 3b)$.

B. $6x^2 + 13x + 6$ $a = 6, b = 13, c = 6, ac = 6(6) = 36$, so both
    $6x^2 + (4 + 9)x + 6$ factors have the same sign. In this case,
    $6x^2 + 4x + 9x + 6$ b is positive, so both factors are positive.
    $(6x^2 + 4x) + (9x + 6)$ $36 = 4(9)$ and $4 + 9 = 13$.
    $2x(3x + 2) + 3(3x + 2)$
    $(3x + 2)(2x + 3)$
    So, $6x^2 + 13x + 6 = (3x + 2)(2x + 3)$.

C. $4x^2 - 35xy - 9y^2$ $a = 4, b = -35, c = -9, ac = -36$, so the
    $4x^2 + (1 - 36)xy - 9y^2$ factors of ac have opposite signs.
    $4x^2 + xy - 36xy - 9y^2$ You need two factors of -36 that sum
    $(4x^2 + xy) + (-36xy - 9y^2)$ to -35. $-36 = 1(-36)$ and $1 - 36 = -35$.
    $x(4x + y) - 9y(4x + y)$
    $(4x + y)(x - 9y)$
    So, $4x^2 - 35xy - 9y^2 = (4x + y)(x - 9y)$.

▬ 10.31 PRACTICE PROBLEMS

1. Reduce these fractions to lowest terms.

   A. $\dfrac{35a^5b^4c^5}{7a^2b^8c^4}$ B. $\dfrac{3x - 6}{12x - 15}$ C. $\dfrac{2x + 5xy}{x^2y^2}$

2. Reduce these fractions to lowest terms.

   A. $\dfrac{xy - 5y}{5x - 25}$ B. $\dfrac{2x + 4}{7x + 14}$ C. $\dfrac{x^2 - x}{xy - y}$

3. Factor these expressions.

   A. $25x^2 - 9y^2$ B. $5x^2 - 10xy + 30y$
   C. $x^4 - 1$ D. $ax + bx - ay - by$
   E. $12x^2 - 30x^5$

4. Factor and reduce these fractions.

   A. $\dfrac{7x^2 - 5xy}{49x^3 - 25xy^2}$ B. $\dfrac{3x^2 + 7x + 4}{3x^2 + x - 4}$

   C. $\dfrac{y^2 + 4y}{y^2 + 6y + 8}$ D. $\dfrac{4x^2 - 9}{6x^2 - 9x}$

■ 10.32 SOLUTIONS

1. A. $\dfrac{35a^5b^4c^5}{7a^2b^8c^4} = \dfrac{35}{7} \cdot \dfrac{a^5}{a^2} \cdot \dfrac{b^4}{b^8} \cdot \dfrac{c^5}{c^4} = \dfrac{5}{1} \cdot \dfrac{a^3}{1} \cdot \dfrac{1}{b^4} \cdot \dfrac{c}{1} =$

   $\dfrac{5a^3c}{b^4}$

   B. $\dfrac{3x-6}{12x-15} = \dfrac{3(x-2)}{3(4x-5)} = \dfrac{x-2}{4x-5}$

   C. $\dfrac{2x+5xy}{x^2y^2} = \dfrac{x(2+5y)}{x^2y^2} = \dfrac{2+5y}{xy^2}$

2. A. $\dfrac{xy-5y}{5x-25} = \dfrac{y(x-5)}{5(x-5)} = \dfrac{y}{5}$

   B. $\dfrac{2x+4}{7x+14} = \dfrac{2(x+2)}{7(x+2)} = \dfrac{2}{7}$

   C. $\dfrac{x^2-x}{xy-y} = \dfrac{x(x-1)}{y(x-1)} = \dfrac{x}{y}$

3. A. $25x^2 - 9y^2 = (5x)^2 - (3y)^2 = (5x+3y)(5x-3y)$

   B. $5x^2 - 10xy + 30y = 5(x^2 - 2xy + 6y)$

   C. $x^4 - 1 = (x^2) - 1^2 = (x^2 + 1)(x^2 - 1) =$
   $(x^2 + 1)(x + 1)(x - 1)$

   D. $ax + bx - ay - by = (ax + bx) + (-ay - by) =$
   $x(a + b) - y(a + b) = (a + b)(x - y)$

   E. $12x^2 - 30x^5 = 6x^2(2 - 5x^3)$

4. A. $\dfrac{7x^2 - 5xy}{49x^3 - 25xy^2} = \dfrac{x(7x - 5y)}{x(49x^2 - 25y^2)} =$

   $\dfrac{x(7x - 5y)}{x(7x + 5y)(7x - 5y)} = \dfrac{1}{7x + 5y}$

   B. $\dfrac{3x^2 + 7x + 4}{3x^2 + x - 4} = \dfrac{(3x + 4)(x + 1)}{(3x + 4)(x - 1)} = \dfrac{x + 1}{x - 1}$

   C. $\dfrac{y^2 + 4y}{y^2 + 6y + 8} = \dfrac{y(y + 4)}{(y + 2)(y + 4)} = \dfrac{y}{y + 2}$

   D. $\dfrac{4x^2 - 9}{6x^2 - 9x} = \dfrac{(2x + 3)(2x - 3)}{3x(2x - 3)} = \dfrac{2x + 3}{3x}$

■ 10.33 OPERATIONS WITH ALGEBRAIC FRACTIONS

In addition and subtraction of algebraic fractions, use factoring to help find the least common denominator for the fractions and to reduce the answer to lowest terms.

Example:

Combine these fractions and reduce to lowest terms.

A. $\dfrac{7}{10x} + \dfrac{8}{15x}$ 　　　B. $\dfrac{3}{4a} + \dfrac{5}{6ab}$ 　　　C. $\dfrac{5}{x^2y} - \dfrac{3}{xy^3}$

D. $\dfrac{2}{x^2 - x} - \dfrac{5}{x^2 - 1}$ 　　　E. $\dfrac{3}{a^2 - 2a - 8} + \dfrac{4}{a^2 + 2a}$

Solution:

A. $\text{LCD}(10x, 15x) = 30x$

   $\dfrac{7}{10x} + \dfrac{8}{15x} = \dfrac{21}{30x} + \dfrac{16}{30x} = \dfrac{37}{30x}$

B. $\text{LCD}(4a, 6ab) = 12ab$

   $\dfrac{3}{4a} + \dfrac{5}{6ab} = \dfrac{9b}{12ab} + \dfrac{10}{12ab} = \dfrac{9b + 10}{12ab}$

C. $\text{LCD}(x^2y, xy^3) = x^2y^3$

   $\dfrac{5}{x^2y} - \dfrac{3}{xy^3} = \dfrac{5y^2}{x^2y^3} - \dfrac{3x}{x^2y^3} = \dfrac{5y^2 - 3x}{x^2y^3}$

D. $LCD = x(x-1)(x+1)$

$$\frac{2}{x^2-x} - \frac{5}{x^2-1} = \frac{2}{x(x-1)} - \frac{5}{(x+1)(x-1)}$$

$$= \frac{2(x+1)}{x(x+1)(x-1)} - \frac{5x}{x(x+1)(x-1)}$$

$$= \frac{2x+2-5x}{x(x+1)(x-1)} = \frac{-3x+2}{x(x+1)(x-1)}$$

E. $LCD = a(a+2)(a-4)$

$$\frac{3}{a^2-2a-8} + \frac{4}{a^2+2a} = \frac{3}{(a-4)(a+2)} + \frac{4}{a(a+2)}$$

$$= \frac{3a}{a(a-4)(a+2)} + \frac{4(a-4)}{a(a-4)(a+2)}$$

$$= \frac{3a+4a-16}{a(a-4)(a+2)} = \frac{7a-16}{a(a-4)(a+2)}$$

When you multiply and divide fractions, factor the numerator and denominator of each fraction and reduce if possible. Then write the indicated product of the numerator factors over the indicated product of the denominator factors. In a product, you may cancel a numerator factor with a common factor in any denominator.

Example:

Multiply these fractions.

A. $\dfrac{x^3y^2}{x^2y} \cdot \dfrac{xy^3}{x^3y^2}$

B. $\dfrac{a}{a-1} \cdot \dfrac{a^2-1}{a^3}$

C. $\dfrac{6x-6}{x^2+2x} \cdot \dfrac{x^2+4x+4}{2x^2+2x-4}$

Solution:

A. $\dfrac{x^3y^2}{x^2y} \cdot \dfrac{xy^3}{x^3y^2} = \dfrac{x \cdot y}{1 \cdot 1} \cdot \dfrac{1 \cdot y}{x^2 \cdot 1} = \dfrac{xy^2}{x^2} = \dfrac{y^2}{x}$

B. $\dfrac{a}{a-1} \cdot \dfrac{a^2-1}{a^3} = \dfrac{a(a+1)(a-1)}{(a-1)a^3} = \dfrac{a(a-1)(a+1)}{a(a-1)a^2} = \dfrac{a+1}{a^2}$

C. $\dfrac{6x-6}{x^2+2x} \cdot \dfrac{x^2+4x+4}{2x^2+2x-4} = \dfrac{6(x-1)}{x(x+2)} \cdot \dfrac{(x+2)(x+2)}{2(x+2)(x-1)}$

$$= \dfrac{6(x-1)(x+2)(x+2)}{2x(x-1)(x+2)(x+2)} = \dfrac{3}{x}$$

Example:

Divide these fractions.

A. $\dfrac{3x^2}{5y} \div \dfrac{2x^3}{6y^3}$

B. $\dfrac{a+b}{4} \div \dfrac{(a+b)^3}{12}$

C. $\dfrac{y^2+5y+6}{y^2-4} \div \dfrac{y^2+4y+4}{y^2-4y+4}$

Solution:

A. $\dfrac{3x^2}{5y} \div \dfrac{2x^3}{6y^3} = \dfrac{3x^2}{5y} \cdot \dfrac{6y^3}{2x^3} = \dfrac{3x^2 \cdot 6y^3}{5y \cdot 2x^3} = \dfrac{2x^2 \cdot y \cdot 3 \cdot 3 \cdot y^2}{2 \cdot x^2 \cdot y \cdot 5 \cdot x} = \dfrac{9y^2}{5x}$

B. $\dfrac{a+b}{4} \div \dfrac{(a+b)^3}{12} = \dfrac{a+b}{4} \cdot \dfrac{12}{(a+b)^3} = \dfrac{4(a+b)3}{4(a+b)(a+b)^2} = \dfrac{3}{(a+b)^2}$

C. $\dfrac{y^2+5y+6}{y^2-4} \div \dfrac{y^2+4y+4}{y^2-4y+4} = \dfrac{(y+2)(y+3)}{(y+2)(y-2)} \cdot \dfrac{(y-2)^2}{(y+2)^2} = \dfrac{(y+3)(y-2)}{(y+2)^2}$

Example:

Combine these fractions as indicated.

A. $\dfrac{x^2-x}{y^2-1} \cdot \dfrac{y-1}{x^2y-xy} \div \dfrac{y-1}{y^2}$ B. $\dfrac{2a-1}{2a^2+2a} \div \dfrac{6a^2-6}{4a^2+a-3} \cdot \dfrac{4a^3-4a}{8a^2-10a+3}$

Solution:

A. $\dfrac{x^2-x}{y^2-1} \cdot \dfrac{y-1}{x^2y-xy} \div \dfrac{y-1}{y^2} = \dfrac{x(x-1)}{(y+1)(y-1)} \cdot \dfrac{y-1}{xy(x-1)} \cdot \dfrac{y^2}{y-1} =$

$\dfrac{xy^2(x-1)(y-1)}{xy(y+1)(y-1)^2(x-1)} = \dfrac{y}{(y-1)(y+1)}$ or $\dfrac{y}{y^2-1}$

B. $\dfrac{2a-1}{2a^2+2a} \div \dfrac{6a^2-6}{4a^2+a-3} \cdot \dfrac{4a^3-4a}{8a^2-10a+3}$

$= \dfrac{2a-1}{2a(a+1)} \cdot \dfrac{(4a-3)(a+1)}{6(a+1)(a-1)} \cdot \dfrac{4a(a+1)(a-1)}{(4a-3)(2a-1)}$

$= \dfrac{4a(2a-1)(4a-3)(a+1)(a+1)(a-1)}{12a(2a-1)(4a-3)(a+1)(a+1)(a-1)} = \dfrac{1}{3}$

▬▬ 10.34 PRACTICE PROBLEMS

1. Add or subtract as indicated and simplify the results.

A. $\dfrac{11}{18ab} + \dfrac{1}{18ab}$ B. $\dfrac{3x}{x-4} + \dfrac{2}{x+1}$

C. $\dfrac{3}{2a^2-a-1} - \dfrac{2}{a^2+a-2}$

D. $\dfrac{2y}{2y^2+5y+2} + \dfrac{y}{3y^2+5y-2} - \dfrac{1}{6y^2+y-1}$

2. Multiply or divide as indicated and simplify the results.

A. $\dfrac{x^2+xy}{xy} \cdot \dfrac{5y}{x^2-y^2}$

B. $\dfrac{3x^2-12}{3x^2-3} \cdot \dfrac{x-1}{2x+4}$

C. $\dfrac{a^2+7a+10}{a^2+10a+25} \div \dfrac{a+2}{a+5}$

D. $\dfrac{a^3b^2-a^2b^3}{2a} \div \dfrac{a^2-b^2}{4a}$

E. $\dfrac{12-6a}{7a-21} \div \dfrac{2a-4}{a^2-4}$

10.35 SOLUTIONS

1. A. $\dfrac{11}{18ab} + \dfrac{1}{18ab} = \dfrac{12}{18ab} = \dfrac{6 \cdot 2}{6 \cdot 3ab} = \dfrac{2}{3ab}$

   B. $\dfrac{3x}{x-4} + \dfrac{2}{x+1} = \dfrac{3x(x+1)}{(x-4)(x+1)} +$

   $\dfrac{2(x-4)}{(x-4)(x+1)} = \dfrac{3x^2 + 3x + 2x - 8}{(x-4)(x+1)} =$

   $\dfrac{3x^2 + 5x - 8}{(x-4)(x+1)}$

   C. $\dfrac{3}{2a^2 - a - 1} - \dfrac{2}{a^2 + a - 2} = \dfrac{3}{(2a+1)(a-1)} -$

   $\dfrac{2}{(a-1)(a+2)} = \dfrac{3(a+2)}{(2a+1)(a-1)(a+2)} -$

   $\dfrac{2(2a+1)}{(2a+1)(a-1)(a+2)} = \dfrac{3a + 6 - 4a - 2}{(2a+1)(a-1)(a+2)}$

   $= \dfrac{-a+4}{(2a+1)(a-1)(a+2)}$

   D. $\dfrac{2y}{2y^2 + 5y + 2} + \dfrac{y}{3y^2 + 5y - 2} - \dfrac{1}{6y^2 + y - 1}$

   $= \dfrac{2y}{(2y+1)(y+2)} + \dfrac{y}{(3y-1)(y+2)} - \dfrac{1}{(3y-1)(2y+1)}$

   $= \dfrac{2y(3y-1) + y(2y+1) - 1(y+2)}{(3y-1)(2y+1)(y+2)}$

   $= \dfrac{6y^2 - 2y + 2y^2 + y - y - 2}{(3y-1)(2y+1)(y+2)}$

   $= \dfrac{8y^2 - 2y - 2}{(3y-1)(2y+1)(y+2)}$

2. A. $\dfrac{x^2 + xy}{xy} \cdot \dfrac{5y}{x^2 - y^2} = \dfrac{x(x+y) \cdot 5y}{xy(x+y)(x-y)} = \dfrac{5}{x-y}$

   B. $\dfrac{3x^2 - 12}{3x^2 - 3} \cdot \dfrac{x-1}{2x+4} = \dfrac{3(x^2 - 4)}{3(x^2 - 1)} \cdot \dfrac{x-1}{2(x+2)}$

   $= \dfrac{3(x+2)(x-2)(x-1)}{6(x+2)(x+1)(x-1)} = \dfrac{x-2}{2(x+1)}$

   C. $\dfrac{a^2 + 7a + 10}{a^2 + 10a + 25} \div \dfrac{a+2}{a+5} = \dfrac{(a+2)(a+5)}{(a+5)(a+5)} \cdot \dfrac{a+5}{a+2} = 1$

   D. $\dfrac{a^3 b^2 - a^2 b^3}{2a} \div \dfrac{a^2 - b^2}{4a}$

   $= \dfrac{a^2 b^2 (a-b)}{2a} \cdot \dfrac{4a}{(a+b)(a-b)} = \dfrac{2a^2 b^2}{a+b}$

   E. $\dfrac{12 - 6a}{7a - 21} \div \dfrac{2a - 4}{a^2 - 4} = \dfrac{-6(a-2)}{7(a-3)} \cdot$

   $\dfrac{(a+2)(a-2)}{2(a-2)} = \dfrac{-3(a-2)(a+2)}{7(a-3)}$

10.36 ALGEBRA TEST 2

Use the following test to assess how well you have mastered the material in this chapter so far. Mark your answers by blackening the corresponding answer oval in each question. An answer key and solutions are provided at the end of the test.

1. Which is the value of C when $F = 68$, using the formula $F = 1.8C + 32$?

   (A) 2
   (B) 18
   (C) 20
   (D) 33.8
   (E) 36

2. Which is the value of $\dfrac{a-b}{a+b}$ when $a = 5$ and $b = 10$?

(A) 3

(B) $\dfrac{1}{3}$

(C) $-\dfrac{1}{3}$

(D) $-\dfrac{1}{2}$

(E) -3

3. Which is the value of $3x^2y - 5xy^3$ when $x = -2$ and $y = -3$?

(A) -306

(B) -126

(C) -54

(D) 234

(E) 306

4. What is the value of v when $V = 45$, $g = 32$, and $t = 5$ using the formula $v = V + gt$?

(A) 82

(B) 205

(C) 257

(D) 385

(E) 7,200

5. $\begin{aligned} 5x^2 + 3x - 2y^2 \\ + \underline{-2x^2 - 8x + 2y^2} \end{aligned}$ What is the sum?

(A) $3x^4 - 5x^2$

(B) $3x^4 - 5x^2 + y^4$

(C) $7x^2 + 1x - 4y^2$

(D) $3x^2 - 5x$

(E) $3x^2 - 5x + y^2$

6. $\begin{aligned} 3a^2 - 6ab - 2b^2 \\ - \underline{-4a^2 + 8ab - 8b^2} \end{aligned}$ What is the difference?

(A) $-a^2 + 2ab - 10b^2$

(B) $-a^2 - 14ab - 10b^2$

(C) $7a^2 - 14ab - 10b^2$

(D) $7a^2 + 2ab - 10b^2$

(E) $7a^2 - 14ab + 6b^2$

7. $3x^4 - 5x^3$
     $\times \underline{-5x + 4}$ What is the product?

  (A) $-15^5 - 20x^3$
  (B) $25x^3 + 12x^4$
  (C) $-15x^5 + 37x^4 - 20x^3$
  (D) $-15x^5 + 25x^4 + 4$
  (E) $3x^4 - 5x^3 - 5x + 4$

8. $\dfrac{12a^2b^3 - 28a^4b^4}{4ab^2}$ What is the value of the quotient?

  (A) $3ab - 28a^4b^4$
  (B) $12a^2b^3 - 7a^3b^2$
  (C) $3ab - 7a^3b^2$
  (D) $8a^2b - 22a^3b^2$
  (E) $-4a^4b^3$

9. Which is equal to $(3x - 5)^2$?

  (A) $9x^2 + 25$
  (B) $9x^2 - 25$
  (C) $9x^2 - 15x + 25$
  (D) $9x^2 - 30x + 25$
  (E) $9x^2 + 30x + 25$

10. Which expression is not defined when $x = 4$?

  (A) $\dfrac{x - 4}{x + 4}$

  (B) $\dfrac{3}{x^2 + 4}$

  (C) $\dfrac{2x^2}{x^2 - 4}$

  (D) $x - 4$

  (E) $\dfrac{3}{x - 4}$

11. Which is $25x^2 - 49$ factored completely?

  (A) $(25x - 1)(x + 49)$
  (B) $(5x - 7)(5x + 7)$
  (C) $(5x - 7)(7x - 5)$
  (D) $(5x + 7)(5x + 7)$
  (E) $(5x - 7)(5x - 7)$

12. Which is $3x^2 + 10x + 3$ factored completely?

   (A) $(3x + 3)(x + 1)$
   (B) $(3x + 1)(x + 3)$
   (C) $(3x - 1)(x + 3)$
   (D) $(3x + 1)(x - 3)$
   (E) $(3x - 1)(x - 3)$

13. Which is $10x^2 + 11x - 6$ factored completely?

   (A) $(5x + 2)(2x - 3)$
   (B) $(10x - 1)(x + 6)$
   (C) $(5x + 2)(2x + 3)$
   (D) $(10x + 1)(x - 6)$
   (E) $(5x - 2)(2x + 3)$

14. Which is $\dfrac{x^2 + 2x - 8}{x^2 - x - 2}$ reduced to lowest terms?

   (A) $\dfrac{x + 4}{x + 1}$

   (B) $\dfrac{x - 4}{x - 1}$

   (C) $\dfrac{x - 4}{x + 1}$

   (D) $\dfrac{x + 4}{x - 1}$

   (E) 4

15. Which is equal to $\dfrac{4}{x^2 - x} - \dfrac{2}{x^2 - 1}$?

   (A) $\dfrac{x + 2}{x(x - 1)(x + 1)}$

   (B) $\dfrac{2}{x(x - 1)}$

   (C) $\dfrac{2(x + 2)}{x(x - 1)(x + 1)}$

   (D) $\dfrac{2}{x(x + 1)}$

   (E) $\dfrac{4}{x}$

16. Which is $\dfrac{a^2}{a+1} \cdot \dfrac{a^2-1}{a^3}$ in simplest form?

    (A) $\dfrac{a+1}{a^2}$

    (B) $a(a+1)$

    (C) $\dfrac{a-1}{a}$

    (D) $\dfrac{a+1}{a}$

    (E) $\dfrac{a-1}{a^2}$

17. Which is equal to $\dfrac{3}{a+2} + \dfrac{5}{a-2}$?

    (A) $\dfrac{4}{a}$

    (B) $\dfrac{4}{a-2}$

    (C) $\dfrac{2a+1}{(a+1)(a-1)}$

    (D) $\dfrac{4(2a+1)}{(a+2)(a-2)}$

    (E) $\dfrac{8a}{(a+2)(a-2)}$

18. Which is $\dfrac{1}{a^2-5a+6} \div \dfrac{a-3}{a-2}$ in simplest form?

    (A) 1

    (B) $\dfrac{1}{(a-3)^2}$

    (C) $\dfrac{1}{(a-2)^2}$

    (D) $\dfrac{3}{2(a^2-5a+6)}$

    (E) $\dfrac{2}{3(a^2-5a+6)}$

19. Which is $9x^2 - 42xy + 49y^2$ factored completely?

    (A) $(3x-7)^2$

    (B) $(3x+7)^2$

    (C) $(3x-7y)^2$

    (D) $(3x+7y)^2$

    (E) $(9x-49y)^2$

20. Which is $x^2 - x - 30$ factored completely?

   (A) $(x - 5)(x + 6)$
   (B) $(x + 5)(x - 6)$
   (C) $(x - 3)(x - 10)$
   (D) $(x - 5)(x - 6)$
   (E) $(x + 5)(x + 6)$

ALGEBRA TEST 2
Answer Key

1. C	6. E	11. B	16. C
2. C	7. C	12. B	17. D
3. A	8. C	13. E	18. B
4. B	9. D	14. A	19. C
5. D	10. E	15. C	20. B

▰ 10.37 SOLUTIONS

1. **C** 20

   $F = 1.8C + 32$, $F = 68$, $68 = 1.8C + 32$, $36 = 1.8C$, $C = 20$

2. **C** $-\dfrac{1}{3}$

   $\dfrac{a-b}{a+b} = \dfrac{5-10}{5+10} = \dfrac{-5}{15} = \dfrac{-1}{3}$

3. **A** -306

   $3x^2y - 5xy^3 = 3(-2)^2(-3) - 5(-2)(-3)^3 = -36 - 270 = -306$

4. **B** 205

   $v = V + gt$, $V = 45$, $g = 32$, $t = 5$, $v = 45 + 32(5) = 45 + 160 = 205$

5. **D** $3x^2 - 5x$

   $$\begin{array}{r} 5x^2 + 3x - 2y^2 \\ + \underline{-2x^2 - 8x + 2y^2} \\ 3x^2 - 5x \end{array}$$

6. **E** $7a^2 - 14ab + 6b^2$

   $$\begin{array}{r} 3a^2 - 6ab - 2b^2 \\ - \underline{-4a^2 + 8ab - 8b^2} \\ 7a^2 - 14ab + 6b^2 \end{array}$$

7. **C** $-15x^5 + 37x^4 - 20x^3$

   $$\begin{array}{r} 3x^4 - 5x^3 \\ \times \underline{- 5x + 4} \\ -15x^5 + 25x^4 \\ + 12x^4 - 20x^3 \\ \hline -15x^5 + 37x^4 - 20x^3 \end{array}$$

8. **C** $3ab - 7a^3b^2$

   $\dfrac{12a^2b^3 - 28a^4b^4}{4ab^2} = 3ab - 7a^3b^2$

9. **D** $9x^2 - 30x + 25$

$$(3x - 5)^2 = (3x)^2 - 2(3x)(5) + 5^2 = 9x^2 - 30x + 25$$

10. **E** $\dfrac{3}{x - 4}$

$$\frac{3}{x - 4} = \frac{3}{4 - 4} = \frac{3}{0} \text{ Not defined.}$$

11. **B** $(5x - 7)(5x + 7)$

$$25x^2 - 49 = (5x)^2 - (7)^2 = (5x - 7)(5x + 7)$$

12. **B** $(3x + 1)(x + 3)$

$$3x^2 + 10x + 3 = (3x + 1)(x + 3)$$

13. **E** $(5x - 2)(2x + 3)$

$$10x^2 + 11x - 6, 10(-6) = -60, 15(-4) = 60 \text{ and } +15 - 4 = 11$$
$$10x^2 + 15x - 4x - 6 = 5x(2x + 3) - 2(2x + 3) = (5x - 2)(2x + 3)$$

14. **A** $\dfrac{x + 4}{x + 1}$

$$\frac{x^2 + 2x - 8}{x^2 - x - 2} = \frac{(x + 4)(x - 2)}{(x + 1)(x - 2)} = \frac{(x + 4)}{(x + 1)}$$

15. **C** $\dfrac{2(x + 2)}{x(x + 1)(x - 1)}$

$$\frac{4}{x^2 - x} - \frac{2}{x^2 - 1} = \frac{4}{x(x - 1)} - \frac{2}{(x + 1)(x - 1)} = \frac{4(x + 1) - 2x}{x(x + 1)(x - 1)}$$

$$= \frac{4x + 4 - 2x}{x(x + 1)(x - 1)} = \frac{2x + 4}{x(x + 1)(x - 1)} = \frac{2(x + 2)}{x(x - 1)(x + 1)}$$

16. **C** $\dfrac{a - 1}{a}$

$$\frac{a^2}{a + 1} \times \frac{a^2 - 1}{a^3} = \frac{a^2(a + 1)(a - 1)}{a^3(a + 1)} = \frac{a - 1}{a}$$

17. **D** $\dfrac{4(2a + 1)}{(a + 2)(a - 2)}$

$$\frac{3}{a + 2} + \frac{5}{a - 2} = \frac{3(a - 2) + 5(a + 2)}{(a + 2)(a - 2)} = \frac{3a - 6 + 5a + 10}{(a + 2)(a - 2)} =$$

$$\frac{8a + 4}{(a + 2)(a - 2)} = \frac{4(2a + 1)}{(a + 2)(a - 2)}$$

18. **B** $\dfrac{1}{(a - 3)^2}$

$$\frac{1}{a^2 - 5a + 6} \div \frac{a - 3}{a - 2} = \frac{1}{(a - 3)(a - 2)} \cdot \frac{a - 2}{a - 3} = \frac{1}{(a - 3)(a - 3)} = \frac{1}{(a - 3)^2}$$

19. **C** $(3x - 7y)^2$

$$9x^2 - 42xy + 49y^2 = (3x)^2 + 2(3x)(-7y) + (-7y)^2 = (3x - 7y)^2$$

20. **B** $(x + 5)(x - 6)$

Because $a = 1$, you need two factors of $c = -30$ that add up to $b = -1$. Because $5(-6) = -30$ and $5 + (-6) = -1$, $x^2 - x - 30 = (x + 5)(x - 6)$.

▬▬ 10.38 SOLVED GRE PROBLEMS

For each question, select the best answer unless otherwise instructed.

	Quantity A	Quantity B
1.	$10 - 5 \times 2$	$(10 - 5) \times 2$

- Ⓐ Quantity A is greater.
- Ⓑ Quantity B is greater.
- Ⓒ The two quantities are equal.
- Ⓓ The relationship cannot be determined from the given information.

2. **Which illustrates the distributive property?**

- Ⓐ $13 \times 19 = 19 \times 13$
- Ⓑ $x \times (8 \times y) = (x \times 8) \times y$
- Ⓒ $5 \times (8 + 2) = (8 + 2) \times 5$
- Ⓓ $5 + 8 \times w = 5 \times w + 8$
- Ⓔ $(3 + w) \times 7 = 3 \times 7 + w \times 7$

For this question, enter your answer in the box.

3. **What is the value of $2x^3 - x$ when $x = -3$?**

4. **Which of these equations has -4 as its solution? Select all that apply.**

- Ⓐ $-x = -4$
- Ⓑ $16x = -4$
- Ⓒ $8 - x = 4$
- Ⓓ $2x - 5 = -13$
- Ⓔ $-2x = 8$

	Quantity A	Quantity B
5.	x^2 when $x = -2$	y^{-3} when $y = 5$

- Ⓐ Quantity A is greater.
- Ⓑ Quantity B is greater.
- Ⓒ The two quantities are equal.
- Ⓓ The relationship cannot be determined from the given information.

10.39 SOLUTIONS

1. **B** Quantity B is greater.
   $10 - 5 \times 2 = 10 - 10 = 0$ and $(10 - 5) \times 2 = 5 \times 2 = 10$.
2. **E** $(3 + w) \times 7 = 3 \times 7 + w \times 7$
   The multiplication by 7 is distributed over the quantity $(3 + w)$. This is an example of the distributive property.
3. $\boxed{-51}$
   $2x^3 - x = 2(-3)^3 - (-3) = 2(-27) + 3 = -54 + 3 = -51$.
4. **D and E** $2x - 5 = -13$ and $-2x = 8$
   $2(-4) - 5 = -8 - 5 = -13$ and $-2(-4) = 8$, so both D and E are correct.
5. **A** Quantity A is greater.
   If $x = -2$, then $x^2 = (-2)^2 = 4$, and if $y = 5$, then $y^{-3} = (5)^{-3} = \dfrac{1}{125}$.

10.40 GRE PRACTICE PROBLEMS

For each question, select the best answer unless otherwise instructed.

1. What is the sum of $2r^2 - 5rs + 6s^2$ and $-3r^2 + 5rs + 11s^2$?

   (A) $5r^2 - 10rs + 17s^2$
   (B) $5r^2 - 10rs - 5s^2$
   (C) $-r^2 + 17s^2$
   (D) $-r^2 + rs + 17s^2$
   (E) $-r^4 + 17s^4$

2. $\begin{array}{r} 3x^3 - 2x^2 \\ \times\ x^2 - 3 \end{array}$ What is the product?

   (A) $3x^5 + 6x^2$
   (B) $3x^5 - 6x^2$
   (C) $3x^5 - 2x^4 - 9x^3 - 6x^2$
   (D) $3x^5 - 2x^4 - 9x^3 + 6x^2$
   (E) $3x^5 + 2x^4 - 9x^3 - 6x^2$

Quantity A	Quantity B
3. | x | x^3 |

   (A) Quantity A is greater.
   (B) Quantity B is greater.
   (C) The two quantities are equal.
   (D) The relationship cannot be determined from the given information.

For this question, enter your answer in the box.

4. If $A = 6s^2$ and $s = 4$, what is the value of A?

5. What is the value of i when $i = prt$; $r = 4\%$ per year; $p = \$7,000$; and $t = 2$ years?

   (A) $56
   (B) $140
   (C) $280
   (D) $560
   (E) $5,600

Quantity A	Quantity B
6. $5 + (4 + 8)$	$(5 + 8) + 4$

   (A) Quantity A is greater.
   (B) Quantity B is greater.
   (C) The two quantities are equal.
   (D) The relationship cannot be determined from the given information.

Quantity A	Quantity B
7. The product of 8 and 4 decreased by 3	The product of 8 decreased by 3 and 4 decreased by 3

   (A) Quantity A is greater.
   (B) Quantity B is greater.
   (C) The two quantities are equal.
   (D) The relationship cannot be determined from the given information.

8. Which is the value of $3x^3y^2$ when $x = -3$ and $y = -2$?

   (A) -972
   (B) -324
   (C) 108
   (D) 324
   (E) 972

Quantity A	Quantity B
9. $1.8C + 32$ when $C = 20$	$\dfrac{9C + 160}{5}$ when $C = 20$

   (A) Quantity A is greater.
   (B) Quantity B is greater.
   (C) The two quantities are equal.
   (D) The relationship cannot be determined from the given information.

10. Which is equal to $(5x - 7y)^2$?

   (A) $25x^2 - 49y^2$
   (B) $25x^2 + 49y^2$
   (C) $25x^2 - 35xy + 49y^2$
   (D) $25x^2 - 70xy + 49y^2$
   (E) $25x^2 + 70xy - 49y^2$

ANSWER KEY

1. C	6. C
2. D	7. A
3. D	8. B
4. 96	9. C
5. D	10. D

10.41 LINEAR EQUATIONS

A **linear equation in one variable** has the form $ax + b = 0$ if $a \neq 0$ and its solution is $x = -\dfrac{b}{a}$. It is linear since the variable is raised to the first power.

The general procedure for solving linear equations is:

1. If there are parentheses in the equation, perform the operations needed to remove them.
2. Combine like terms, if possible, on each side of the equation.
3. Get all terms containing the variables on one side of the equation.
4. If there is a constant term on the variable side of the equation, undo the operation so that the constant is zero.
5. If the coefficient of the variable is not a positive one, divide each side of the equation by the coefficient.
6. Check the solution in the original equation.

$2(x + 3) = 3(x - 1) + 4$	
$2x + 6 = 3x - 3 + 4$	Remove parentheses.
$2x + 6 = 3x + 1$	Combine like terms.
$2x + 6 - 3x = 3x + 1 - 3x$	To get all variable terms on left, add $-3x$.
$-x + 6 = 1$	Combine like terms.
$-x + 6 - 6 = 1 - 6$	To get the variable on one side by itself, add -6.
$-x = -5$	Combine like terms.
$\dfrac{-x}{-1} = \dfrac{-5}{-1}$	Divide each side by -1.
$x = 5$	

Check: Left side: $2(x + 3) = 2(5 + 3) = 2(8) = 16$
       Right side: $3(x - 1) + 4 = 3(5 - 1) + 4 = 3(4) + 4 = 12 + 4 = 16$
       Since $16 = 16$, $x = 5$ is a solution of the equation.

Example:

Solve each equation for x.

A. $x + 8 - 2(x + 1) = 3x - 6$ B. $3x + 2 = 6x - 4$
C. $2(x + 3) = 5(x - 1) - 7(x - 3)$ D. $3x + 4(x - 2) = x - 5 + 3(2x - 1)$

Solution:

A. $x + 8 - 2(x + 1) = 3x - 6$
$x + 8 - 2x - 2 = 3x - 6$
$-x + 6 = 3x - 6$
$-x + 6 - 3x = 3x - 6 - 3x$
$-4x + 6 = -6$
$-4x + 6 - 6 = -6 - 6$
$-4x = -12$
$\dfrac{-4x}{-4} = \dfrac{-12}{-4}$
$x = 3$

B. $3x + 2 = 6x - 4$
$3x + 2 - 6x = 6x - 4 - 6x$
$-3x + 2 = -4$
$-3x + 2 - 2 = -4 - 2$
$-3x = -6$
$\dfrac{-3x}{-3} = \dfrac{-6}{-3}$
$x = 2$

C. $2(x + 3) = 5(x - 1) - 7(x - 3)$
$2x + 6 = 5x - 5 - 7x + 21$
$2x + 6 = -2x + 16$
$4x + 6 = 16$
$4x = 10$
$\dfrac{4x}{4} = \dfrac{10}{4}$
$x = \dfrac{5}{2}$

D. $3x + 4(x - 2) = x - 5 + 3(2x - 1)$
$3x + 4x - 8 = x - 5 + 6x - 3$
$7x - 8 = 7x - 8$
This is an identity equation, so the equation is true for all real numbers.

10.42 LITERAL EQUATIONS

Literal equations are equations in which some of the constants are letters, not specific numbers. The literal equations you encounter most often are formulas.

For example, $d = rt$ can be solved for r to get $r = \dfrac{d}{t}$.

Example:

Solve each equation for the letter indicated.

A. $A = 0.5(h + b)$ for h

B. $6k = 3(l + w) - 2h$ for h

C. $\dfrac{2a - 3b}{c} = \dfrac{3a - 2c}{b}$ for a

D. $A = P + irt$ for P

Solution:

A. $A = 0.5(h + b)$ for h
$2(A) = 2[0.5(h + b)]$
$2A = 1(h + b)$
$2A = h + b$
$2A - b = h$
$h = 2A - b$

B. $6k = 3(l + w) - 2h$ for h
$6k - 3(l + w) = -2h$
$-6k + 3(l + w) = 2h$
$\dfrac{3(l + w) - 6k}{2} = h$

$h = \dfrac{3(l + w) - 6k}{2}$

C. $\dfrac{2a - 3b}{c} = \dfrac{3a - 2c}{b}$ for a
$b(2a - 3b) = c(3a - 2c)$
$2ab - 3b^2 = 3ac - 2c^2$
$2ab - 3b^2 - 3ac = -2c^2$
$2ab - 3ac = 3b^2 - 2c^2$
$a(2b - 3c) = 3b^2 - 2c^2$

$a = \dfrac{3b^2 - 2c^2}{2b - 3c}$

D. $A = P + irt$ for P
$A - irt = P$
$P = A - irt$

10.43 EQUATIONS WITH FRACTIONS

A **fractional equation** is an algebraic equation with the variable in the denominator of a fraction. You have to be sure that any answer that you get will not make the denominator zero, which would cause the problem to be undefined.

For example, if $\dfrac{4}{x} = 2$, then $4 = 2x$ and $x = 2$. You have to check to see if $x = 2$ makes the denominator zero, and in this case it does not. So, $x = 2$ is the solution.

In $\dfrac{4}{x} = \dfrac{1}{x}$, you get $4x = x$ and $3x = 0$, so $x = 0$. Check to see if $x = 0$ makes the denominator zero, and in this case it does. Thus, $x = 0$ is not a solution.

Example:

Solve these equations for x.

A. $\dfrac{3}{x} - \dfrac{4}{5x} = \dfrac{1}{10}$

B. $\dfrac{2}{x - 1} + \dfrac{6}{x} = \dfrac{5}{x - 1}$

C. $\dfrac{3}{x - 3} + 4 = \dfrac{x}{x - 3}$

D. $\dfrac{1}{2x} + \dfrac{8}{5} = \dfrac{3}{x}$

Solution:

A. $\dfrac{3}{x} - \dfrac{4}{5x} = \dfrac{1}{10}$ LCD $= 10x$

$$10x\left(\dfrac{3}{x} - \dfrac{4}{5x}\right) = 10x\left(\dfrac{1}{10}\right)$$

$30 - 8 = x$
$22 = x$ Solution, since only $x = 0$ is not allowed.

B. $\dfrac{2}{x-1} + \dfrac{6}{x} = \dfrac{5}{x-1}$ LCD $= x(x-1)$

$$x(x-1)\left(\dfrac{2}{x-1} + \dfrac{6}{x}\right) = x(x-1)\dfrac{5}{x-1}$$
$2x + 6(x-1) = 5x$
$2x + 6x - 6 - 5x = 5x - 5x$
$6x - 6 = 0$
$3x - 6 + 6 = 0 + 6$
$3x = 6$
$\dfrac{3x}{3} = \dfrac{6}{3}$
$x = 2$ Solution, since only $x = 0$ is not allowed.

C. $\dfrac{3}{x-3} + 4 = \dfrac{x}{x-3}$

$$x - 3\left(\dfrac{3}{x-3} + 4\right) = (x-3)\dfrac{x}{x-3}$$
$3 + 4(x-3) = x$
$3 + 4x - 12 - x = x - x$
$3x - 9 = 0$
$3x - 9 + 9 = 0 + 9$
$3x = 9$
$\dfrac{3x}{3} = \dfrac{9}{3}$
$x = 3$ Not a solution, since $x = 3$ is not allowed.

D. $\dfrac{1}{2x} + \dfrac{8}{5} = \dfrac{3}{x}$ LCD $= 10x$

$$10x\left(\dfrac{1}{2x} + \dfrac{8}{5}\right) = 10x \cdot \dfrac{3}{x}$$
$5 + 8(2x) = 10 \cdot 3$
$5 + 16x = 30$
$5 + 16x - 5 = 30 - 5$
$16x = 25$
$\dfrac{16x}{16} = \dfrac{25}{16}$
$x = \dfrac{25}{16}$ Solution, since only $x = 0$ is not allowed.

These equations contain fractions, but they do not have variables in the denominator of the fractions. They are equations containing fractions but are not fractional equations.

Example:

Solve for x.

A. $\dfrac{2x}{5} - \dfrac{4}{5} = \dfrac{9}{5}$ 　　B. $\dfrac{x}{6} - \dfrac{1}{2} = \dfrac{2}{3}$ 　　C. $\dfrac{9x+1}{6} = x + \dfrac{1}{3}$ 　　D. $\dfrac{x-4}{3} = \dfrac{x}{5} + 2$

Solution:

A. $\dfrac{2x}{5} - \dfrac{4}{5} = \dfrac{9}{5}$ 　　LCD $= 5$

$$5\left(\dfrac{2x}{5} - \dfrac{4}{5}\right) = 5\left(\dfrac{9}{5}\right)$$

$2x - 4 = 9$
$2x - 4 + 4 = 9 + 4$
$2x = 9 + 4$
$2x = 13$

$$\dfrac{2x}{2} = \dfrac{13}{2}$$

$$x = \dfrac{13}{2}$$

B. $\dfrac{x}{6} - \dfrac{1}{2} = \dfrac{2}{3}$ 　　LCD $= 6$

$$6\left(\dfrac{x}{6} - \dfrac{1}{2}\right) = 6\left(\dfrac{2}{3}\right)$$

$x - 3 = 4$
$x - 3 + 3 = 4 + 3$
$x = 7$

C. $\dfrac{9x+1}{6} = x + \dfrac{1}{3}$ 　　LCD $= 6$

$$6\left(\dfrac{9x+1}{6}\right) = 6\left(x + \dfrac{1}{3}\right)$$

$9x + 1 = 6x + 2$
$9x + 1 - 6x = 6x + 2 - 6x$
$3x + 1 = 2$
$3x + 1 - 1 = 2 - 1$
$3x = 1$

$$\dfrac{3x}{3} = \dfrac{1}{3}$$

$$x = \dfrac{1}{3}$$

D. $\dfrac{x-4}{3} = \dfrac{x}{5} + 2$ LCD = 15

$$15\left(\dfrac{x-4}{3}\right) = 15\left(\dfrac{x}{5} + 2\right)$$

$5(x-4) = 3x + 15(2)$
$5x - 20 = 3x + 30$
$5x - 20 - 3x = 3x + 30 - 3x$
$2x - 20 = 30$
$2x - 20 + 20 = 30 + 20$
$2x = 50$

$$\dfrac{2x}{2} = \dfrac{50}{2}$$

$x = 25$

▇▇ 10.44 EQUATIONS THAT ARE PROPORTIONS

Proportions are equations with two fractions equal to each other. The equation may or may not have a variable in the denominator of a fraction.

Example:

Solve for x.

A. $(5 - x) : (x + 1) = 2 : 1$

B. $(x + 3) : 10 = (3x - 2) : 8$

C. $\dfrac{4}{x} = \dfrac{2}{7}$

D. $\dfrac{x+3}{x-2} = \dfrac{3}{2}$

Solution:

A. $(5 - x) : (x + 1) = 2 : 1$

$\dfrac{5 - x}{x + 1} = \dfrac{2}{1}$

$1(5 - x) = 2(x + 1)$
$5 - x = 2x + 2$
$5 - x + x = 2x + 2 + x$
$5 = 3x + 2$
$5 - 2 = 3x + 2 - 2$
$3 = 3x$
$1 = x$
$x = 1$

Solution, since only $x = -1$ is not allowed.

B. $(x + 3) : 10 = (3x - 2) : 8$

$\dfrac{x+3}{10} = \dfrac{3x-2}{8}$

$8(x + 3) = 10(3x - 2)$
$8x + 24 = 30x - 20$
$8x + 24 - 8x = 30x - 20 - 8x$
$24 = 22x - 20$
$24 + 20 = 22x - 20 + 20$

$$44 = 22x$$
$$2 = x$$

Solution, since all real numbers are allowed.

C. $\dfrac{4}{x} = \dfrac{2}{7}$

$$4(7) = 2(x)$$
$$28 = 2x$$
$$14 = x$$

Solution, since only $x = 0$ is not allowed.

D. $\dfrac{x+3}{x-2} = \dfrac{3}{2}$

$$2(x + 3) = 3(x - 2)$$
$$2x + 6 = 3x - 6$$
$$2x + 6 + 6 = 3x - 6 + 6$$
$$2x + 12 = 3x$$
$$2x + 12 - 2x = 3x - 2x$$
$$12 = x$$

Solution, since only $x = 2$ is not allowed.

10.45 EQUATIONS WITH RADICALS

When an equation contains a radical, remove the radical by squaring each side of the equation after you get the radical on one side alone. When there are two radicals, get one radical on each side of the equation before squaring it.

Example:

Solve these radical equations for x.

A. $\sqrt{8x} - 7 = 17$ B. $\sqrt{10x} + 10 = 2$ C. $\sqrt{-2x} = 8$
D. $\sqrt{10 - 2x} = \sqrt{3x + 25}$ E. $\sqrt{5x - 1} = \sqrt{2x + 8}$

Solution:

A. $\sqrt{8x} - 7 = 17$
   $\sqrt{8x} = 24$
   $\left(\sqrt{8x}\right)^2 = (24)^2$
   $8x = 576$
   $x = 72$

B. $\sqrt{10x} + 10 = 2$
   $\sqrt{10x} = -8$
   No solution because the principal square root is never a negative number.

C. $\sqrt{-2x} = 8$
   $\left(\sqrt{-2x}\right)^2 = 8^2$
   $-2x = 64$
   $x = -32$

D. $\sqrt{10 - 2x} = \sqrt{3x + 25}$

$\left(\sqrt{10 - 2x}\right)^2 = \left(\sqrt{3x + 25}\right)^2$

$10 - 2x = 3x + 25$

$10 - 5x = 25$

$-5x = 15$

$x = -3$

You have to check to be sure each radicand is non-negative.

$10 - 2x = 10 - 2(-3) = 10 + 6 = 16 \geq 0$

$3x + 25 = 3(-3) + 25 = -9 + 25 = 16 \geq 0$

$x = -3$ is a solution.

E. $\sqrt{5x - 1} = \sqrt{2x + 8}$

$\left(\sqrt{5x - 1}\right)^2 = \left(\sqrt{2x + 8}\right)^2$

$5x - 1 = 2x + 8$

$3x - 1 = 8$

$3x = 9$

$x = 3$

▬▬ 10.46 PRACTICE PROBLEMS

1. Solve the equations for x.

   A. $7x + 8 = 1$ B. $8 - 8x = -16$

   C. $5x - 4 = 21$ D. $9 - 2x = 15$

   E. $9x - 5x + 13 = 3x + 6$

2. Solve these equations for y.

   A. $2(y + 3) = -4(y + 2) + 2$

   B. $3(2y + 1) - 2(y - 2) = 5$

   C. $4(y - 9) = 8(y - 3)$

   D. $2y + 3(y - 4) = 2(y - 3)$

   E. $6y - 3(5y + 2) = 4(1 - y)$

3. Solve for the indicated letter.

   A. $P = 2l + 2w$ for l B. $P = a + b + c$ for c

   C. $A = lw$ for w D. $A = \dfrac{1}{2}(B + b)h$ for B

   E. $C = 2\pi r$ for r

4. Solve these equations for x.

   A. $\dfrac{3x}{4} + \dfrac{5x}{2} = 13$

   B. $\dfrac{8x}{3} - \dfrac{2x}{4} = -13$

   C. $\dfrac{x - 8}{5} + \dfrac{8}{5} = \dfrac{-x}{3}$

5. Solve these equations for y.

   A. $\dfrac{1}{y} + \dfrac{2}{y} = 3 - \dfrac{3}{y}$

   B. $\dfrac{3}{4y} - \dfrac{1}{6} = \dfrac{4}{8y} + \dfrac{1}{2}$

   C. $\dfrac{1}{y} + \dfrac{1}{y - 1} = \dfrac{5}{y - 1}$

6. Solve these equations for x.

   A. $\dfrac{9}{x} = \dfrac{3}{38}$ B. $\dfrac{16}{x} = \dfrac{2}{7}$

   C. $\dfrac{4}{x - 2} = \dfrac{10}{25}$ D. $\dfrac{x - 2}{3} = \dfrac{x + 6}{5}$

7. Solve these equations for y.

   A. $\sqrt{12y + 1} - \sqrt{25 - 12y}$

   B. $\sqrt{7y} - 5 = 23$

   C. $\sqrt{y + 1} = \sqrt{21 + 5y}$

   D. $\sqrt{3y} + 21 = 33$

   E. $\sqrt{5y} + 8 = -2$

10.47 SOLUTIONS

1. Solve the equations for x.

   A. $7x + 8 = 1$
     $7x = -7$
     $x = -1$

   B. $8 - 8x = -16$
     $-8x = -24$
     $x = 3$

   C. $5x - 4 = 21$
     $5x = 25$
     $x = 5$

   D. $9 - 2x = 15$
     $-2x = 6$
     $x = -3$

   E. $9x - 5x + 13 = 3x + 6$
     $4x + 13 = 3x + 6$
     $x + 13 = 6$
     $x = -7$

2. Solve these equations for y.

   A. $2(y + 3) = -4(y + 2) + 2$
     $2y + 6 = -4y - 8 + 2$
     $2y + 6 = -4y - 8 + 2$
     $2y + 6 = -4y - 6$
     $6y + 6 = -6$
     $6y = -12$
     $y = -2$

   B. $3(2y + 1) - 2(y - 2) = 5$
     $6y + 3 - 2y + 4 = 5$
     $4y + 7 = 5$
     $4y = -2$
     $y = -\dfrac{1}{2}$

   C. $4(y - 9) = 8(y - 3)$
     $4y - 36 = 8y - 24$
     $-4y - 36 = -24$
     $-4y = 12$
     $y = -3$

   D. $2y + 3(y - 4) = 2(y - 3)$
     $2y + 3y - 12 = 2y - 6$
     $5y - 12 = 2y - 6$
     $3y - 12 = -6$
     $3y = 6$
     $y = 2$

   E. $6y - 3(5y + 2) = 4(1 - y)$
     $6y - 15y - 6 = 4 - 4y$
     $-9y - 6 = 4 - 4y$
     $-5y - 6 = 4$
     $-5y = 10$
     $y = -2$

3. Solve for the indicated letter.

   A. $P = 2l + 2w$ for l
     $P - 2w = 2l$
     $\dfrac{P - 2w}{2} = l$

   B. $P = a + b + c$ for c
     $P - a - b = c$

   C. $A = lw$ for w
     $\dfrac{A}{l} = w$

   D. $A = \dfrac{1}{2}(B + b)h$ for B
     $2A = (B + b)h$
     $2A = Bh + bh$
     $2A - bh = Bh$
     $\dfrac{2A - bh}{h} = B$

   E. $C = 2\pi r$ for r
     $\dfrac{C}{2\pi} = r$

4. Solve these equations for x.

   A. $\dfrac{3x}{4} + \dfrac{5x}{2} = 13$
     $3x + 2(5x) = 4(13)$
     $3x + 10x = 52$
     $13x = 52$
     $x = 4$

   B. $\dfrac{8x}{3} - \dfrac{2x}{4} = -13$
     $4(8x) + 3(-2x) = 12(-13)$
     $32x - 6x = -156$
     $26x = -156$
     $x = -6$

   C. $\dfrac{x - 8}{5} + \dfrac{8}{5} = \dfrac{-x}{3}$
     $3(x - 8) + 3(8) = 5(-x)$
     $3x - 24 + 24 = -5x$
     $3x = -5x$
     $8x = 0$
     $x = 0$

5. Solve these equations for y.

   A. $\dfrac{1}{y} + \dfrac{2}{y} = 3 - \dfrac{3}{y}$
     $1 + 2 = 3y - 3$
     $3 = 3y - 3$
     $6 = 3y$
     $2 = y$

B. $\dfrac{3}{4y} - \dfrac{1}{6} = \dfrac{4}{8y} + \dfrac{1}{2}$ LCD $= 24y$

$24y\left(\dfrac{3}{4y} - \dfrac{1}{6}\right) = 24y\left(\dfrac{4}{8y} + \dfrac{1}{2}\right)$

$6(3) + 4y(-1) = 3(4) + 12y(1)$

$18 - 4y = 12 + 12y$

$18 = 12 + 16y$

$6 = 16y$

$\dfrac{6}{16} = y$

$\dfrac{3}{8} = y$

C. $\dfrac{1}{y} + \dfrac{1}{y-1} = \dfrac{5}{y-1}$ LCD $= y(y-1)$

$y - 1 + y = 5y$

$2y - 1 = 5y$

$-1 = 3y$

$-\dfrac{1}{3} = y$

6. Solve these equations for x.

A. $\dfrac{9}{x} = \dfrac{3}{38}$

$3x = 342$

$x = 114$

B. $\dfrac{16}{x} = \dfrac{2}{7}$

$2x = 112$

$x = 56$

C. $\dfrac{4}{x-2} = \dfrac{10}{25}$

$10(x - 2) = 25(4)$

$10x - 20 = 100$

$10x = 120$

$x = 12$

D. $\dfrac{x-2}{3} = \dfrac{x+6}{5}$

$5(x - 2) = 3(x + 6)$

$5x - 10 = 3x + 18$

$2x - 10 = 18$

$2x = 28$

$x = 14$

7. Solve these equations for y.

A. $\sqrt{12y + 1} = \sqrt{25 - 12y}$

$\left(\sqrt{12y + 1}\right)^2 = \left(\sqrt{25 - 12y}\right)^2$

$12y + 1 = 25 - 12y$

$24y + 1 = 25$

$24y = 24$

$y = 1$

B. $\sqrt{7y} - 5 = 23$

$\sqrt{7y} = 28$

$\left(\sqrt{7y}\right)^2 = (28)^2$

$7y = 784$

$y = 112$

C. $\sqrt{y + 1} = \sqrt{21 + 5y}$

$\left(\sqrt{y+1}\right)^2 = \left(\sqrt{21 + 5y}\right)^2$

$y + 1 = 21 + 5y$

$1 = 21 + 4y$

$-20 = 4y$

$-5 = y$

Since each radicand becomes negative when $y = -5$, -5 is not a solution. No solution.

D. $\sqrt{3y} + 21 = 33$

$\sqrt{3y} = 12$

$\left(\sqrt{3y}\right)^2 = (12)^2$

$3y = 144$

$y = 48$

E. $\sqrt{5y} + 8 = -2$

$\sqrt{5y} = -10$

No solution, because the principal square root is never a negative number.

▬▬ 10.48 SYSTEMS OF LINEAR EQUATIONS

A system of two linear equations in two variables has a solution when there is a point (x, y) that makes each of the equations a true statement. There are graphical and algebraic methods for finding the point if one exists.

Example:

Graph each system and find the solution.

A. $2x - y = 4$ and $x + y = 5$

B. $3x - y = -6$ and $2x + 3y = 7$

Solution:

A. (1) for $2x - y = 4$

x	-1	0	1
y	-6	-4	-2

(2) for $x + y = 5$

x	-1	0	1
y	6	5	4

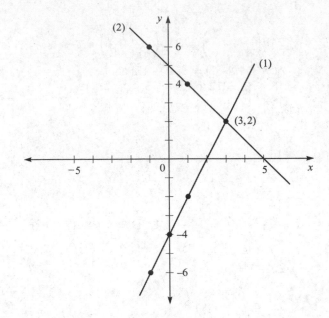

Figure 10.1

B. (1) $3x - y = -6$

x	-2	-1	0
y	0	3	6

(2) $2x + 3y = 7$

x	-1	2	5
y	3	1	-1

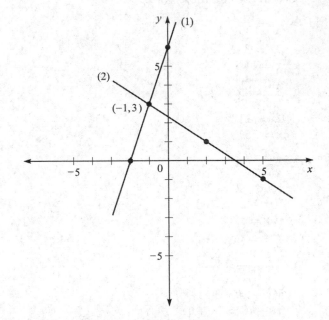

Figure 10.2

When you use the **elimination method**, you choose a variable to elimi-nate and multiply each equation as needed so that the variables have the same coefficient but with the opposite signs. Then add the equations and the chosen variable should be eliminated. After that, solve the resulting equation and then substitute into a given equation to find the value of the second variable.

Example:

Solve these systems of equation by elimination.

A. $5x + 2y = 3$ and $2x + 3y = -1$ 　　　　　　B. $2x - 3y = 7$ and $3x + y = 5$

Solution:

A. (1) $5x + 2y = 3$ (2) $2x + 3y = -1$ Eliminate x, LCM $(5, 2) = 10$.
   $2 \times (1)$ $10x + 4y = 6$ Substitute $y = -1$ into (1).
   $-5 \times (2)$ $\underline{-10x - 15y = 5}$ $5x + 2(-1) = 3$
              $-11y = 11$ $5x - 2 = 3$
                 $y = -1$ $5x = 5$
                                                 $x = 1$

   $(x, y) = (1, -1)$

B. (1) $2x - 3y = 7$ (2) $3x + y = 5$ Eliminate y, LCM $(3, 1) = 3$.
   $1 \times (1)$ $2x - 3y = 7$ Substitute $x = 2$ into (1).
   $3 \times (2)$ $\underline{9x + 3y = 15}$ $2(2) - 3y = 7$
          $11x$ $= 22$ $4 - 3y = 7$
              $x = 2$ $-3y = 3$
                                        $y = -1$

   $(x, y) = (2, -1)$

The **substitution method** of solving a system of equations requires you to solve one equation for one variable in terms of the other and then substitute that value into the other equation.

Example:

Solve these systems of equations by substitution.

A. $2x + y + 1 = 0$ and $3x - 2y + 5 = 0$ B. $3x + 2y = 13$ and $4x - y = -1$

Solution:

A. (1) $2x + y + 1 = 0$ (2) $3x - 2y + 5 = 0$ Solve (1) for y
   (1) $2x + y + 1 = 0$
      $y = -2x - 1$ Substitute $x = -1$ into $y = -2x - 1$.
   (2) $3x - 2y + 5 = 0$
      $3x - 2(-2x - 1) + 5 = 0$ $y = -2(-1) - 1 = 2 - 1 = 1$
      $3x + 4x + 2 + 5 = 0$ $y = 1$
      $7x + 7 = 0$
      $7x = -7$
       $x = -1$
     $(x, y) = (-1, 1)$

B. $3x + 2y = 13$ and $4x - y = -1$
   (1) $3x + 2y = 13$
   (2) $4x - y = -1$ (2) $4x - y = -1$ Solve (2) for y.
     $4x + 1 = y$
   (1) $3x + 2y = 13$ Substitute $x = 1$ into $y = 4x + 1$.
     $3x + 2(4x + 1) = 13$ $y = 4(1) + 1 = 4 + 1 = 5$
     $3x + 8x + 2 = 13$ $y = 5$
     $11x + 2 = 13$
     $11x = 11$
      $x = 1$
     $(x, y) = (1, 5)$

▬▬ 10.49 PRACTICE PROBLEMS

1. Solve these systems by graphing.
   A. $3x + y = 3$ and $x + 2y = -4$
   B. $3x + y = 3$ and $3x - 2y = 3$

2. Solve these systems by elimination.
   A. $2x + 3y = 2$ and $4x - 2y = -12$
   B. $2x + y = 3$ and $2x + 2y = 6$

3. Solve these systems by substitution.
   A. $2x + y = 6$ and $3x - 6y = 24$
   B. $x + y = 3$ and $x - y = 4$

▬▬ 10.50 SOLUTIONS

1. A. $3x + y = 3$ and $x + 2y = -4$
   (1) $3x + y = 3$

x	−1	0	1
y	6	3	0

   (2) $x + 2y = -4$

x	−2	0	2
y	−1	−2	−3

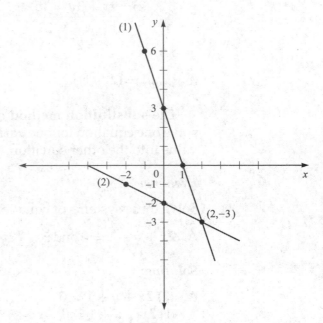

Figure 10.3

B. $3x + y = 3$ and $3x - 2y = 3$
   (1) $3x + y = 3$

x	−1	0	1
y	6	3	0

   (2) $3x - 2y = 3$

x	−1	1	3
y	−3	0	3

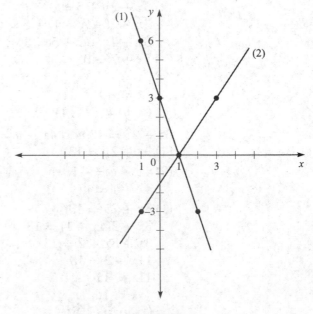

Figure 10.4

2. Solve these systems by elimination.

A. (1) $2x + 3y = 2$
    (2) $4x - 2y = -12$ Eliminate y.

$2 \times (1)$ $4x + 6y = 4$ $2x + 3y = 2$
$3 \times (2)$ $12x - 6y = -36$ $2(-2) + 3y = 2$
$\overline{16x = -32}$ $-4 + 3y = 2$
$ x = -2$ $3y = 6$
$ y = 2$

$(x, y) = (-2, 2)$

B. (1) $2x + y = 3$ (2) $2x + 2y = 6$
                   Eliminate x.

(1) $2x + y = 3$ $2x + y = 3$
$-1 \times (2)$ $\underline{-2x - 2y = -6}$ $2x + 3 = 3$
$ -y = -3$ $2x = 0$
$ y = 3$ $x = 0$

$(x, y) = (0, 3)$

3. A. (1) $2x + y = 6$ (2) $3x - 6y = 24$
Solve (1) for y.
$y = -2x + 6$ $y = -2x + 6$
Substitute into (2).
$3x - 6(-2x + 6) = 24$ $y = -2(4) + 6$
$3x + 12x - 36 = 24$ $y = -8 + 6$
$15x = 60$ $y = -2$
$x = 4$
$(x, y) = (4, -2)$

B. (1) $x + y = 3$ (2) $x - y = 4$
Solve (2) for x. $x = y + 4$ $x = y + 4$
Substitute
into (1). $y + 4 + y = 3$ $x = -.5 + 4$
                     $2y = -1$ $x = 3.5$
                       $y = -.5$

$(x, y) = (3.5, -.5)$

▇▇ 10.51 LINEAR INEQUALITIES

An **inequality** is a statement that two quantities are related by less than, $<$, greater than, $>$, less than or equal to, $\le$, or greater than or equal to, $\ge$. Linear inequalities in one variable are solved by methods similar to those for linear equations, with one major difference. When multiplying or dividing both sides of an inequality by a negative, you must change the direction of the inequality symbol.

Example:

Solve these inequalities for x.

A. $x + 2 > 7$ B. $3x < 9$ C. $x - 6 < -5$ D. $-2x > 6$

Solution:

A. $x + 2 > 7$ B. $3x < 9$ C. $x - 6 < -5$
      $x + 2 - 2 > 7 - 2$ $\dfrac{3x}{3} < \dfrac{9}{3}$ $x - 6 + 6 < -5 + 6$
      $x > 5$ $x < 3$ $x < 1$

D. $-2x > 6$
      $\dfrac{-2x}{-2} < \dfrac{6}{-2}$ Since you divide by negative 2, change the
                         direction of the sign.
      $x < -3$

▇▇ 10.52 PRACTICE PROBLEMS

Solve each inequality for x.

1. $2x + 7 < 15$

2. $7x - 5 \ge 9$

3. $-6x + 5 < 17$

4. $2x + 13 > -41$

5. $4(3 - x) \ge 2(x - 3)$

6. $0 < 2x + 7 - 9x$

7. $23 - 8x \le 5 + x$

8. $3x - 4 > 2x - 9$

9. $3x - 2 + x \le 5x$

10. $4(-x + 2) - (1 - 5x) > -8$

11. $6(10 - x) + 3(7 - 2x) < 45$

▬▬ 10.53 SOLUTIONS

1. $2x + 7 < 15$
   $2x + 7 - 7 < 15 - 7$
   $2x < 8$
   $x < 4$

2. $7x - 5 \ge 9$
   $7x \ge 14$
   $x \ge 2$

3. $-6x + 5 < 17$
   $-6x < 12$
   $x > -2$

4. $2x + 13 > -41$
   $2x > -54$
   $x > -27$

5. $4(3 - x) \ge 2(x - 3)$
   $12 - 4x \ge 2x - 6$
   $12 - 6x \ge -6$
   $-6x \ge -18$
   $x \le 3$

6. $0 < 2x + 7 - 9x$
   $0 < -7x + 7$
   $7x < 7$
   $x < 1$

7. $23 - 8x \le 5 + x$
   $23 - 9x \le 5$
   $-9x \le -18$
   $x \ge 2$

8. $3x - 4 > 2x - 9$
   $x - 4 > -9$
   $x > -5$

9. $3x - 2 + x \le 5x$
   $4x - 2 \le 5x$
   $-x - 2 \le 0$
   $-x \le 0 + 2$
   $x \ge -2$

10. $4(-x + 2) - (1 - 5x) > -8$
    $-4x + 8 - 1 + 5x > -8$
    $x + 7 > -8$
    $x > -15$

11. $6(10 - x) + 3(7 - 2x) < 45$
    $60 - 6x + 21 - 6x < 45$
    $81 - 12x < 45$
    $-12x < -36$
    $x > 3$

▬▬ 10.54 QUADRATIC EQUATIONS AND INEQUALITIES

A **quadratic equation in one variable** is an equation of the form $ax^2 + bx + c = 0$ where a, b, and c are real numbers and $a \ne 0$. Quadratic equations are generally solved by two methods: factoring and the quadratic formula. Some quadratic equations cannot be factored, but all quadratic equations can be solved by the quadratic formula.

The zero product principle is the basis for factoring. If a and b are real numbers and $ab = 0$, then $a = 0$ or $b = 0$. To solve a quadratic equation using this property, you must have one side of the equation equal to zero.

Example:

Solve each equation by factoring.

A. $x^2 + 4x = 21$

B. $2y^2 - y = 1$

C. $5x^2 + 15x = 0$

D. $9y^2 = 7(6y - 7)$

E. $(x + 6)(x - 2) = -7$

Solution:

A. $x^2 + 4x = 21$
   $x^2 + 4x - 21 = 0$
   $(x + 7)(x - 3) = 0$
   $x + 7 = 0$ or $x - 3 = 0$
   $x = -7$ or $x = 3$
   The solutions are -7 and 3.

B. $2y^2 - y = 1$
   $2y^2 - y - 1 = 0$
   $(2y + 1)(y - 1) = 0$
   $2y + 1 = 0$ or $y - 1 = 0$
   $2y = -1$ or $y = 1$
   $y = -\dfrac{1}{2}$

   The solutions are $-\dfrac{1}{2}$ and 1.

C. $5x^2 + 15x = 0$
   $5x(x + 3) = 0$
   $5x = 0$ or $x + 3 = 0$
   $x = 0$ or $x = -3$
   The solutions are 0 and -3.

D. $9y^2 = 7(6y - 7)$
   $9y^2 = 42y - 49$
   $9y^2 - 42y + 49 = 0$
   $(3y - 7)(3y - 7) = 0$
   $3y - 7 = 0$ or $3y - 7 = 0$
   $3y = 7$ or $3y = 7$
   $y = \dfrac{7}{3}$ or $y = \dfrac{7}{3}$

   The solutions are both $\dfrac{7}{3}$.

E. $(x + 6)(x - 2) = -7$ (Note that the equation does not equal zero.
   $x^2 + 4x - 12 + 7 = 0$ Multiply out and get the equation to equal zero.)
   $x^2 + 4x - 5 = 0$
   $(x + 5)(x - 1) = 0$
   $x + 5 = 0$ or $x - 1 = 0$
   $x = -5$ or $x = 1$
   The solutions are -5 and 1.

When $ax^2 + bx + c = 0$ and $a \neq 0$, then $x = \dfrac{-b \pm \sqrt{b^2 - 4ac}}{2a}$.

If $b^2 - 4ac$ is negative, the quadratic equation has no real roots.

Example:

Solve each equation using the quadratic formula.

A. $3x^2 + 4x - 4 = 0$ B. $y^2 - 4y + 3 = -y^2 + 2y$
C. $(x - 1)(x + 3) = -5$ D. $4y^2 - 8y + 3 = 0$

Solution:

A. $3x^2 + 4x - 4 = 0$
$a = 3, b = 4, c = -4$

$$x = \frac{-4 \pm \sqrt{4^2 - 4(3)(-4)}}{2(3)}$$

$$x = \frac{-4 \pm \sqrt{64}}{6}$$

$$x = \frac{-4 \pm 8}{6}$$

$$x = \frac{-4 + 8}{6} \text{ or } x = \frac{-4 - 8}{6}$$

$$x = \frac{4}{6} \text{ or } x = \frac{-12}{6}$$

$$x = \frac{2}{3} \text{ or } x = -2$$

The solutions are $\frac{2}{3}$ and -2.

B. $y^2 - 4y + 3 = -y^2 + 2y$
$2y^2 - 6y + 3 = 0$
$a = 2, b = -6, c = 3$

$$y = \frac{-(-6) \pm \sqrt{(-6)^2 - 4(2)(3)}}{2(2)}$$

$$y = \frac{6 \pm \sqrt{12}}{4}$$

$$y = \frac{6 \pm 2\sqrt{3}}{4}$$

$$y = \frac{2(3 \pm \sqrt{3})}{4}$$

$$y = \frac{3 \pm \sqrt{3}}{2}$$

The solutions are $\frac{3 + \sqrt{3}}{2}$ and $\frac{3 - \sqrt{3}}{2}$.

C. $(x - 1)(x + 3) = -5$
$x^2 + 2x - 3 = -5$
$x^2 + 2x + 2 = 0$
$a = 1, b = 2, \text{ and } c = 2$

$$x = \frac{-2 \pm \sqrt{2^2 - 4(1)(2)}}{2(1)}$$

$$x = \frac{-2 \pm \sqrt{-4}}{2}$$

Since $b^2 - 4ac < 0$, there are no real solutions.

D. $4y^2 - 8y + 3 = 0$
$a = 4, b = -8, c = 3$

$$y = \frac{-(-8) \pm \sqrt{(-8)^2 - 4\,(4)\,(3)}}{2\,(4)}$$

$$y = \frac{8 \pm \sqrt{16}}{8}$$

$$y = \frac{8 \pm 4}{8}$$

$$y = \frac{8+4}{8} \text{ or } y = \frac{8-4}{8}$$

$$y = \frac{12}{8} \text{ or } y = \frac{4}{8}$$

$$y = 1.5 \text{ or } y = 0.5$$

The solutions are 1.5 and 0.5.

Quadratic inequalities that can be factored can be solved using the properties of the product of signed numbers. If $ab > 0$, then $a > 0$ and $b > 0$, or $a < 0$ and $b < 0$; and if $ab < 0$, then $a > 0$ and $b < 0$, or $a < 0$ and $b > 0$.

Example:

Solve these quadratic inequalities.

A. $6y^2 - 7y - 5 > 0$
C. $2y^2 - 5y - 3 \geq 0$

B. $6x^2 - 7x + 2 < 0$
D. $6x^2 + x - 5 \leq 0$

Solution:

A. $6y^2 - 7y - 5 > 0$
$(3y - 5)(2y + 1) > 0$

Case 1: $3y - 5 > 0$ and $2y + 1 > 0$ or Case 2: $3y - 5 < 0$ and $2y + 1 < 0$

$\qquad\quad 3y > 5$ and $2y > -1$ $\qquad\qquad\qquad\quad 3y < 5$ and $2y < -1$

$\qquad\quad y > \dfrac{5}{3}$ and $y > -\dfrac{1}{2}$ $\qquad$ or $\qquad y < \dfrac{5}{3}$ and $y < \dfrac{-1}{2}$

$\qquad\qquad\quad y > \dfrac{5}{3}$ $\qquad\qquad\qquad$ or $\qquad\qquad y < -\dfrac{1}{2}$

The solution is $y > \dfrac{5}{3}$ or $y < -\dfrac{1}{2}$.

B. $6x^2 - 7x + 2 < 0$
$(3x - 2)(2x - 1) < 0$

Case 1: $3x - 2 < 0$ and $2x - 1 > 0$ or Case 2: $3x - 2 > 0$ and $2x - 1 < 0$

$\qquad\quad 3x < 2$ and $2x > 1$ $\qquad\qquad\qquad\quad 3x > 2$ and $2x < 1$

$\qquad\quad x < \dfrac{2}{3}$ and $x > \dfrac{1}{2}$ $\qquad\qquad\qquad x > \dfrac{2}{3}$ and $x < \dfrac{1}{2}$

$\qquad\quad \dfrac{1}{2} < x < \dfrac{2}{3}$ $\qquad\qquad$ or $\qquad\qquad$ no solution.

The solution is $\dfrac{1}{2} < x < \dfrac{2}{3}$.

C. $2y^2 - 5y - 3 \geq 0$
$(2y + 1)(y - 3) \geq 0$

Case 1: $2y + 1 \geq 0$ and $y - 3 \geq 0$ or   Case 2: $2y + 1 \leq 0$ and $y - 3 \leq 0$

$2y \geq -1$ and $y \geq 3$ or $2y \leq -1$ and $y \leq 3$

$y \geq -\dfrac{1}{2}$ and $y \geq 3$ or $y \leq \dfrac{-1}{2}$ and $y \leq 3$

$y \geq 3$ or $y \leq -\dfrac{1}{2}$

The solution is $y \geq 3$ or $y \leq -\dfrac{1}{2}$.

D. $6x^2 + x - 5 \leq 0$
$(6x - 5)(x + 1) \leq 0$

Case 1: $6x - 5 \leq 0$ and $x + 1 \geq 0$ or   Case 2: $6x - 5 \geq 0$ and $x + 1 \leq 0$

$6x \leq 5$ and $x \geq -1$ or $6x \geq 5$ and $x \leq -1$

$x \leq \dfrac{5}{6}$ and $x \geq -1$ or $x \geq \dfrac{5}{6}$ and $x \leq -1$

$-1 \leq x \leq \dfrac{5}{6}$ or   no solution.

The solution is $-1 \leq x \leq \dfrac{5}{6}$.

▬▬ 10.55 PRACTICE PROBLEMS

1. Solve these equations by factoring.
   A. $x^2 - 5x + 6 = 0$ B. $y^2 = 4y$
   C. $x^2 + 3x = 28$ D. $5y - 2y^2 = 2$
   E. $5x^2 + 40 = 33x$

2. Solve these equations using the quadratic formula.
   A. $x^2 - 6x + 8 = 0$ B. $y^2 = 4 - 3y$
   C. $3x^2 + 8x + 5 = 0$ D. $x^2 + 4x + 1 = 0$
   E. $2y^2 + 3y = 5$

3. Solve these quadratic inequalities.
   A. $x^2 - 7x > -12$ B. $2y^2 + 2 < 5y$
   C. $9x^2 < 9x - 2$ D. $x^2 + x \leq 6$
   E. $y^2 \geq 5y + 24$

▬▬ 10.56 SOLUTIONS

1. Solve these equations by factoring.
   A. $x^2 - 5x + 6 = 0$
   $(x - 3)(x - 2) = 0$
   $x - 3 = 0$ or $x - 2 = 0$
   $x = 3$ or $x = 2$

   B. $y^2 = 4y$
   $y^2 - 4y = 0$
   $y(y - 4) = 0$
   $y = 0$ or $y - 4 = 0$
   $y = 0$ or $y = 4$

   C. $x^2 + 3x = 28$
   $x^2 + 3x - 28 = 0$
   $(x + 7)(x - 4) = 0$

   $x + 7 = 0$ or $x - 4 = 0$
   $x = -7$ or $x = 4$

   D. $5y - 2y^2 = 2$
   $-2y^2 + 5y - 2 = 0$
   $2y^2 - 5y + 2 = 0$
   $(2y - 1)(y - 2) = 0$
   $2y - 1 = 0$ or $y - 2 = 0$
   $2y = 1$ or $y = 2$
   $y = \dfrac{1}{2}$ or $y = 2$

E. $5x^2 + 40 = 33x$
$5x^2 - 33x + 40 = 0$
$(5x - 8)(x - 5) = 0$
$5x - 8 = 0$ or $x - 5 = 0$
$5x = 8$ or $x = 5$
$x = 1.6$ or $x = 5$

2. Solve these equations using the quadratic formula.
A. $x^2 - 6x + 8 = 0$
$a = 1, b = -6, c = 8$

$$x = \frac{-(-6) \pm \sqrt{(-6)^2 - 4(1)(8)}}{2(1)}$$

$$x = \frac{6 \pm \sqrt{36 - 32}}{2}$$

$$x = \frac{6 \pm \sqrt{4}}{2}$$

$$x = \frac{6 \pm 2}{2}$$

$$x = \frac{6 + 2}{2} \text{ or } x = \frac{6 - 2}{2}$$

$$x = \frac{8}{2} \text{ or } x = \frac{4}{2}$$

$x = 4$ or $x = 2$

B. $y^2 = 4 - 3y$
$y^2 + 3y - 4 = 0$
$a = 1, b = 3, c = -4$

$$y = \frac{-3 \pm \sqrt{3^2 - 4(1)(-4)}}{2(1)}$$

$$y = \frac{-3 \pm \sqrt{25}}{2}$$

$$y = \frac{-3 \pm 5}{2}$$

$$y = \frac{-3 + 5}{2} \text{ or } y = \frac{-3 - 5}{2}$$

$y = 1$ or $y = -4$

C. $3x^2 + 8x + 5 = 0$
$a = 3, b = 8, c = 5$

$$x = \frac{-8 \pm \sqrt{(8)^2 - 4(3)(5)}}{2(3)}$$

$$x = \frac{-8 \pm \sqrt{4}}{6}$$

$$x = \frac{-8 \pm 2}{6}$$

$$x = \frac{-8 + 2}{6} \text{ or } x = \frac{-8 - 2}{6}$$

$$x = -1 \text{ or } x = -\frac{5}{3}$$

D. $x^2 + 4x + 1 = 0$
$a = 1, b = 4, c = 1$

$$x = \frac{-4 \pm \sqrt{4^2 - 4(1)(1)}}{2(1)}$$

$$x = \frac{-4 \pm \sqrt{12}}{2}$$

$$x = \frac{-4 \pm 2\sqrt{3}}{2}$$

$x = -2 \pm \sqrt{3}$

E. $2y^2 + 3y = 5$
$2y^2 + 3y - 5 = 0$
$a = 2, b = 3, c = -5$

$$y = \frac{-3 \pm \sqrt{3^2 - 4(2)(-5)}}{2(2)}$$

$$y = \frac{-3 \pm \sqrt{49}}{4}$$

$$y = \frac{-3 \pm 7}{4}$$

$$y = \frac{-3 + 7}{4} \text{ or } y = \frac{-3 - 7}{4}$$

$y = 1$ or $y = -2.5$

3. Solve these quadratic inequalities.
A. $x^2 - 7x > -12$
$x^2 - 7x + 12 > 0$
$(x - 3)(x - 4) > 0$

Case 1: $x - 3 > 0$ and $x - 4 > 0$
$x > 3$ and $x > 4$
Solution: $x > 4$
or
Case 2: $x - 3 < 0$ and $x - 4 < 0$
$x < 3$ and $x < 4$
Solution: $x < 3$
The solution is $x > 4$ or $x < 3$.

B. $2y^2 + 2 < 5y$
$2y^2 - 5y + 2 < 0$
$(2y - 1)(y - 2) < 0$

Case 1: $2y - 1 < 0$ and $y - 2 > 0$
$2y < 1$ and $y > 2$
$y < 1/2$ and $y > 2$
No solution.
or
Case 2: $2y - 1 > 0$ and $y - 2 < 0$
$2y > 1$ and $y < 2$
$y > 1/2$ and $y < 2$
$1/2 < y < 2$
The only solution is $1/2 < y < 2$.

C. $9x^2 < 9x - 2$
$9x^2 - 9x + 2 < 0$
$(3x - 2)(3x - 1) < 0$

Case 1: $3x - 2 < 0$ and $3x - 1 > 0$
$3x < 2$ and $3x > 1$
$x < 2/3$ and $x > 1/3$
$1/3 < x < 2/3$
or
Case 2: $3x - 2 > 0$ and $3x - 1 < 0$
$3x > 2$ and $3x < 1$
$x > 2/3$ and $x < 1/3$
No solution.
The only solution is $1/3 < x < 2/3$.

D. $x^2 + x \leq 6$
$x^2 + x - 6 \leq 0$
$(x + 3)(x - 2) \leq 0$

Case 1: $x + 3 \leq 0$ and $x - 2 \geq 0$
$x \leq -3$ and $x \geq 2$

No solution.
or
Case 2: $x + 3 \geq 0$ and $x - 2 \leq 0$
$x \geq -3$ and $x \leq 2$
$-3 \leq x \leq 2$
The only solution is $-3 \leq x \leq 2$.
The solution is $-3 \leq x \leq 2$.

E. $y^2 \geq 5y + 24$
$y^2 - 5y - 24 \geq 0$
$(y - 8)(y + 3) \geq 0$

Case 1: $y - 8 \geq 0$ and $y + 3 \geq 0$
$y \geq 8$ and $y \geq -3$
$y \geq 8$
or
Case 2: $y - 8 \leq 0$ and $y + 3 \leq 0$
$y \leq 8$ and $y \leq -3$
$y \leq -3$
The solution is $y \geq 8$ or $y \leq -3$.

10.57 FUNCTIONS

A **relation** is a set of ordered pairs, an equation, or a rule. In the ordered pair, the first element belongs to the domain, and this value is usually represented by x. The second element in the ordered pair belongs to the range, and this value is usually represented as y.

A **function** is a relation for which the first element in the ordered pair is paired with exactly one second element. It can be stated that for each x value, there is exactly one y value. You denote a function in the variable x by $f(x)$.

Sometimes a special symbol such as $\diamond$, $\square$, #, or $*$ may be used to denote the rule for a function. For example, $\square$ could be defined so that $3\square = 5(3) - 2$, and # could denote $5\# = 5^2 + 3(5)$, and $x* = x^2 + 3x$.

Example:

Evaluate each function as indicated.

A. $f(x) = 3x^2 - 2x$ when $x = -2, x = 0, x = 3$
B. $f(x) = 7x - 2$ when $x = -3, x = 0, x = 5$
C. $f(x) = x^3 - 5x$ when $x = -1, x = 0, x = 2$

Solution:

A. $f(x) = 3x^2 - 2x$ when $x = -2, x = 0, x = 3$
$f(-2) = 3(-2)^2 - 2(-2) = 3(4) + 4 = 12 + 4 = 16$
$f(0) = 3(0)^2 - 2(0) = 3(0) - 0 = 0 - 0 = 0$
$f(3) = 3(3)^2 - 2(3) = 3(9) - 6 = 27 - 6 = 21$
B. $f(x) = 7x - 2$ when $x = -3, x = 0, x = 5$
$f(-3) = 7(-3) - 2 = -21 - 2 = -23$
$f(0) = 7(0) - 2 = 0 - 2 = -2$
$f(5) = 7(5) - 2 = 35 - 2 = 33$

C. $f(x) = x^3 - 5x$ when $x = -1$, $x = 0$, $x = 2$
$f(-1) = (-1)^3 - 5(-1) = -1 + 5 = 4$
$f(0) = 0^3 - 5(0) = 0 - 0 = 0$
$f(2) = 2^3 - 5(2) = 8 - 10 = -2$

Example:

Evaluate each function as indicated.

A. $x \diamond = 5x^2 - 3x + 10$ when $x = 2$, $x = -3$

B. $x \square = \dfrac{x^2 - 5x}{x + 2}$, if $x \neq -2$ when $x = 0$, $x = 3$

C. $x \square y = \dfrac{x^2 + xy - y^2}{xy}$, if $x \neq 0$ and $y \neq 0$ when $x = 2$ and $y = 3$ and when $x = 3$ and $y = -1$

Solution:

A. $x \diamond = 5x^2 - 3x + 10$ when $x = 2$, $x = -3$
$2 \diamond = 5(2)^2 - 3(2) + 10 = 5(4) - 6 + 10 = 20 - 6 + 10 = 24$
$-3 \diamond = 5(-3)^2 - 3(-3) + 10 = 5(9) + 9 + 10 = 45 + 9 + 10 = 64$

B. $x \square = \dfrac{x^2 - 5x}{x + 2}$, if $x \neq -2$ when $x = 0$, $x = 3$

$0 \square = \dfrac{0^2 - 5(0)}{0 + 2} = \dfrac{0 - 0}{2} = 0$

$3 \square = \dfrac{3^2 - 5(3)}{3 + 2} = \dfrac{9 - 15}{5} = \dfrac{-6}{5}$

C. $x \square y = \dfrac{x^2 + xy - y^2}{xy}$, if $x \neq 0$ and $y \neq 0$ when $x = 2$ and $y = 3$ and when $x = 3$ and $y = -1$

$2 \square 3 = \dfrac{2^2 + 2(3) - 3^2}{2(3)} = \dfrac{4 + 6 - 9}{6} = \dfrac{1}{6}$

$3 \square -1 = \dfrac{3^2 + 3(-1) - (-1)^2}{3(-1)} = \dfrac{9 - 3 - 1}{-3} = \dfrac{5}{-3} = -\dfrac{5}{3}$

▆▆ 10.58 PRACTICE PROBLEMS

1. Demonstrate that each expression is not a function by evaluating it for the given x value.

   A. $x^2 + y^2 = 25$ for $x = 3$
   B. $y^2 = 5x + 11$ for $x = 5$

2. Evaluate each function as indicated.

   A. $f(x) = \dfrac{3}{x + 4}$, $x \neq -4$ for $x = -3$

   B. $f(x) = x^2 - 9x$ for $x = 6$
   C. $f(x) = 9 - 2x^2$ for $x = -2$

   D. $f(x) = \dfrac{x^2 - 3}{2x}$, $x \neq 0$, for $x = 5$

   E. $f(x) = 2(x + 5)^2$ for $x = 2$

3. Evaluate each function as indicated.

   A. $x \diamond = 3x^2 - 2x$ for $x = 2$

   B. $x \square = \dfrac{x}{x^2 + 2}$ for $x = 6$

   C. $x \square = \dfrac{5x - 2}{7 - 4x}$ for $x = -3$

   D. $x \square y = 2x^2 - 5y^2$ for $x = 3$ and $y = -2$
   E. $x \square y = y^2 - 5xy$ for $x = 5$ and $y = 3$

▆▆ 10.59 SOLUTIONS

1. Demonstrate that each expression is not a function by evaluating it for the given x value.

   A. $x^2 + y^2 = 25$ for $x = 3$
   $3^2 + y^2 = 25$
   $9 + y^2 = 25$
   $y^2 = 16$
   $y = \pm 4$
   Since $(3, 4)$ and $(3, -4)$ are values for the expression, it is not a function.

   B. $y^2 = 5x + 11$ for $x = 5$
   $y^2 = 5(5) + 11$
   $y^2 = 25 + 11$
   $y^2 = 36$
   $y = \pm 6$
   Since $(5, 6)$ and $(5, -6)$ are values for the expression, it is not a function.

2. Evaluate each function as indicated.

   A. $f(x) = \dfrac{3}{x+4}$, $x \ne -4$ for $x = -3$

   $f(-3) = \dfrac{3}{-3+4} = \dfrac{3}{1} = 3$

   B. $f(x) = x^2 - 9x$ for $x = 6$
   $f(6) = 6^2 - 9(6)$
   $f(6) = 36 - 54 = -18$

   C. $f(x) = 9 - 2x^2$ for $x = -2$
   $f(-2) = 9 - 2(-2)^2 = 9 - 2(4) = 9 - 8 = 1$

   D. $f(x) = \dfrac{x^2 - 3}{2x}$, $x \ne 0$, for $x = 5$

   $f(x) = \dfrac{5^2 - 3}{2(5)} = \dfrac{25 - 3}{10} = \dfrac{22}{10} = 2.2$

   E. $f(x) = 2(x + 5)^2$ for $x = 2$
   $f(x) = 2(2 + 5)^2 = 2(7)^2 = 2(49) = 98$

3. Evaluate each function as indicated.

   A. $x \diamond = 3x^2 - 2x$ for $x = 2$
   $2 \diamond = 3(2)^2 - 2(2) = 3(4) - 4 = 12 - 4 = 8$

   B. $x \square = \dfrac{x}{x^2 + 2}$ for $x = 6$

   $6 \square = \dfrac{6}{6^2 + 2} = \dfrac{6}{36 + 2} = \dfrac{6}{38} = \dfrac{3}{19}$

   C. $x \square = \dfrac{5x - 2}{7 - 4x}$ for $x = -3$

   $3 \square = \dfrac{5(-3) - 2}{7 - 4(-3)} = \dfrac{-15 - 2}{7 + 12} = \dfrac{-17}{19} = -\dfrac{17}{19}$

   D. $x \square y = 2x^2 - 5y^2$ for $x = 3$ and $y = -2$
   $3 \square 2 = 2(3)^2 - 5(2)^2 = 2(9) - 5(4) = 18 - 20$
   $= -2$

   E. $x \square y = y^2 - 5xy$ for $x = 5$ and $y = 3$
   $5 \square 3 = 3^2 - 5(5)3 = 9 - 75 = -66$

▆▆ 10.60 ALGEBRAIC WORD PROBLEMS

Solving word problems combines two skills: translating word statements into algebraic expressions and solving equations. A general procedure for solving word problems allows you to solve problems over a wide variety of applications.

Solving Word Problems

1. Read the problem carefully, looking for key terms and concepts. Identify the question you must answer. A diagram may help you interpret the given information.
2. List all the unknown quantities in the problem and represent them in terms of one variable if possible, such as x or x and y.
3. Use the information identified in Step 1 to write algebraic relationships among the quantities identified in Step 2.
4. Combine the algebraic relationships into equations.
5. Solve the equation or system of equations.
6. Verify your results by checking against the facts in the problem.

Example:

Solve these number word problems.

A. The sum of two numbers is 94, and the larger number is 5 less than twice the smaller number. Find the numbers.
B. Two numbers have a sum of 18. Find the numbers if one number is 8 larger than the other.
C. Find three consecutive integers if their sum is 21.
D. Find three consecutive even integers such that the first plus twice the second plus four times the third equals 174.

Solution:

A. $x =$ the smaller number   Represent the smaller number by using x.
  $2x - 5 =$ the larger number   Represent the larger number with respect
                                        to the smaller.

$x + 2x - 5 = 94$ Add the two numbers to get the sum.
$3x - 5 = 94$ Combine like terms to simplify.
$x = 33$ Solve the equation.
$2x - 5 = 2(33) - 5 = 66 - 5 = 61.$ Find the larger number.
The two numbers are 33 and 61.

B. $x =$ the smaller number   Represent the smaller number with x.
  $y =$ the larger number   Represent the larger number with y.
    $x + y = 18$ Add the variables to get the sum.
    $x + 8 = y$ Add 8 to smaller number to get larger number.

$$\begin{aligned} x + y &= 18 \\ + \; x - y &= -8 \\ \hline 2x &= 10 \\ x &= 5 \end{aligned}$$
$$\begin{aligned} x + y &= 18 \\ 5 + y &= 18 \\ y &= 13 \end{aligned}$$

Solve the system for x.

Find the larger number.

The two numbers are 5 and 13.

C. $x =$ the first integer
  $x + 1 =$ the second integer
  $x + 2 =$ the third integer
  $x + x + 1 + x + 2 = 21$
  $3x + 3 = 21$
  $3x = 18$
  $x = 6$
  $x + 1 = 7$
  $x + 2 = 8$
The three consecutive integers are 6, 7, and 8.

D. $x =$ the first even integer
  $x + 2 =$ the second even integer
  $x + 4 =$ the third even integer
  $x + 2(x + 2) + 4(x + 4) = 174$
  $x + 2x + 4 + 4x + 16 = 174$
  $7x + 20 = 174$
  $7x = 154$
  $x = 22$
  $x + 2 = 24$
  $x + 4 = 26$
The three consecutive even integers are 22, 24, and 26.

Example:

Solve these age word problems.

A. Carlos is 3 years older than his brother Jose. In 4 years from now, the sum of their ages will be 33 years. How old is each now?
B. Kia is 5 years younger than her sister Yvette. Three years ago, the sum of their ages was 23. How old is each now?

Solution:

A. x = the age of Jose now   Represent the younger person's age by x.
   $x + 3$ = the age of Carlos now   Using x, represent the older person's age.

$x + 4 + x + 3 + 4 = 33$ Add their ages in 4 years to get that sum.
$2x + 11 = 33$ Combine like terms to simplify.
$2x = 22$ Solve the equation.
$x = 11$ Find the second person's age.
$x + 3 = 14$ Write an answer to the question asked.
Jose is 11 years old now and Carlos is 14 years old.

B. x = the age of Yvette now
   $x - 5$ = the age of Kia now
$x - 3 + x - 5 - 3 = 23$
$2x - 11 = 23$
$2x = 34$
$x = 17$
$x - 5 = 12$
Kia is 12 years old now and Yvette is 17 years old.

Example:

Solve these statistical word problems.

A. Ken's percent grades on five tests were 84, 72, 91, 64, and 83. Find the average (arithmetic mean) grade.
B. Von had an average percent score on the first four tests of 84, and his average percent score on the next six tests was 92. What was Von's average score for all of the tests?
C. Marie's monthly school expenses were $64, $82, $51, $90, $67, $71, $58, $94, and $63. What is Marie's median monthly school expense?
D. Tim's monthly food costs were $412, $408, $410, $408, $401, $410, and $408. What is the mode for the monthly food costs?

Solution:

A. Sum = $84 + 72 + 91 + 64 + 83 = 394$, $N = 5$
AVE = SUM $\div N = 394 \div 5 = 78.8$
Ken's average percent score is 78.8.
B. Average of four tests 84, average of six tests 92
$$\text{AVE} = \frac{4(84) + 6(92)}{4 + 6} = \frac{336 + 552}{10} = \frac{888}{10} = 88.8.$$
Von's average score on the ten tests was 88.8.
C. Sequenced expenses: $51, $58, $63, $64, $67, $71, $82, $90, $94. Since there is an odd number of values, the median is the middle value, the fifth one. The median of Marie's monthly school expense is $67.

D. Sequenced costs: $401, $401, $408, $408, $408, $410, $410, $412, $420. The most frequent number is $408. The mode monthly food cost for Tim is $408.

Example:

Solve these mixture word problems.

A. Beth has one solution that is 16% acid and a second solution that is 26% acid. How many ounces of each is needed to make 30 ounces of a solution that is 18% acid?

B. Lynn has 100 pounds of candy worth $4.80 per pound. How many pounds of a second candy worth $5.20 per pound should she mix with the 100 pounds in order to obtain a mixture worth $5.00 per pound?

Solution:

A. x = number of ounces of 16% acid solution   Represent one quantity.
    $30 - x$ = number of ounces of 26% acid solution   Represent second quantity.

$0.16x + 0.26(30 - x) = 0.18(30)$ Represent the number of ounces of acid.

$0.16x + 7.8 - 0.26x = 5.4$ Simplify the equation.
$-0.10x + 7.8 = 5.4$
$-0.10x = -2.4$ Solve the equation.
$x = 24$ Find the second quantity.
$30 - x = 6$ Write an answer to the question asked.
Beth needs to use 24 ounces of the 16% acid solution and 6 ounces of the 26% acid solution.

B. x = number of pounds of $5.20 a pound candy needed
$100(\$4.80) + x(\$5.20) = (100 + x)(\$5.00)$
$\$480 + \$5.20x = \$500 + \$5.00x$
$\$480 + \$0.20x = \$500$
$\$0.20x = \$20$
$x = 100$
Lynn needs to use 100 pounds of the $5.20 candy.

Example:

Solve these rate word problems.

A. Tyrone rides his bicycle 5 miles from his home to the city bus stop at the rate of 8 miles per hour. He arrives in time to catch the bus to work, which travels at 25 miles per hour. If he spends 1.5 hours traveling from home to work, how far does he travel on the bus?

B. Maria jogs 15 miles to her sister's house to get her bicycle and then bicycles home. The total trip takes 3 hours. If she bicycles twice as fast as she jogs, how fast does she bicycle?

C. Two drivers, Brenda and Julie, are 300 miles apart. Brenda drives at 30 miles per hour and Julie drives at 45 miles per hour. If they drive toward each other, how far will each have traveled when they meet?

D. At what rate must Eden travel to overtake Rex, who is traveling at a rate of 20 miles per hour slower than Eden, if Eden starts 2 hours after Rex and wants to overtake him in 4 hours?

E. Two planes start from Minneapolis at the same time and fly in opposite directions, one averaging 40 miles per hour faster than the other. If they are 2,000 miles apart in 5 hours, what is the average speed for each plane?

Solution:

A. $d = rt$ where d is the distance traveled, r is the rate and t is the time.

$x = $ distance traveled on the bus   Represent the quantities.

$\dfrac{5}{8} = $ time spent riding bicycle

$\dfrac{x}{25} = $ time spent riding bus

$\dfrac{5}{8} + \dfrac{x}{25} = \dfrac{3}{2}$ $1.5 \text{ hours} = \dfrac{3}{2} \text{ hours}$ Add the times to get total time.

$\text{LCM}(8, 25, 2) = 200$ Clear equation of fractions.

$$200\left(\dfrac{5}{8}\right) + 200\left(\dfrac{x}{25}\right) = 200\left(\dfrac{3}{2}\right)$$

$25(5) + 8(x) = 100(3)$ Simplify the equation.

$125 + 8(x) = 300$ Solve the equation.

$8x = 175$

$x = 21.875$ Write the answer to the question.

Tyrone traveled 21.875 miles on the bus.

B. $x = $ miles per hour for jogging

$2x = $ miles per hour for bicycling

$\dfrac{15}{x} + \dfrac{15}{2x} = 3$

$30 + 15 = 6x$

$45 = 6x$

$7.5 = x$

$15 = 2x$

Maria bicycles at the rate of 15 miles per hour.

C. $x = $ time they traveled

$30x + 45x = 300$

$75x = 300$

$x = 4$

$30x = 120$

$45x = 180$

Brenda traveled 120 miles and Julie traveled 180 miles.

D. $x = $ miles per hour for Eden

$x - 20 = $ miles per hour for Rex

$6(x - 20) = 4x$

$6x - 120 = 4x$

$2x - 120 = 0$

$2x = 120$

$x = 60$

Eden needs to travel at 60 miles per hour.

E. x = rate of slower plane
$x + 40$ = rate of faster plane
$5x + 5(x + 40) = 2,000$
$5x + 5x + 200 = 2,000$
$10x = 1,800$
$x = 180$
$x + 40 = 220$
The planes travel at 180 and 220 miles per hour.

Example:

Solve these work word problems.

A. Jamal can fill the vending machine in 45 minutes. When his sister, Violet, helps him, it takes them 20 minutes. How long would it take Violet to fill the machine by herself?
B. One pipe can fill a pool in 18 hours. Another pipe can fill it in 24 hours. The drainpipe can empty the tank in 12 hours. With all three pipes open, how long will it take to fill the pool?
C. One work crew can do a job in 8 days. After the first crew worked 3 days, a second crew joins them, and together, the two crews finish the job in 3 more days. How long would it take the second crew to do the job alone?
D. One machine can wrap 200 boxes per hour. A newer machine can wrap 250 boxes per hour. How long would it take the two machines working together to wrap 4,950 boxes?
E. Barbara and Wesley can paint a room together in 6 hours. If Barbara can paint the room alone in 10 hours, how long would it take Wesley working alone to paint the room?

Solution:

A. x = number of minutes for Violet alone   Represent each person's time.

$$\frac{20}{x} + \frac{20}{45} = 1$$ Add work done by each person.

$900 + 20x = 45x$ Solve equation.
$900 = 25x$
$36 = x$
It would take Violet 36 minutes working alone.   Write an answer to the question.

B. x = the number of hours together

$$\frac{x}{18} + \frac{x}{24} - \frac{x}{12} = 1 \quad \text{LCM}(18, 24, 12) = 72$$

$$72\left(\frac{x}{18} + \frac{x}{24} - \frac{x}{12}\right) = 72(1)$$

$4x + 3x - 6x = 72$
$x = 72$
With all three pipes open, it would take 72 hours to fill the pool.

C. $x =$ the number of days for second crew

$$\frac{6}{8} + \frac{3}{x} = 1$$

$6x + 24 = 8x$

$24 = 2x$

$12 = x$

The second crew could do the job alone in 12 days.

D. $x =$ number of hours together

$200x + 250x = 4,950$

$450x = 4,950$

$x = 11$

It would take the two machines together 11 hours to do the job.

E. $x =$ the number of hours for Wesley

$$\frac{6}{10} + \frac{6}{x} = 1$$

$6x + 60 = 10x$

$60 = 4x$

$15 = x$

Wesley can paint the room alone in 15 hours.

Sometimes you will use data from a survey to answer questions. A problem with the data occurs when you allow an object to have multiple attributes. For example, a person might own both a desktop computer and a laptop computer, so that person would belong to two categories.

Use **Venn diagrams** to represent the data in distinct categories while showing relationships among the categories. The overlapping circles allow you to visually analyze the data. For each of the given groups, use one circle. With two groups there are four regions: group one only, group two only, both groups, and neither group.

Example:

Solve these problems using Venn diagrams.

A. One hundred people were asked about the computers that they owned. Laptop computers were owned by 55 people, desktop computers were owned by 68 people, and 28 people owned both a laptop and a desktop computer. How many people did not own a computer? How many people owned just one computer?

B. At a wine tasting party, 150 people were asked which of the three wines served that they liked.
68 like Boone's Farm Strawberry Hill
68 like Ripple
80 like Thunderbird
35 like Boone's and Thunderbird
30 like Ripple and Thunderbird
28 like Boone's and Ripple
20 like all three
How many people like exactly two of these wines?
How many people like exactly one of these wines?

Solution:

A. Two groups: L = Laptops and D = Desktops
Draw a rectangle enclosing two intersecting circles.

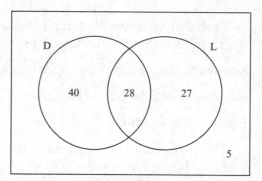

Figure 10.5

$N = 100$, so the sum of the numbers in the regions must total 100.

Since 28 people owned both a laptop and a desktop, put 28 in the overlap of the circles labeled D and L. There are 55 laptop owners, so $55 - 28 = 27$ in the part of L not shared with D. There are 68 desktop owners, so $68 - 28 = 40$ in the part of D not in L. You have accounted for $40 + 28 + 27 = 95$ of the 100 people, so 5 are outside the circles for D and L. Thus, there are 5 people who do not own a computer. There are 40 people with desktops only and 27 people with laptops only, so there are $40 + 27 = 67$ people who own one computer.

B. Three groups: B = Boone's, R = Ripple, T = Thunderbird
Draw a rectangle enclosing three intersecting circles.

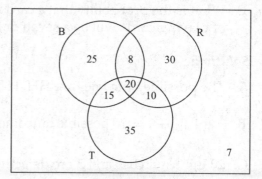

Figure 10.6

$N = 150$, so the numbers put in the eight regions must total 150. Since 20 people like all three wines, place 20 in the region common to B, R, and T. There are $28 - 20 = 8$ in the overlap of B and R not in all three, $30 - 20 = 10$ in the overlap of R and T not in all three, and $35 - 20 = 15$ in the overlap of B and T and not in all three. There are $68 - 15 - 20 - 8 = 25$ in B only, $68 - 8 - 20 - 10 = 30$ in R only, and $80 - 15 - 20 - 10 = 35$ in T only. There are $25 + 15 + 20 + 8 + 30 + 10 + 35 = 143$ accounted for by those who like at least one of the wines. There are $150 - 143 = 7$ people who did not like any of the wines.

There are 15 who like Boone's and Thunderbird only, 10 who like Ripple and Thunderbird only, and 8 who like Boone's and Ripple only, so 33 people like exactly two of these wines. There are 25 who like Boone's only, 30 who like Ripple only, and 35 who like Thunderbird only, so 90 people like exactly one of these wines.

Probability is the likelihood that a given event will occur. It is the ratio of the number of favorable outcomes to the total number of outcomes. There are two ways a coin can come up when it is flipped. $P(\text{heads}) = \frac{1}{2}$. There are six outcomes for a die when it is rolled: 1, 2, 3, 4, 5, or 6. Thus $P(1) = P(2) = P(3) = P(4) = P(5) = P(6) = \frac{1}{6}$.

The probability of an event is $0 \leq P(\text{event}) \leq 1$. There is a probability of 0 only when the event cannot occur, and a probability of 1 when the event is certain to occur. If $P(E)$ is the probability event E will occur and $P(\text{not E})$ is the probability event E will not occur, $P(\text{not E}) = 1 - P(E)$.

Example:

Find the probability of the event.

A. Two coins are flipped once. What is the probability of getting two tails?
B. A jar contains 3 yellow marbles, 4 green marbles, 2 black marbles, and 1 white marble. One marble is selected without looking. What is the probability that the marble will be green?
C. A box contains 20 light switches, and 3 of them are defective. If you select one switch from the box at random, what is the probability that it will not be defective?
D. What is the probability that a card selected at random from a standard deck of playing cards is a ten?
E. What is the probability that a card selected at random from a standard deck of playing cards will be a seven or a club?

Solution:

A. The possible outcomes are HH, HT, TH, and TT. $P(\text{two tails}) = \frac{1}{4}$.

B. $3 + 4 + 2 + 1 = 10$ possible outcomes, 4 outcomes green. $P(\text{green}) = \frac{4}{10} = \frac{2}{5}$.

C. 20 light switches and 3 are defective. $P(\text{defective}) = \frac{3}{20}$, $P(\text{not defective}) = 1 - P(\text{defective})$.

$$P(\text{not defective}) = 1 - \frac{3}{20} = \frac{17}{20}$$

D. There are 52 cards in a standard deck. There are 4 tens: 10 of hearts, 10 of clubs, 10 of spades, and 10 of diamonds. $P(\text{ten}) = \frac{4}{52} = \frac{1}{13}$

E. There are 52 cards in a deck. There are 13 each in hearts, clubs, spades, and diamonds. There are 4 sevens, one each in hearts, clubs, spades, and diamonds. Thus, the seven of clubs is counted twice, once as a club and once as a seven. There are 13 clubs plus 3 sevens that are not clubs. $P(7 \text{ or club}) = \frac{16}{52} = \frac{4}{13}$.

10.61 PRACTICE PROBLEMS

1. Find two numbers such that twice the first plus five times the second is 20, and four times the first less three times the second is 14.

2. Three more than twice a certain number is 57. Find the number.

3. Wendy's mother is three times as old as Wendy. In 14 years, she will be twice as old as Wendy is then. How old is each now?

4. Joan is 3 years older than Susan. Eight years ago, Joan was four times the age of Susan. How old is each now?

5. Michelle scored 95, 91, 98, 90, 96, and 100. What is her average test score?

6. David weighed himself each week. In May, he weighed himself five times with an average weight of 185 pounds. In June, he weighed himself four times with an average weight of 180 pounds. What is his average weight for May and June?

7. John's cell phone bills were $87, $81, $88, $87, $84, $87, $89, $80, $78, $79, $81, and $82 for the past year. What was the median for his bills? What is the mode for his bills?

8. An 80% acid solution is mixed with a 20% acid solution to get 3 gallons of a solution that is $\frac{1}{3}$ acid. How much of each acid solution was used?

9. A mixture of 40 pounds of mixed nuts worth $1.80 a pound is to be made from peanuts costing $1.35 a pound and fancy mixed nuts costing $2.55 a pound. How many pounds of each kind of nuts should be used?

10. In her motorboat, Ruth can go downstream in 1 hour less time than she can go upstream. If the current is 5 miles per hour, how fast can she travel in still water if it takes her 2 hours to travel upstream the given distance?

11. Two drivers started toward each other from towns 255 miles apart. One driver traveled at 40 miles per hour, and the other traveled at 45 miles per hour. How long did the people drive until they met?

12. Amanda can mow a lawn in 1 hour, 20 minutes. Kim can mow the same lawn in 2 hours. How long would it take them, working together, to mow the lawn?

13. One computer can do a payroll in 12 hours. A second computer can do the payroll in 6 hours. How long will it take to do the payroll if both computers work on the payroll at the same time?

14. A survey was taken of 52 students at Macon High School. They were asked which amusement parks they would like to visit on a class trip: Six Flags, Disney World, and Opryland. The data are summarized as follows.
    28 preferred Six Flags
    25 preferred Disney World
    26 preferred Opryland
    10 preferred Opryland and Disney World
    11 preferred Disney World and Six Flags
    14 preferred Six Flags and Opryland
    6 preferred all three
    How many students prefer both Disney World and Six Flags and did not prefer Opryland? How many did not prefer any of these three sites?

15. Three coins are flipped once. What is the probability that exactly two coins will be heads?

16. A card is selected at random from a standard deck of playing cards. What is the probability that the card selected is a red card or a jack?

10.62 SOLUTIONS

1. $x =$ first number
   $y =$ second number
   (1) $2x + 5y = 20$
   (2) $4x - 3y = 14$

   $$\begin{array}{ll} (2)\ 4x - 3y = 14 & 4x - 3y = 14 \\ -2 \times (1)+\ \underline{-4x - 10y = -40} & 4x - 3(2) = 14 \\ \qquad\quad -13y = -26 & 4x - 6 = 14 \\ \qquad\qquad\quad y = 2 & 4x = 20 \\ & x = 5 \end{array}$$

   The first number is 5 and the second number is 2.

2. $x =$ the number
   $2x + 3 = 57$
   $2x = 54$
   $x = 27$
   The number is 27.

3. $x =$ Wendy's age now
   $3x =$ mother's age now
   $3x + 14 = 2(x + 14)$
   $3x + 14 = 2x + 28$
   $x + 14 = 28$
   $x = 14$
   $3x = 42$
   Wendy is 14 years old, and her mother is 42 years old now.

4. $x =$ Susan's age now
   $x + 3 =$ Joan's age now
   $x + 3 - 8 = 4(x - 8)$
   $x - 5 = 4x - 32$
   $-3x - 5 = -32$
   $-3x = -27$
   $x = 9$
   $x + 3 = 12$
   Joan's age is 12 years, and Susan's age is 9 years now.

5. $SUM = 95 + 91 + 98 + 90 + 96 + 100 = 570$,
   $N = 6$
   $AVE = SUM \div N = 570 \div 6 = 95$
   Michelle's average test score is 95.

6. May: average is $185 - 5$ weighs. June: average is $180 - 4$ weighs.
   $$AVE = \frac{5(185) + 4(180)}{5 + 4} = \frac{925 + 720}{9} = \frac{1645}{9} =$$
   182.7777
   $AVE = 182.8$
   David's average weight is 182.8 pounds to the nearest tenth of a pound.

7. Sequenced bills: $78, $79, $80, $81, $81, $82, $84, $87, $87, $87, $88, $89. There are 12 bills, and the middle two values are $82 and $84.
   The median is ($82 + $84) $\div$ 2 = $83.
   The bill of $87 occurred the most, three times.
   The mode is $87.

8. $x =$ gallons of 80% acid used
   $3 - x =$ gallons of 20% acid used
   Since $\frac{1}{3}$ of the 3 gallons is acid, there is one gallon of acid.
   $0.80x + 0.20(3 - x) = 1$
   $0.80x + 0.20(3 - x) = 1$
   $0.80x + 0.60 - 0.20x = 1$
   $0.60x + 0.60 = 1$
   $0.60x = 0.40$
   $x = \dfrac{0.40}{0.60} = \dfrac{2}{3}$
   $3 - x = 2\dfrac{1}{3}$
   $\frac{2}{3}$ gallon of the 80% acid solution and $2\frac{1}{3}$ gallons of the 20% solution were used.

9. $x =$ number of pounds of peanuts used
   $40 - x =$ number of pounds of fancy mixed nuts used
   $\$1.35x + \$2.55(40 - x) = \$1.80(40)$
   $\$1.35x + \$102.00 - \$2.55x = \$72.00$
   $-\$1.20x + \$102.00 = \$72.00$
   $-\$1.20x = -\$30.00$
   $x = 25$
   $40 - x = 15$
   25 pounds of the peanuts and 15 pounds of the fancy mixed nuts should be used.

10. $x =$ the still water rate of Ruth's boat
    $x + 5 =$ the downstream rate of the boat
    $x - 5 =$ the upstream rate of the boat
    $2(x - 5) = 1(x + 5)$
    $2x - 10 = x + 5$
    $x = 15$
    The still water rate of Ruth's motorboat is 15 miles per hour.

11. $x =$ time each drove until they met
    $40x + 45x = 255$
    $85x = 255$
    $x = 3$
    It took the drivers 3 hours to meet.

12. $x =$ the number of minutes working together
$$\frac{x}{80} + \frac{x}{120} = 1$$
$$3x + 2x = 240$$
$$5x = 240$$
$$x = 48$$
They can mow the lawn together in 48 minutes.

13. $x =$ the number of hours for the two computers to do the payroll together
$$\frac{x}{12} + \frac{x}{6} = 1$$
$$x + 2x = 12$$
$$3x = 12$$
$$x = 4$$
It would take the two computers 4 hours to do the payroll together.

14. 3 categories: S = Six Flags, O = Opryland, D = Disney World

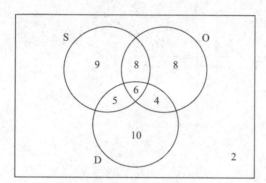

Figure 10.7

Place the number who prefer all three in the diagram first, then determine the regions for two. $10 - 6 = 4$, $11 - 6 = 5$, and $14 - 6 = 8$. From the completed Venn diagram, 5 people prefer 6 Flags and Disney World and do not prefer Opryland. Two did not prefer any of the three sites.

15. Three coins are flipped with possible outcomes of HHH, HHT, HTH, HTT, THH, THT, TTH, TTT. Of the 8 outcomes, 3 have exactly 2 heads.
$$P(\text{exactly 2 heads}) = \frac{3}{8}$$

16. There are 52 cards in a standard deck of playing cards, of which 26 are red cards. Of the 4 jacks, 2 are red and 2 are black. Thus, there are $26 + 2$ red cards and non-red jacks.
$$P(\text{red or jack}) = \frac{28}{52} = \frac{7}{13}$$

▬▬ 10.63 ALGEBRA TEST 3

Use the following test to assess how well you have mastered the material in this chapter. Mark your answers by blackening the corresponding answer oval in each question. An answer key and solutions are provided at the end of the test.

1. Which is the solution of $x - 3 - 2(6 - 2x) = 2(2x - 5)$?

   (A) -2

   (B) -1

   (C) $\dfrac{20}{7}$

   (D) 5

   (E) 10

2. Which is the solution of $(2x + 1)^2 = (x - 4)^2 + 3x(x + 3)$?

  (A) $-\dfrac{17}{3}$

  (B) -5

  (C) 5

  (D) $\dfrac{17}{3}$

  (E) 6

3. If $A = 0.5h(a + b)$ is solved for b, what is the value of b?

  (A) $b = 2A - h - a$

  (B) $b = \dfrac{A - 0.05h}{a}$

  (C) $b = A - 0.5ah$

  (D) $b = A - 0.5h - a$

  (E) $b = \dfrac{2A - ah}{h}$

4. Which is the solution of $\dfrac{2}{x - 3} - \dfrac{4}{x + 3} = \dfrac{16}{x^2 - 9}$?

  (A) 1

  (B) 4

  (C) -0.5

  (D) -6.5

  (E) -8

5. Which is the solution of $\dfrac{5}{6} = \dfrac{n}{30}$?

  (A) 150

  (B) 36

  (C) 29

  (D) 25

  (E) 20

6. Which is the solution of $\sqrt{5x - 3} = 4$?

  (A) 1

  (B) 1.4

  (C) 3.8

  (D) 4.2

  (E) 5

7. Which is a solution for the system of equations $3x - y = -2$ and $x + y = 6$?

    (A) $(3, 1)$
    (B) $(2, 4)$
    (C) $(-4, -10)$
    (D) $(1, 5)$
    (E) $(3, 3)$

8. Which is a solution for the system of equations $13x + 11y = 21$ and $7x + 6y = -3$?

    (A) $(159, -186)$
    (B) $(3, -4)$
    (C) $(-56, 49)$
    (D) $(-4, 5)$
    (E) $(-43, 38)$

9. Which is the solution for $7(x - 3) \leq 4(x + 5) - 47$?

    (A) $x \leq 2$
    (B) $x > -2$
    (C) $x \geq 2$
    (D) $x \leq -2$
    (E) $x \geq 0$

10. Which are the solutions for $9x^2 - 2x - 11 = 0$?

    (A) $\dfrac{11}{9}, -1$

    (B) $-\dfrac{11}{9}, 1$

    (C) $-\dfrac{11}{9}, -1$

    (D) $\dfrac{11}{9}, 1$

    (E) $\dfrac{11}{3}, -\dfrac{1}{3}$

11. Which are the solutions for $(x + 1)(2x - 1) = 2$?

    (A) $\dfrac{3}{2}, -1$

    (B) $1, 1$

    (C) $\dfrac{2}{3}, -1$

    (D) $\dfrac{3}{2}, 0$

    (E) $-\dfrac{3}{2}, 1$

12. Which are the solutions for $4x^2 = -16x$? Select all that apply.

   - [A] -4
   - [B] 0
   - [C] 2
   - [D] 4
   - [E] 16

13. Which is the solution for $6x^2 - 8x = 3$?

   (A) $\dfrac{4 \pm \sqrt{34}}{6}$

   (B) $\dfrac{-4 \pm \sqrt{34}}{6}$

   (C) $\dfrac{2 \pm \sqrt{34}}{3}$

   (D) $\dfrac{4 \pm \sqrt{2}}{6}$

   (E) No real solutions

14. Which is the solution for $x^2 + 3x < 28$?

   (A) $x < -7$
   (B) $x > 4$
   (C) $-4 < x < 7$
   (D) $-7 < x < 4$
   (E) No real solutions

15. If $x\,\square = 2x^2 - 4x - 3$, what is $-2\,\square$?

   (A) -3
   (B) -1
   (C) 11
   (D) 13
   (E) 21

16. Two bicyclists travel in opposite directions. One travels at 5 miles per hour faster than the other. In 2 hours they are 50 miles apart. What is the rate of the faster bicyclist?

   (A) 10 mph
   (B) 11.25 mph
   (C) 15 mph
   (D) 20 mph
   (E) 22.5 mph

17. Two sisters were born in consecutive years. How old are they now if the product of their present ages is 1,056?

    (A) 12, 13
    (B) 31, 36
    (C) 16, 66
    (D) 32, 33
    (E) 42, 43

18. A bicycle rider travels 8 miles per hour faster than a jogger. It takes the bicyclist half as long as it takes the jogger to travel 16 miles. What is the jogger's speed?

    (A) 8 mph
    (B) 16 mph
    (C) 24 mph
    (D) 32 mph
    (E) 40 mph

19. Anthony and Ben can paint a water tower in 6 days when they work together. If Anthony works twice as fast as Ben, how long would it take Ben to paint the water tower working alone?

    (A) 1.5 days
    (B) 3 days
    (C) 4.5 days
    (D) 9 days
    (E) 18 days

20. Pipe A can fill a tank in 20 minutes. Pipe B can empty the tank in 30 minutes. Pipe C can fill the tank in 60 minutes. If all three pipes work together, how long would it take to fill the tank?

    (A) 10 minutes
    (B) 20 minutes
    (C) 30 minutes
    (D) 45 minutes
    (E) 60 minutes

21. What is the average (mean) of these values: 18, 16, 24, 16, 16, 24?

    (A) 16
    (B) 17
    (C) 18
    (D) 19
    (E) 24

22. What is the median for this set of values: 62, 72, 62, 83, 79, 68, 72, 62?

    (A) 62
    (B) 68
    (C) 70
    (D) 72
    (E) 83

23. A survey of 50 people at a shopping center provided the following information:

    25 like country music.
    21 like rap music.
    10 like both country music and rap music.

    How many people in the survey did not like either country music or rap music?

    (A) 24
    (B) 16
    (C) 14
    (D) 6
    (E) 4

24. What is the probability of drawing at random 1 green marble from a bag containing 24 green marbles and 32 yellow marbles?

    (A) $\dfrac{1}{24}$

    (B) $\dfrac{3}{7}$

    (C) $\dfrac{4}{7}$

    (D) $\dfrac{2}{3}$

    (E) $\dfrac{3}{4}$

25. Two hundred tickets were sold for a charity raffle. Isabella buys five tickets. What is the probability that she will not win the raffle?

    (A) 0.02
    (B) 0.025
    (C) 0.5
    (D) 0.975
    (E) 0.98

ALGEBRA TEST 3
Answer Key

1. D	6. C	11. E	16. C	21. D
2. C	7. D	12. A and B	17. D	22. C
3. E	8. A	13. A	18. A	23. C
4. A	9. D	14. D	19. E	24. B
5. D	10. A	15. D	20. C	25. D

▩ 10.64 SOLUTIONS

1. **D** 5

$$x - 3 - 2(6 - 2x) = 2(2x - 5)$$
$$x - 3 - 12 + 4x = 4x - 10$$
$$x - 15 = -10$$
$$x = 5$$

2. **C** 5

$$(2x + 1)^2 = (x - 4)^2 + 3x(x + 3)$$
$$4x^2 + 4x + 1 = x^2 - 8x + 16 + 3x^2 + 9x$$
$$4x + 1 = -8x + 16 + 9x$$
$$3x + 1 = 16$$
$$3x = 15$$
$$x = 5$$

3. **E** $b = \dfrac{2A - ah}{h}$

$$A = 0.5h(a + b)$$
$$2A = h(a + b)$$
$$2A = ha + hb$$
$$2A - ha = hb$$
$$\frac{2A - ah}{h} = b$$

4. **A** 1

$$\frac{2}{x - 3} - \frac{4}{x + 3} = \frac{16}{x^2 - 9}$$
$$2(x + 3) - 4(x - 3) = 16$$
$$2x + 6 - 4x + 12 = 16$$
$$-2x + 18 = 16$$
$$-2x = -2$$
$$x = 1$$

5. **D** 25

$$\frac{5}{6} = \frac{n}{30}, 6n = 150, n = 25$$

6. **C** 3.8

$$\sqrt{5x - 3} = 4$$

$$\left(\sqrt{5x - 3}\right)^2 = (4)^2$$

$$5x - 3 = 16$$
$$5x = 19$$
$$x = 3.8$$

7. **D** (1, 5)

$$\begin{array}{ll} 3x - y = -2 & \quad 3(1) - y = -2 \\ +\ \underline{x + y = \ 6} & \qquad\quad -y = -5 \\ \ \ 4x\ \ \ \ \ \ = \ 4 & \qquad\ \ \ y = 5 \\ \qquad\ x = 1 & \qquad\ \ \ y = 5 \end{array}$$

$$(x, y) = (1, 5)$$

8. **A** (159, −186)

$$\begin{array}{llll} (1)\ \ 13x + 11y = 21 & \quad 6 \times (1) & \quad 78x + 66y = 126 \\ (2)\ \ \ 7x + \ 6y = -3 & \ -11 \times (2) & +\ \underline{-77x - 66y = \ 33} \\ & & \qquad x \qquad\ \ = 159 \end{array}$$

$$7(159) + 6y = -3,\ 1113 + 6y = -3,\ 6y = -1,116,\ y = -186$$

$$(x, y) = (159, -186)$$

9. **D** $x \leq -2$

$$7(x - 3) \leq 4(x + 5) - 47$$
$$7x - 21 \leq 4x + 20 - 47$$
$$7x - 21 \leq 4x - 27$$
$$3x - 21 \leq -27$$
$$3x \leq -6$$
$$x \leq -2$$

10. **A** $\dfrac{11}{9}, -1$

$$9x^2 - 2x - 11 = 0$$
$$(9x - 11)(x + 1) = 0$$
$$9x - 11 = 0 \text{ or } x + 1 = 0$$
$$9x = 11 \text{ or } x = -1$$
$$x = \frac{11}{9} \text{ or } x = -1$$

11. **E** $-\dfrac{3}{2}, 1$

$$(x + 1)(2x - 1) = 2$$
$$2x^2 + x - 1 = 2$$
$$2x^2 + x - 3 = 0$$
$$(2x + 3)(x - 1) = 0$$
$$2x + 3 = 0 \text{ or } x - 1 = 0$$
$$2x = -3 \text{ or } x = 1$$
$$x = -\frac{3}{2} \text{ or } x = 1$$

12. **A and B** 0, −4

$$4x^2 = -16x$$
$$4x^2 + 16x = 0$$
$$4x(x + 4) = 0$$
$$4x = 0 \text{ or } x + 4 = 0$$
$$x = 0 \text{ or } x = -4$$

13. **A** $\dfrac{4 \pm \sqrt{34}}{6}$

$$6x^2 - 8x = 3$$
$$6x^2 - 8x - 3 = 0$$
$$a = 6, b = -8, c = -3$$

$$x = \frac{-b \pm \sqrt{b^2 - 4ac}}{2a}$$

$$x = \frac{-(-8) \pm \sqrt{(-8)^2 - 4(6)(-3)}}{2(6)}$$

$$x = \frac{8 \pm \sqrt{136}}{12}$$

$$x = \frac{8 \pm 2\sqrt{34}}{12}$$

$$x = \frac{4 \pm \sqrt{34}}{6}$$

14. **D** $-7 < x < 4$

$$x^2 + 3x < 28$$
$$x^2 + 3x - 28 < 0$$
$$(x + 7)(x - 4) < 0$$
Case 1: $x + 7 < 0$ and $x - 4 > 0$ or Case 2: $x + 7 > 0$ and $x - 4 < 0$
$x < -7$ and $x > 4$ or $x > -7$ and $x < 4$
No solution or $-7 < x < 4$
The solution is $-7 < x < 4$.

15. **D** 13

$$x\square = 2x^2 - 4x - 3$$
$$-2\square = 2(-2)^2 - 4(-2) - 3 = 8 + 8 - 3 = 13$$

16. **C** 15 mph

$x = $ slower rate
$x + 5 = $ faster rate
$$2x + 2(x + 5) = 50$$
$$2x + 2x + 10 = 50$$
$$4x = 40$$
$$x = 10$$
$$x + 5 = 15$$

17. **D** 32, 33

    $x = 1^{st}$ age
    $x + 1 = 2^{nd}$ age
    $x(x + 1) = 1,056$
    $x^2 + x - 1,056 = 0$
    $(x + 33)(x - 32) = 0$
    $x + 33 = 0$ or $x - 32 = 0$
    $x = -33$ or $x = 32$
    No solution or $x + 1 = 33$
    The ages are 32 and 33.

18. **A** 8 mph

    $x =$ rate for jogger
    $x + 8 =$ rate for bicyclist

    $$\frac{16}{x} = 2\left(\frac{16}{x + 8}\right)$$

    $$\frac{16}{x} = \frac{32}{x + 8}$$

    $16(x + 8) = 32x$
    $16x + 128 = 32x$
    $128 = 16x$
    $8 = x$

19. **E** 18 days

    $x =$ time for Anthony alone
    $2x =$ time for Ben alone

    $$\frac{6}{x} + \frac{6}{2x} = 1$$

    $12 + 6 = 2x$
    $18 = 2x$
    $9 = x$
    Ben's time alone is 18 days.

20. **C** 30 minutes

    $x =$ time to complete job together

    $$\frac{x}{20} - \frac{x}{30} + \frac{x}{60} = 1$$

    LCM(20, 30, 60) = 60

    $3x - 2x + x = 60$
    $2x = 60$
    $x = 30$ minutes

21. **D** 19

    SUM $= 18 + 16 + 24 + 16 + 16 + 24 = 114, N = 6$
    AVE $=$ SUM $\div N = 114 \div 6 = 19$

22. **C** 70

Sequential list: 62, 62, 62, 68, 72, 72, 79, 83
There are eight values in list, so the average of the middle two values is the median.
Median = $(68 + 72) \div 2 = 140 \div 2 = 70$

23. **C** 14

Two categories: C = country music, R = rap music

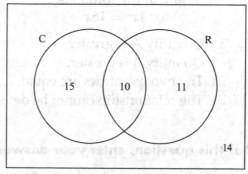

Figure 10.8

$50 - (15 + 10 + 11) = 50 - 36 = 14$ not in any category.
There are 14 people who do not like either.

24. **B** $\dfrac{3}{7}$

24 green marbles, 32 yellow marbles, $24 + 32 = 56$ marbles
$P(\text{green}) = \dfrac{24}{56} = \dfrac{3}{7}$

25. **D** 0.975

$P(\text{win}) = \dfrac{5}{200} = 0.025$
$P(\text{lose}) = 1 - P(\text{win}) = 1 - 0.025 = 0.975$

10.65 SOLVED GRE PROBLEMS

For each question, select the best answer unless otherwise instructed.

1. Which is the solution of $x - 5 - 3(2 - 3x) = 9$?

   (A) -7.5
   (B) -2.5
   (C) -0.2
   (D) 2
   (E) 2.5

	Quantity A	Quantity B
2.	The value of x when $3x - 2(x - 6) = 8$	The value of y when $2 - 5(3 - y) = 2$

    (A) Quantity A is greater.
    (B) Quantity B is greater.
    (C) The two quantities are equal.
    (D) The relationship cannot be determined from the given information.

	Quantity A	Quantity B
3.	Sum of the solutions for $4x^2 = 16x$	Product of the solutions for $4x^2 = 16x$

    (A) Quantity A is greater.
    (B) Quantity B is greater.
    (C) The two quantities are equal.
    (D) The relationship cannot be determined from the given information.

For this question, enter your answer in the box.

4. **What is the solution for n in $\dfrac{3}{5} = \dfrac{n}{30}$?**

5. **Which is the solution for $\sqrt{2x - 5} = 3$?**

    (A) -1
    (B) 2
    (C) 4
    (D) 7
    (E) 9

▬ 10.66 SOLUTIONS

1. **D 2**
$x - 5 - 3(2 - 3x) = 9$
$x - 5 - 6 + 9x = 9$
$10x - 11 = 9$
$10x = 20$
$x = 2$

2. **B** Quantity B is greater.
$$
\begin{array}{ll}
3x - 2(x - 6) = 8 & 2 - 5(3 - y) = 2 \\
3x - 2x + 12 = 8 & 2 - 15 + 5y = 2 \\
x + 12 = 8 & -13 + 5y = 2 \\
x = -4 & y = 3
\end{array}
$$

3. **A** Quantity A is greater.
$$4x^2 = 16$$
$$4x^2 - 16 = 0$$
$$4x(x - 4) = 0$$
$$4x = 0 \text{ or } x - 4 = 0$$
$$x = 0 \text{ or } x = 4$$
The sum of the solutions is $0 + 4 = 4$. The product of the solutions is $0(4) = 0$.

4. $\boxed{18}$

$\dfrac{3}{5} = \dfrac{n}{30}.$ $5n = 90.\ n = 18.$

5. **D** 7
$\sqrt{2x - 5} = 3.$ $2x - 5 = 9.\ 2x = 14.\ x = 7.$

■ 10.67   GRE PRACTICE PROBLEMS

For each question, select the best answer unless otherwise instructed.

1. **Which are the solutions of $3x^2 - 2x = 6$?**

   (A) $3, -2$

   (B) $6, -1$

   (C) $\dfrac{1 \pm \sqrt{76}}{3}$

   (D) $\dfrac{2 \pm \sqrt{19}}{3}$

   (E) $\dfrac{1 \pm \sqrt{19}}{3}$

Quantity A	Quantity B

2. The value of x when $3(x - 1) = 4$ The value of y when $2(y + 3) = 7$

   (A) Quantity A is greater.

   (B) Quantity B is greater.

   (C) The two quantities are equal.

   (D) The relationship cannot be determined from the given information.

3. **What is the value of b when $A = \dfrac{1}{2}bh$?**

   (A) $\dfrac{2A}{h}$

   (B) $\dfrac{A}{2h}$

   (C) $\dfrac{1}{2}Ah$

   (D) $A - \dfrac{1}{2}h$

   (E) $2A - h$

4. **What is the solution for the system $2x - 3y = -12$ and $4x + y = -10$?**

   (A) $(2, -3)$
   (B) $(-3, 2)$
   (C) $(-6, 0)$
   (D) $(2, -12)$
   (E) $(3, 6)$

	Quantity A	**Quantity B**
5.	The value of x when $\sqrt{2x - 1} = 3$	The value of y when $2 - 4(3 - y) = 10$

   (A) Quantity A is greater.
   (B) Quantity B is greater.
   (C) The two quantities are equal.
   (D) The relationship cannot be determined from the given information.

For this question, enter your answer in the box.

6. **What is the value of y when $\dfrac{2}{y} - \dfrac{3}{y} = 2 - \dfrac{5}{y}$?**

7. **What is the solution for $3x - 4 < 5(x + 2)$?**

   (A) $x > 7$
   (B) $x < 7$
   (C) $x > -3$
   (D) $x > -7$
   (E) $x < -7$

8. **Jane is 3 years older than her brother Sean. In 14 years, the sum of their ages will be 63 years. How old is Jane now?**

   (A) 15 years
   (B) 16 years
   (C) 19 years
   (D) 30 years
   (E) 33 years

	Quantity A	**Quantity B**
9.	The product of the solutions for $x^2 - 5x + 6 = 0$.	The sum of the solutions for $x^2 - 5x + 6 = 0$.

   (A) Quantity A is greater.
   (B) Quantity B is greater.
   (C) The two quantities are equal.
   (D) The relationship cannot be determined from the given information.

Quantity A	Quantity B

10. The product of the solutions for $x^2 = 121$.   The product of the solutions for $y^2 = 8y$.

- (A) Quantity A is greater.
- (B) Quantity B is greater.
- (C) The two quantities are equal.
- (D) The relationship cannot be determined from the given information.

ANSWER KEY

1.	E	6.	2
2.	A	7.	D
3.	A	8.	C
4.	B	9.	A
5.	C	10.	B

CHAPTER 11
GEOMETRY

11.1 SYMBOLS

Symbol	Meaning of Symbol
π	pi (about 3.14)
$\parallel$	is parallel to
$\perp$	is perpendicular to
$\angle$	angle
$\triangle$	triangle
$a°$	a degrees
$\ulcorner$	right angle

11.2 POINTS, LINES, AND ANGLES

Points are represented with dots and are named by capital letters. A line extends indefinitely in two directions and can be named by naming two points on it. Because a line has infinitely many points, a line can have many names.

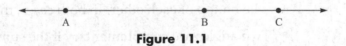

A   B   C

Figure 11.1

The line in Figure 11.1 can be named line AB, line AC or line BC.

➤ A **line segment** is a part of a line between two points. The two points are the endpoints of the line segment.

In Figure 11.1, there is a line segment AB, a line segment AC, and a line segment BC. The line segment AB and the line segment BA are two names for the same segments. $AB = BA$.

➤ A **ray** is a part of a line with exactly one endpoint. Ray AB is a ray with endpoint A going through point B. Ray BA is a ray with endpoint B going through point A. Ray $AB \neq$ Ray BA.

➤ An **angle** is the figure formed by two rays with a common endpoint. The common endpoint of the rays is called the **vertex** of the angle, and the rays are called the **sides** of the angle.

The symbol for an angle is $\angle$. An angle can be named by stating only its vertex if no other angle has the same vertex. In general, an angle is named by naming a point on one side of the angle, then its vertex, and finally a point on the other side of the angle. In Figure 11.2, the angle can be named as $\angle A$, $\angle BAC$, $\angle CAB$, $\angle BAD$, $\angle DAB$, and $\angle EAC$, among many other names.

251

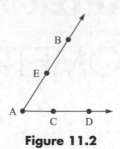

Figure 11.2

Angles are measured in degrees. There are 360° about a point and 180° about the point staying on one side of a line through that point. The measure of an angle is the number of degrees of rotation it takes to get from one side of the angle to the other.

➤ An **acute angle** is an angle whose measure is greater than 0° but less than 90°.
➤ A **right angle** is an angle whose measure is exactly 90°.
➤ An **obtuse angle** is an angle whose measure is greater than 90° but less than 180°.
➤ A **straight angle** is an angle whose sides are a pair of opposite rays and whose measure is exactly 180°.
➤ A **reflex angle** is an angle whose measure is greater than 180° but less than 360°.
➤ **Adjacent angles** are two angles in the same plane that have a common vertex and a common side that separates the two angles.

If the rotation from one side of the angle to the other is 0°, the rays lie on top of one another. Also, if the rotation between the sides of the angle is 360°, the rays lie on top of each other. In both cases, the angle looks like a single ray.

➤ Two angles are **complementary** if the sum of the measures is 90°.
➤ Two angles are **supplementary** if the sum of the measures is 180°.

To be complementary or supplementary, the two angles do not have to be adjacent. The sum of the measures of the two angles is all that matters in deciding whether the pair of angles is either supplementary or complementary. A way to remember which pair has which sum is to put the terms in alphabetical order, complementary then supplementary, and put the sums in numerical order, 90° then 180°. The results match up in the orders shown.

Example:

Find the complement of each angle.

A. 17°   B. 24°   C. 87°   D. 60°

Solution:

A. $90° - 17° = 73°$
B. $90° - 24° = 66°$
C. $90° - 87° = 3°$
D. $90° - 60° = 30°$

Example:

Find the supplement of each angle.

A. 156°   B. 95°   C. 103°   D. 135°

Solution:

A. 180° − 156° = 24°
B. 180° − 95° = 85°
C. 180° − 103° = 77°
D. 180° − 135° = 45°

➤ **Vertical angles** are two nonadjacent angles formed when two lines intersect.
➤ **Two lines are perpendicular** when they intersect to form right angles. This can be written as line $a \perp$ line b.
➤ **Two lines** in same plane **are parallel** if they do not intersect. This can be written as line $a \parallel$ line b.

When two lines intersect, the angles in each pair of vertical angles formed have equal measures. When two parallel lines are intersected by a third line, several special types of angles are formed. In Figure 11.3, lines a and b are parallel; line t is called a transversal.

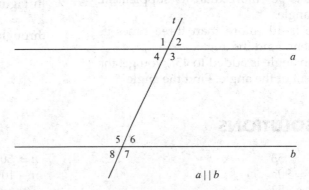

Figure 11.3

In Figure 11.3, ∠1 and ∠5 are a pair of **corresponding angles.** The other pairs of corresponding angles are ∠2 and ∠6, ∠3 and ∠7, and ∠4 and ∠8. Because $a \parallel b$, ∠1 = ∠5, ∠2 = ∠6, ∠3 = ∠7, and ∠4 = ∠8. Another pair of special angles is ∠3 and ∠5, which are **alternate interior angles**, as are ∠4 and ∠6. Line $a \parallel b$, so ∠3 = ∠5 and ∠4 = ∠6. The **alternate exterior angles** are ∠1 and ∠7, and ∠2 and ∠8, which are equal since the lines are parallel. Angles 4 and 5 and angles 3 and 6 are interior angles on the same side of the transversal. Since $a \parallel b$, ∠4 is supplementary to ∠5, and ∠3 is supplementary to ∠6. Thus, we have ∠3 + ∠6 = 180° and ∠4 + ∠5 = 180°.

If the measure of ∠1 is 140°, we can find the measure of all of the other angles in Figure 11.3.

∠1 and ∠3 are a pair of vertical angles, so ∠1 = ∠3; then ∠3 = 140°.

∠1 and ∠5 are corresponding angles, so ∠1 = ∠5; then ∠5 = 140°.

∠5 and ∠7 are vertical angles, so ∠5 = ∠7; then ∠7 = 140°.

∠3 and ∠6 are interior angles on the same side of the transversal t, so ∠3 supplements ∠6, ∠3 + ∠6 = 180°, 140° + ∠6 = 180°, and ∠6 = 180° − 140° = 40°.

$\angle 6$ and $\angle 2$ are corresponding angles, so $\angle 6 = \angle 2$, $\angle 2 = 40°$.
$\angle 6$ and $\angle 4$ are alternate interior angles, so $\angle 6 = \angle 4$, $\angle 4 = 40°$.
$\angle 4$ and $\angle 8$ are corresponding angles, so $\angle 4 = \angle 8$, $\angle 8 = 40°$.
When $a \parallel b$ and $\angle 1 = 140°$, $\angle 3 = \angle 5 = \angle 7 = 140°$ and $\angle 2 = \angle 4 = \angle 6 = \angle 8 = 40°$.

Angles 1 and 2 are adjacent angles and the noncommon sides form a straight line; thus, $\angle 1$ and $\angle 2$ are a linear pair and are supplementary: $\angle 1$ supp $\angle 2$.

▬▬ 11.3 PRACTICE PROBLEMS

1. Find the supplements of these angles.
   A. 127°   B. 90°   C. 28°   D. 30°   E. 57°

2. Find the complements of these angles.
   A. 71°   B. 12°   C. 76°   D. 30°   E. 45°

3. Find the angles requested.
   A. An angle is eight times its supplement. Find the angle.
   B. An angle is 10° less than its complement. Find the angle.
   C. An angle is 30° more than its supplement. Find the angle.
   D. An angle is 10° more than three times its complement. Find the angle.
   E. If twice an angle is added to 45°, you get the supplement of the angle. Find the angle.

4. Using Figure 11.3 with $a \parallel b$, find all the angles 2 through 8.
   A. $\angle 1 = 72°$ B. $\angle 1 = 90°$ C. $\angle 1 = 125°$

5. In Figure 11.3 with $a \parallel b$, $\angle 1 = 4x°$ and $\angle 4 = 4x° + 20°$, find the measures of angles 1 through 8.

6. In Figure 11.3 with $a \parallel b$, $\angle 3 = 5x°$ and $\angle 4 = 3x° - 20°$, find the measures of angles 1 through 8.

7. In Figure 11.3 with $a \parallel b$, $\angle 2 = 3x° + 10°$ and $\angle 7 = x° - 30°$, find the measures of angles 1 through 8.

▬▬ 11.4 SOLUTIONS

1. A. $180° - 127° = 53°$
   B. $180° - 90° = 90°$
   C. $180° - 28° = 152°$
   D. $180° - 30° = 150°$
   E. $180° - 57° = 123°$

2. A. $90° - 71° = 19°$
   B. $90° - 12° = 78°$
   C. $90° - 76° = 14°$
   D. $90° - 30° = 60°$
   E. $90° - 45° = 45°$

3. A. $n =$ the complement, $8n =$ the angle
   $n + 8n = 180°$
   $9n = 180°$
   $n = 20°$
   $8n = 160°$
   The angle is 160°.
   B. $n =$ the complement, $n - 10° =$ the angle
   $n + n - 10° = 90°$
   $2n - 10° = 90°$
   $2n = 100°$

$n = 50°$
$n - 10° = 40°$
The angle is 40°.
C. $n =$ the supplement, $n + 30° =$ the angle
$n + n + 30° = 180°$
$2n = 150°$
$n = 75°$
$n + 30° = 105°$
The angle is 105°.
D. $n =$ the complement, $3n + 10° =$ the angle
$n + 3n + 10° = 90°$
$4n + 10° = 90°$
$4n = 80°$
$n = 20°$
$3n + 10° = 70°$
The angle is 70°.
E. $n =$ the angle, $2n + 45° =$ the supplement
$n + 2n + 45° = 180°$
$3n = 135°$
$n = 45°$
The angle is 45°.

4. A. $\angle 1 = 72°$, $\angle 2$ supp $\angle 1$ so $\angle 2 + \angle 1 = 180°$,
   $\angle 2 = 180° - 72° = 108°$.
   Vertical angles $\angle 1 = \angle 3$, $\angle 3 = 72°$;
   $\angle 4 = \angle 2$, $\angle 4 = 108°$
   Corresponding angles $\angle 1 = \angle 5$, $\angle 4 = \angle 8$,
   $\angle 2 = \angle 6$, $\angle 3 = \angle 7$,
   $\angle 5 = 72°$, $\angle 8 = 108°$, $\angle 6 = 108°$, $\angle 7 = 72°$
   B. $\angle 1 = 90°$, $\angle 2$ supp $\angle 1$, so $\angle 2 + \angle 1 = 180°$,
   $\angle 2 = 180° - 90° = 90°$
   Vertical angles $\angle 1 = \angle 3$, $\angle 3 = 90°$; $\angle 4 = \angle 2$,
   $\angle 4 = 90°$
   Corresponding angles $\angle 1 = \angle 5$, $\angle 4 = \angle 8$,
   $\angle 2 = \angle 6$, $\angle 3 = \angle 7$
   $\angle 5 = 90°$, $\angle 8 = 90°$, $\angle 6 = 90°$, $\angle 7 = 90°$
   C. $\angle 1 = 125°$, $\angle 2$ supp $\angle 1$ so $\angle 2 + \angle 1 = 180°$,
   $\angle 2 = 180° - 125° = 55°$
   Vertical angles $\angle 1 = \angle 3$, $\angle 3 = 125°$; $\angle 4 = \angle 2$,
   $\angle 4 = 55°$
   Corresponding angles $\angle 1 = \angle 5$, $\angle 2 = \angle 6$,
   $\angle 3 = \angle 7$, $\angle 4 = \angle 8$
   $\angle 5 = 125°$, $\angle 6 = 55°$, $\angle 7 = 125°$, $\angle 8 = 55°$

5. $\angle 1$ and $\angle 4$ are supplementary
   $\angle 1 + \angle 4 = 180°$
   $4x° + 4x° + 20° = 180°$
   $8x° + 20° = 180°$

$8x° = 160°$
$x° = 20°$
$\angle 1 = 4x° = 80°$, $\angle 4 = 100°$
$\angle 1 = \angle 3 = \angle 5 = \angle 7 = 80°$,
$\angle 2 = \angle 4 = \angle 6 = \angle 8 = 100°$

6. $\angle 3$ and $\angle 4$ are supplementary
   $\angle 3 + \angle 4 = 180°$
   $5x° + 3x° - 20 = 180°$
   $8x° - 20° = 180°$
   $8x° = 200°$
   $x° = 25°$
   $\angle 3 = 5x° = 125°$, $\angle 4 = 3x° - 20° = 55°$
   $\angle 1 = \angle 3 = \angle 5 = \angle 7 = 125°$,
   $\angle 2 = \angle 4 = \angle 6 = \angle 8 = 55°$

7. $\angle 2$ and $\angle 7$ are supplementary angles
   $\angle 7$ supp $\angle 2$
   $\angle 7 + \angle 2 = 180°$
   $x° - 30° + 3x° + 10° = 180°$
   $4x° - 20° = 180°$
   $4x° = 200°$
   $x° = 50°$
   $\angle 2 = 3x° + 10° = 160°$, $\angle 7 = x° - 30° = 20°$
   $\angle 1 = \angle 3 = \angle 5 = \angle 7 = 20°$,
   $\angle 2 = \angle 4 = \angle 6 = \angle 8 = 160°$

■ 11.5 POLYGONS

A **polygon** is a closed figure whose sides are line segments.

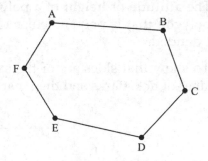

Figure 11.4

Two consecutive sides of a polygon share a common endpoint. The common endpoints are called the **vertices** of the polygon. A polygon is named by naming the vertices as you go around the polygon. In Figure 11.4, the polygon has six sides and can be named *ABCDEF*.

➤ A **diagonal of a polygon** is a line segment that joins two nonconsecutive vertices.
➤ An **equiangular polygon** is a polygon with all angles having the same measure.
➤ An **equilateral polygon** is a polygon with all sides having the same length.
➤ A **regular polygon** is a polygon that is both equiangular and equilateral.

Types of Polygon

Number of Sides	Name
3	Triangle
4	Quadrilateral
5	Pentagon
6	Hexagon
8	Octagon

The sum of the interior angles of a polygon with n sides is $S = (n - 2)180°$.

Example:

Find the sum of the interior angles of a polygon with the given number of sides.

A. 3   B. 4   C. 5   D. 10   E. 15

Solution:

A. $n = 3, S = (n - 2)180° = (3 - 2)180° = 1 \times 180° = 180°$
B. $n = 4, S = (n - 2)180° = (4 - 2)180° = 2 \times 180° = 360°$
C. $n = 5, S = (n - 2)180° = (5 - 2)180° = 3 \times 180° = 540°$
D. $n = 10, S = (n - 2)180° = (10 - 2)180° = 8 \times 180° = 1,440°$
E. $n = 15, S = (n - 2)180° = (15 - 2)180° = 13 \times 180° = 2,340°$

➤ Two polygons are **congruent** if the angles of the first polygon are equal to the corresponding angles of the second polygon and the sides of the first polygon are equal to the corresponding sides of the second polygon.

➤ Two polygons are **similar** if the angles of the first polygon are equal to the corresponding angles of the second polygon and the sides of the first polygon are proportional to the corresponding sides of the second polygon.

➤ The **altitude** or **height** of a polygon is a line segment from one vertex of a polygon that is perpendicular to the opposite side, or to the opposite side extended.

To show that sides are of the same length, you mark them with the same number of tick marks and then mark angles with arcs.

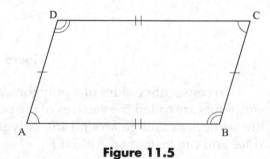

Figure 11.5

In that Figure 11.5, the quadrilateral $ABCD$ has the sides marked to show $AD = BC$ and $AB = CD$. The angles are marked to show that $\angle A = \angle C$ and $\angle B = \angle D$.

11.6 PRACTICE PROBLEMS

1. Find the sum of the interior angles of the following.

   A. A quadrilateral   B. A hexagon   C. An octagon

2. Find the measure of one angle for each polygon.

   A. Equiangular triangle
   B. Equiangular quadrilateral
   C. Regular pentagon
   D. Regular octagon

11.7 SOLUTIONS

1. A. A quadrilateral has four sides, so $n = 4$.
   $S = (n-2)180° = (4-2)180° = 2 \times 180° = 360°$

   B. A hexagon has six sides, so $n = 6$.
   $S = (n-2)180° = (6-2)180° = 4 \times 180° = 720°$

   C. An octagon has eight sides so $n = 8$.
   $S = (n - 2)180° = (8 - 2)180° = 6 \times 180° = 1080°$

2. A. An equiangular triangle is a polygon with three sides and the angles are equal.
   $S = (3 - 2)180° = 180°$, each angle is equal, so each is one-third of the sum. $180° \div 3 = 60°$. Each angle is 60°.

   B. An equiangular quadrilateral is a polygon with all four angles equal.
   $S = (4 - 2)180° = 360°$, $360° \div 4 = 90°$. Each angle is 90°.

   C. A regular pentagon is a polygon with five sides that are equal and five angles that are equal.
   $S = (5 - 2)180° = 540°$, $540° \div 5 = 108°$. Each angle is 108°.

   D. A rectangular octagon is a polygon with eight equal sides and eight equal angles.
   $S = (8 - 2)180° = 1080°$, $1080° \div 8 = 135°$. Each angle is 135°.

11.8 TRIANGLES

A **triangle** is a polygon with three sides. Figure 11.6 shows triangle ABC (written as $\triangle ABC$).

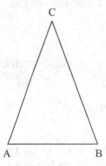

Figure 11.6

Triangles are classified by the number of equal sides they have. If no two sides have the same length, the triangle is a **scalene triangle**. If at least two sides have the same length, it is an **isosceles triangle**. When all three sides have the same length, the triangle is an **equilateral triangle**.

In Figure 11.6, if $AC = BC$, then $\triangle ABC$ is isosceles. The equal sides are called the legs of the isosceles triangle. Side AB, the unequal side, is called its base. The angles A and B are the base angles, and $\angle C$ is the vertex angle. In an isosceles triangle, the base angles are equal: $\angle A = \angle B$.

A triangle can also be classified by the size of its largest angle. If a triangle has an obtuse angle, the triangle is called an **obtuse triangle**. If a triangle has a right angle, the triangle is a **right triangle**. When all three angles of the triangle are acute, the triangle is an **acute triangle**.

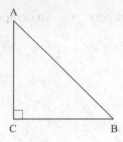

Figure 11.7

In Figure 11.7, △ABC is a right triangle since angle C is marked with the block ⌐. This is the symbol used to indicate right angles. The sides AC and BC are the legs of the right triangle. Side AB is the hypotenuse. In any right triangle, the sides that form the right angle are the legs and the side opposite the right angle is the hypotenuse.

Properties of Triangles

In a triangle, the sum of the angles is 180°. In △ABC, if $AB > BC > AC$, then $\angle C > \angle A > \angle B$. That is, if the sides of a triangle are unequal, then the angles opposite the angles are unequal in the same order. Also, in △ABC, if $\angle A > \angle B > \angle C$, then $BC > AC > AB$. In any △ABC, $AB + BC > AC$, $AC + CB > AB$, and $CA + AB > BC$; that is, in any triangle, the sum of the lengths of any two sides is greater than the third side.

Example:

Can the three given lengths be the sides of a triangle?

A. 3, 8, 12   B. 3, 8, 11   C. 3, 8, 10

Solution:

A. The first check is to add the two smaller numbers to see if they exceed the length of the third side. $3 + 8 = 11$ and $11 < 12$. A triangle cannot have sides of lengths 3, 8, and 12.
B. $3 + 8 = 11$ and $11 = 11$. A triangle cannot have sides of lengths 3, 8, and 11.
C. $3 + 8 = 11$ and $11 > 10$. $3 + 10 = 13$ and $13 > 8$. $8 + 10 = 18$ and $18 > 3$. A triangle can have sides of lengths 3, 8, and 10.

Example:

Can these be the angles of a triangle?

A. 20°, 50°, 130°   B. 30°, 60°, 90°   C. 40°, 40°, 100°
D. 35°, 50°, 95°   E. 40°, 90°, 110°

Solution:

A. $20° + 50° + 130° = 200°$ $200° \neq 180°$ These are not the angles of a triangle.
B. $30° + 60° + 90° = 180°$ $180° = 180°$ These are the angles of a triangle.
C. $40° + 40° + 100° = 180°$ $180° = 180°$ These are the angles of a triangle.
D. $35° + 50° + 95° = 180°$ $180° = 180°$ These are the angles of a triangle.
E. $40° + 90° + 110° = 240°$ $240° \neq 180°$ These are not the angles of a triangle.

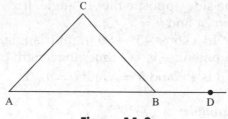

Figure 11.8

➤ An **exterior angle** of a triangle is created when a side of the triangle is extended through a vertex. In Figure 11.8, $\angle CBD$ is an exterior angle for $\triangle ABC$.

The interior and the exterior angles at the same vertex are supplementary. In Figure 11.8, $\angle ABC$ supp $\angle CBD$. The exterior angle at one vertex of a triangle has the same measure as the sum of the measures of the interior angles at the other two vertices. In Figure 11.8, $\angle CBD = \angle A + \angle C$.

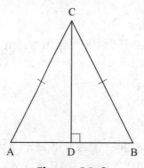

Figure 11.9

In Figure 11.9, $\triangle ABC$ is isosceles with altitude CD. The base is AB. CD bisects the base AB and the vertex $\angle C$. $\angle ACD = \angle BCD$ and $AD = BD$. Since AC and BC are marked as the equal sides, we know that $\angle A = \angle B$.

An equilateral triangle is also an equiangular triangle. The measure of each angle of an equilateral triangle is $60°$. An equilateral triangle is also an isosceles triangle.

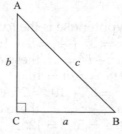

Figure 11.10

The right triangle in Figure 11.10 has legs a and b and hypotenuse c. In a right triangle ABC with right angle at C, $\angle A$ and $\angle B$ are complementary.

➤ **Pythagorean Theorem:** In a right triangle, the square of the hypotenuse is equal to the sum of the square of the legs. For Figure 11.10, $c^2 = a^2 + b^2$.

In a $30° - 60° - 90°$ right triangle, the hypotenuse is twice the length of the side opposite the $30°$ angle, and the side opposite the $60°$ angle is $\sqrt{3}$ multiplied by the side opposite the $30°$ angle. If $\angle A = 30°$ in Figure 11.10, then $\angle B = 60°$ and $c = 2a$ and $b = a\sqrt{3}$.

In a $45° - 45° - 90°$ right triangle, the two legs have the same length and the hypotenuse is $\sqrt{2}$ times the length of the legs. If $\angle A$ is $45°$ in Figure 11.10, then $\angle B$ is $45°$ and $b = a$ and $c = a\sqrt{2}$.

Example:

In a $30° - 60° - 90°$ right triangle, side a is opposite the $30°$ angle. Find the lengths of the other sides.

A. $a = 5$ B. $a = 8$ C. $a = \sqrt{3}$

Solution:

A. $a = 5, b = a\sqrt{3} = 5\sqrt{3}, c = 2a = 10$
B. $a = 8, b = a\sqrt{3} = 8\sqrt{3}, c = 2a = 16$
C. $a = \sqrt{3}, b = a\sqrt{3} = 3, c = 2a = 2\sqrt{3}$

Example:

In a $45° - 45° - 90°$ right triangle, a leg has length a. Find the lengths of the other sides.

A. $a = 8$ B. $a = 10$ C. $a = \sqrt{2}$

Solution:

A. $a = 8, b = a = 8, c = a\sqrt{2} = 8\sqrt{2}$
B. $a = 10, b = a = 10, c = a\sqrt{2} = 10\sqrt{2}$
C. $a = \sqrt{2}, b = a = \sqrt{2}, c = a\sqrt{2} = \sqrt{2} \times \sqrt{2} = 2$

Example:

Could a right triangle have sides of the lengths given?

A. $3, 4, 5$ B. $5, 10, 12$ C. $5, 12, 13$

Solution:

A. The hypotenuse is the longest side, so $c = 5, a = 3, b = 4$.
$a^2 + b^2 = 3^2 + 4^2 = 9 + 16 = 25$ and $c^2 = 5^2 = 25$.
Since $a^2 + b^2 = c^2$, these can be the sides of a right triangle.
B. $c = 12, a = 5, b = 10$.
$a^2 + b^2 = 5^2 + 10^2 = 25 + 100 = 125$ and $c^2 = 12^2 = 144$.
Since $a^2 + b^2 \neq c^2$, these are not the sides of a right triangle.
C. $c = 13, a = 5, b = 12$.
$a^2 + b^2 = 5^2 + 12^2 = 25 + 144 = 169$ and $c^2 = 13^2 = 169$.
Since $a^2 + b^2 = c^2$, these can be the sides of a right triangle.

Two triangles are **similar** when two angles of one triangle are equal to the corresponding two angles of the other triangle.

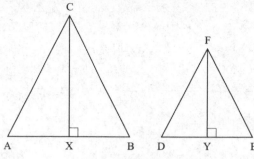

Figure 11.11

In Figure 11.11, the two triangles are similar. $\triangle ABC \sim \triangle DEF$. CX is the altitude to AB, and FY is the altitude to DE. Since $\triangle ABC \sim \triangle DEF$, the corresponding sides are proportional, so

$$\frac{AB}{DE} = \frac{BC}{EF} = \frac{CA}{FD} \text{ and } \frac{AB}{DE} = \frac{CX}{FY}.$$

▉ 11.9 PRACTICE PROBLEMS

1. Which sets of lengths can be the sides of a triangle?
   A. 5, 8, 17   B. 6, 8, 12   C. 5, 12, 17   D. 12, 18, 20

2. Which sets of angles can be the angles of a triangle?
   A. $70°, 80°, 30°$ B. $75°, 45°, 60°$
   C. $18°, 72°, 90°$ D. $70°, 30°, 40°$

3. If triangle ABC is isosceles and $\angle A$ is a base angle, find the measure of the other base angle and the vertex angle. Use Figure 11.8.
   A. $\angle A = 40°$ B. $\angle A = 32°$ C. $\angle A = 75°$

4. If triangle ABC is isosceles and $\angle C$ is the vertex angle, find the measure of the base angles. Use Figure 11.8.
   A. $\angle C = 160°$ B. $\angle C = 90°$ C. $\angle C = 70°$

5. In a $30° - 60° - 90°$ triangle, the hypotenuse has length c. Find the lengths of the legs of the triangle.
   A. $c = 18$ B. $c = 24$ C. $c = 4\sqrt{3}$

6. In a $30° - 60° - 90°$ triangle, the hypotenuse has length c. Find the lengths of the legs of the triangles.
   A. $c = 20$ B. $c = 36$ C. $c = 6\sqrt{3}$

7. In a $45° - 45° - 90°$ triangle, the length of a leg is a. What are the lengths of the other two sides?
   A. $a = 8\sqrt{2}$ B. $a = 10$ C. $a = 16$

8. Could a right triangle have sides of the lengths given?
   A. 7, 24, 25   B. 10, 24, 26   C. 10, 15, 20

9. Find the missing side of the right triangle when the other two sides are as given.
   A. Leg $= 11$, leg $= 60$ B. Leg $= 36$, hypotenuse $= 39$

10.

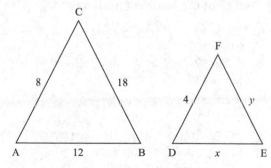

Figure 11.12

Triangle $ABC \sim$ triangle DEF with sides as indicated. Find x and y. Use Figure 11.12.

11.

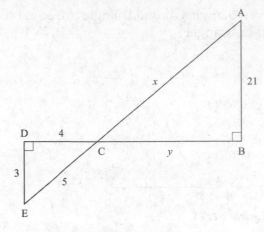

Figure 11.13

In Figure 11.13, triangle $DEC \sim$ triangle BAC with sides as indicated. Find x and y.

12.

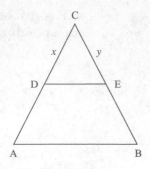

Figure 11.14

In Figure 11.14, triangle $ABC \sim$ triangle DEC, $AB = 10$, $BC = 14$, $AC = 18$, $DE = 5$, $CD = x$, and $CE = y$. Find x and y.

▮▮▮ 11.10 SOLUTIONS

1. A. 17 is the longest side. $17 > 5 + 8$, so 5, 8, and 17 are not the sides of a triangle.
   B. 12 is the longest side. $12 < 6 + 8$, so 6, 8, and 12 are the sides of a triangle.
   C. 17 is the longest side. $17 = 5 + 12$, so 5, 12, and 17 are not the sides of a triangle.
   D. 20 is the longest side. $20 < 12 + 18$, so 12, 18, and 20 are the sides of a triangle.

2. A. $70° + 80° + 30° = 180°$. These are the angles of a triangle.
   B. $75° + 45° + 60° = 180°$. These are the angles of a triangle.
   C. $18° + 72° + 90° = 180°$. These are the angles of a triangle.
   D. $70° + 30° + 40° = 140°$. $140° < 180°$. These are not the angles of a triangle.

3. A. $\angle A = 40°$, $\angle B = 40°$, and $\angle C = 180° - 40° - 40° = 100°$
   B. $\angle A = 32°$, $\angle B = 32°$, and $\angle C = 180° - 32° - 32° = 116°$
   C. $\angle A = 75°$, $\angle B = 75°$, and $\angle C = 180° - 75° - 75° = 30°$

4. A. $\angle C = 160°$, $\angle A + \angle B = 180° - 160° = 20°$, $\angle A = \angle B = 10°$
   B. $\angle C = 90°$, $\angle A + \angle B = 180° - 90° = 90°$, $\angle A = \angle B = 45°$
   C. $\angle C = 70°$, $\angle A + \angle B = 180° - 70° = 110°$, $\angle A = \angle B = 55°$

5. A. $c = 18$, $c = 2a$, $a = 9$, $b = a\sqrt{3} = 9\sqrt{3}$
   B. $c = 24$, $c = 2a$, $a = 12$, $b = a\sqrt{3} = 12\sqrt{3}$

C. $c = 4\sqrt{3}$, $c = 2a$, $a = 2\sqrt{3}$, $b = a\sqrt{3} = 2\sqrt{3} \times \sqrt{3} = 2 \times 3 = 6$

6. A. $c = 20$, $c = 2a$, $a = 10$, $b = a\sqrt{3} = 10\sqrt{3}$
   B. $c = 36$, $c = 2a$, $a = 18$, $b = a\sqrt{3} = 18\sqrt{3}$
   C. $c = 6\sqrt{3}$, $c = 2a$, $a = 3\sqrt{3}$, $b = a\sqrt{3} = 3\sqrt{3} \times \sqrt{3} = 3 \times 3 = 9$

7. $a = b$, $c = a\sqrt{2}$
   A. $a = 8\sqrt{2}$, $b = 8\sqrt{2}$, $c = 8\sqrt{2} \times \sqrt{2} = 8 \times 2 = 16$
   B. $a = 10$, $b = 10$, $c = 10\sqrt{2}$
   C. $a = 16$, $b = 16$, $c = 16\sqrt{2}$

8. A. $7^2 + 24^2 = 49 + 576 = 625$, $25^2 = 625$
      $7^2 + 24^2 = 25^2$, so 7, 24, and 25 are the sides of a right triangle.
   B. $10^2 + 24^2 = 100 + 576 = 676$, $26^2 = 676$
      $10^2 + 24^2 = 26^2$, so 7, 24, and 26 are the sides of a right triangle.
   C. $10^2 + 15^2 = 100 + 225 = 325$, $20^2 = 400$
      $10^2 + 15^2 \neq 20^2$, so 10, 15, and 20 are not the sides of a right triangle.

9. A. $a^2 + b^2 = c^2$, $a = 11$, $b = 60$
      $11^2 + 60^2 = c^2$
      $121 + 3600 = 3721 = c^2$
      $\sqrt{3721} = 61 = c$ (hypotenuse)
   B. $a^2 + b^2 = c^2$
      $36^2 + b^2 = 39^2$
      $1296 + b^2 = 1521$
      $b^2 = 1521 - 1296 = 225$
      $b = \sqrt{225} = 15$ (second leg)

10. Triangle $ABC \sim$ triangle DEF, $AC = 8$, $AB = 12$,
    $BC = 18$, $DE = x$, $EF = y$, $DF = 4$

    $$\frac{AB}{DE} = \frac{BC}{EF} = \frac{AC}{DF}$$

    $$\frac{12}{x} = \frac{8}{4}, \frac{18}{y} = \frac{8}{4}$$

    $8x = 48,\ 8y = 72$
    $x = 6,\ y = 9$

11. Triangle $DEC \sim$ triangle BAC, $DE = 3$, $EC = 5$,
    $DC = 4$, $AB = 21$, $AC = x$, $BC = y$

    $$\frac{DE}{BA} = \frac{EC}{AC} = \frac{DC}{BC}$$

$$\frac{3}{21} = \frac{4}{y}, \frac{3}{21} = \frac{5}{x}$$

$3y = 84,\ 3x = 105$
$y = 28,\ x = 35$

12. Triangle $ABC \sim$ triangle DEC, $AB = 10$,
    $BC = 14$, $AC = 18$, $DE = 5$, $CD = x$,
    $CE = y$

    $$\frac{AB}{DE} = \frac{BC}{EC} = \frac{AC}{DC}$$

    $$\frac{10}{5} = \frac{14}{y}, \frac{10}{5} = \frac{18}{x}$$

    $10y = 70,\ 10x = 90$
    $y = 7,\ x = 9$

11.11 QUADRILATERALS

A **quadrilateral** is a polygon with four sides. The sum of the angles of a quadrilateral is 360°.

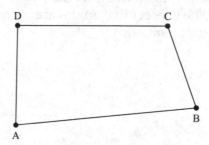

Figure 11.15

In quadrilateral $ABCD$ in Figure 11.15, $\angle A + \angle B + \angle C + \angle D = 360°$.

➤ A **trapezoid** is a quadrilateral with exactly one pair of parallel sides.

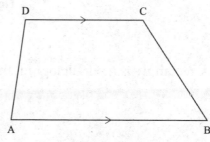

Figure 11.16

Quadrilateral $ABCD$ is a trapezoid since exactly one pair of sides, AB and CD, are parallel. The use of the > indicates which sides are parallel. See Figure 11.16. The parallel sides, AB and CD, are called the bases of the trapezoid. The non-parallel sides, AD and BC, are called the legs.

➤ A **parallelogram** is a quadrilateral with both pairs of opposite sides parallel.

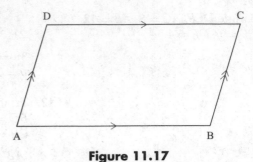

Figure 11.17

The quadrilateral in Figure 11.17 is a parallelogram because the figure is marked with $AB \parallel CD$ and $AD \parallel BC$.

Properties of a Parallelogram See Figure 11.18

1. The opposite sides are equal.
2. The opposite angles are equal.
3. The opposite sides are parallel.
4. The diagonals bisect each other.
5. The consecutive angles are supplementary.

1. $AB = CD$ and $AD = BC$
2. $\angle A = \angle C$ and $\angle B = \angle D$
3. $AB \parallel CD$ and $AD \parallel BC$
4. $AE = EC$ and $DE = EB$
5. $\angle DAB$ supp $\angle ABC$, $\angle ABC$ supp $\angle BCD$, $\angle BCD$ supp $\angle CDA$, $\angle CDA$ supp $\angle DAB$

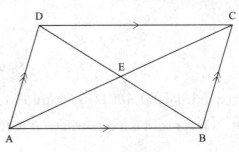

Figure 11.18

➤ A **rectangle** is a parallelogram that has all right angles. See Figure 11.19.

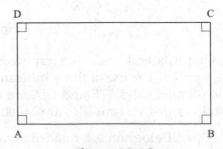

Figure 11.19

Properties of a Rectangle See Figure 11.20

1. Opposite sides are parallel.
2. Opposite sides are equal.
3. All angles are equal.
4. The diagonals are equal.
5. The diagonals bisect each other.

1. $AB \parallel CD$ and $AD \parallel BC$
2. $AB = CD$ and $AD = BC$
3. $\angle A = \angle B = \angle C = \angle D = 90°$
4. $AC = DB$
5. $AE = EC = BE = DE$

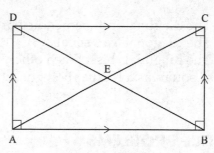

Figure 11.20

➤ A **rhombus** is a parallelogram with all sides equal. See Figure 11.21.

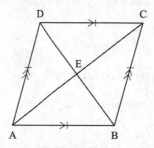

Figure 11.21

Properties of a Rhombus

1. Opposite sides are parallel.
2. Opposite angles are congruent.
3. All sides are equal.
4. The diagonals bisect the angles.

5. The diagonals bisect each other.
6. The diagonals are perpendicular.

1. $AB \parallel CD$ and $AD \parallel BC$
2. $\angle DAB = \angle BCD$ and $\angle ABC = \angle CDA$
3. $AB = BC = CD = DA$
4. $\angle DAC = \angle BAC, \angle DCA = \angle BCA$
   $\angle CBD = \angle ABD$ and $\angle ADB = \angle CDB$
5. $AE = EC$ and $BE = ED$
6. $AC \perp BD$

➤ A **square** is a rectangle with all sides equal. See Figure 11.22.

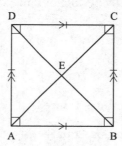

Figure 11.22

Properties of a Square

1. Opposite sides are parallel.
2. All angles are equal.

3. All sides are equal.
4. The diagonals bisect the angles.

5. The diagonals are perpendicular.
6. The diagonals are equal.
7. The diagonals bisect each other.
8. A square is a regular polygon.

1. $AB \parallel CD$ and $AD \parallel BC$
2. $\angle DAB = \angle ABC = \angle BCD = \angle CDA = 90°$
3. $AB = BC = CD = DA$
4. $\angle DAC = \angle BAC, \angle BCA = \angle DCA$ $\angle CDB = \angle ADB$ and $\angle ABD = \angle CBD$
5. $AC \perp BD$
6. $AC = BD$
7. $AE = EC = BE = ED$
8. All sides and angles are equal.

▨▨▨ 11.12 PRACTICE PROBLEMS

1. Which quadrilaterals have the sum of the angles equal to 360°?

2. Which quadrilaterals have only one pair of parallel sides?

3. Which quadrilaterals have two diagonals?

4. Which quadrilaterals have equal diagonals?

5. Which quadrilaterals have perpendicular diagonals?

6. Which quadrilaterals have two pairs of parallel sides?

7. Which quadrilaterals have diagonals that bisect each other?

8. Which quadrilaterals have diagonals that bisect the angles?

9. Which quadrilaterals have the opposite angles equal?

10. Which quadrilaterals are equiangular?

11. Which quadrilaterals are equilateral?

12. Which quadrilaterals have the consecutive angles supplementary?

13. Which quadrilaterals have opposite sides equal?

14. Which quadrilaterals are regular polygons?

▨▨▨ 11.13 SOLUTIONS

1. All quadrilaterals: trapezoid, parallelogram, rectangle, rhombus, square

2. Trapezoid

3. All quadrilaterals: trapezoid, parallelogram, rectangle, rhombus, square

4. Rectangle, square

5. Rhombus, square

6. Parallelogram, rectangle, rhombus, square

7. Parallelogram, rectangle, rhombus, square

8. Rhombus, square

9. Parallelogram, rectangle, rhombus, square

10. Rectangle, square

11. Rhombus, square

12. Parallelogram, rectangle, rhombus, square

13. Parallelogram, rectangle, rhombus, square

14. Square

11.14 PERIMETER AND AREA

Perimeter and area are measurements that are commonly subjects of GRE math problems.

► The **perimeter** of a polygon is the distance around the polygon. Hence, the perimeter is the sum of the lengths of the sides of the polygon.

If a polygon has n sides, the perimeter P of the polygon is $P = s_1 + s_2 + s_3 + \cdots + s_n$, where each s is the length of a side. The perimeter P of a rectangle with length l and width w is $P = 2l + 2w$. The perimeter P of a square with sides of length s is $P = 4s$.

► The **area** of a polygon is the amount of the surface enclosed by the polygon. The area is expressed as the number of square units of surface inside the polygon.

Areas of Polygons

Polygon	Symbols Used in Formula	Area
Triangle	b = base, h = altitude to the base	$A = 0.5\,bh$
Trapezoid	h = distance between parallel sides a and b	$A = 0.5\,h(a + b)$
Parallelogram	b = base side of parallelogram, h = altitude to base	$A = bh$
Rectangle	l = length, w = width	$A = lw$
Square	s = side of square	$A = s^2$
Right triangle	a and b = legs of the triangle	$A = 0.5\,ab$

The area of a general polygon can be found by dividing it into smaller regions, each of which is a polygon whose area can be found.

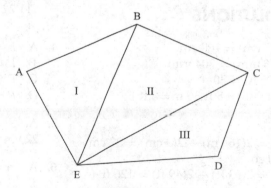

Figure 11.23

In Figure 11.23, polygon *ABCDE* was divided into three nonoverlapping regions by drawing all of the diagonals from one vertex, *E*. The area of polygon *ABCDE* equals the area of triangle I plus the area of triangle II plus the area of triangle III.

11.15 PRACTICE PROBLEMS

1. Find the perimeter of a square with side s.
   A. $s = 25$ cm   B. $s = 76$ mm   C. $s = 7.5$ ft
   D. $s = 2$ ft 9 in

2. Find the perimeter of a rectangle with length l and width w.
   A. $l = 16$ cm, $w = 9$ cm   B. $l = 63$ ft, $w = 49$ ft
   C. $l = 3.5$ m, $w = 9.75$ m

3. Find the perimeter of a triangle with these sides.
   A. 16 m, 8 m, 14 m   B. 20 in, 16 in, 28 in
   C. 6.22 m, 4.7 m, 5.84 m

4. Find the perimeter of an equilateral triangle with side s.
   A. $s = 40$ cm   B. $s = 14$ ft   C. $s = 4.75$ m
   D. $s = 8.4$ in

5. Find the perimeter of a pentagon whose sides are 24 in, 58 in, 32 in, 66 in, and 43 in.

6. Find the area of a parallelogram with the base b and altitude h.
   A. $b = 26$ in, $h = 14$ in   B. $b = 98$ cm, $h = 75$ cm
   C. $b = 25$ in, $h = 32$ in

7. Find the area of a triangle with the base b and altitude h.
   A. $b = 12$ cm, $h = 18$ cm   B. $b = 13$ m, $h = 10$ m
   C. $b = 5$ ft, $h = 7$ ft

8. Find the area of a square with side s.
   A. $s = 18$ ft   B. $s = 10$ ft   C. $s = 15$ cm
   D. $s = 6$ ft   E. $s = 8$ m

9. Find the area of a trapezoid with altitude h and bases a and b.
   A. $h = 8$ cm, $a = 4$ cm, $b = 10$ cm
   B. $h = 5$ mm, $a = 9$ mm, $b = 13$ mm
   C. $h = 18$ in, $a = 29$ in, $b = 36$ in
   D. $h = 7$ ft, $a = 8$ ft, $b = 14$ ft

10. Find the area of a right triangle with the given sides.
    A. 11 in, 60 in, 61 in   B. 7 cm, 24 cm, 25 cm

11.

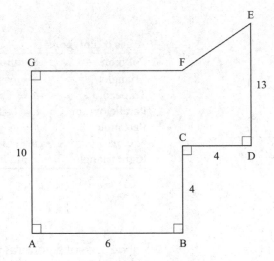

Figure 11.24

Find the area of figure $ABCDEFG$ using Figure 11.24.

11.16 SOLUTIONS

1. A. $P = 4s = 4(25$ cm$) = 100$ cm
   B. $P = 4s = 4(76$ mm$) = 304$ mm
   C. $P = 4s = 4(7.5$ ft$) = 30$ ft
   D. $P = 4s = 4(2$ ft 9 in$) = 8$ ft 36 in $= 8$ ft $+$ 3 ft $= 11$ ft

2. A. $P = 2l + 2w = 2(16$ cm$) + 2(9$ cm$) = 32$ cm $+ 18$ cm $= 50$ cm
   B. $P = 2l + 2w = 2(63$ ft$) + 2(49$ ft$) = 126$ ft $+$ 98 ft $= 224$ ft
   C. $P = 2l + 2w = 2(3.5$ m$) + 2(9.75$ m$) = 7.0$ m $+ 19.5$ m $= 26.5$ m

3. A. $P = s_1 + s_2 + s_3 = 16$ m $+ 8$ m $+ 14$ m $= 38$ m
   B. $P = 20$ in $+ 16$ in $+ 28$ in $= 64$ in
   C. $P = 6.22$ m $+ 4.7$ m $+ 5.84$ m $= 16.76$ m

4. A. $P = 3s = 3(40$ cm$) = 120$ cm
   B. $P = 3s = 3(14$ ft$) = 42$ ft
   C. $P = 3s = 3(4.75$ m$) = 14.25$ m
   D. $P = 3s = 3(8.4$ in$) = 25.2$ in

5. A. $P = 24$ in $+ 58$ in $+ 32$ in $+ 66$ in $+ 43$ in $=$ 223 in

6. A. $A = bh = (26$ in$)(14$ in$) = 364$ in^2
   B. $A = bh = (98$ cm$)(75$ cm$) = 7,350$ cm^2
   C. $A = bh = (25$ in$)(32$ in$) = 800$ in^2

7. A. $A = 0.5\,bh = 0.5(12$ cm$)(18$ cm$) = 108$ cm^2
   B. $A = 0.5\,bh = 0.5(13$ m$)(10$ m$) = 65$ m^2
   C. $A = 0.5\,bh = 0.5(5$ ft$)(7$ ft$) = 17.5$ ft^2

8. A. $A = s^2 = (18 \text{ ft})^2 = 324 \text{ ft}^2$
    B. $A = s^2 = (10 \text{ ft})^2 = 100 \text{ ft}^2$
    C. $A = s^2 = (15 \text{ cm})^2 = 225 \text{ cm}^2$
    D. $A = s^2 = (6 \text{ ft})^2 = 36 \text{ ft}^2$
    E. $A = s^2 = (8 \text{ m})^2 = 64 \text{ m}^2$

9. A. $A = 0.5h(a + b) = 0.5(8 \text{ cm})(4 \text{ cm} + 10 \text{ cm}) = 0.5(8 \text{ cm})(14 \text{ cm}) = 56 \text{ cm}^2$
    B. $A = 0.5h(a + b) = 0.5(5 \text{ mm})(9 \text{ mm} + 13 \text{ mm}) = 0.5(5 \text{ mm})(22 \text{ mm}) = 55 \text{ mm}^2$
    C. $A = 0.5h(a + b) = 0.5(18 \text{ in})(29 \text{ in} + 36 \text{ in}) = 0.5(18 \text{ in})(65 \text{ in}) = 585 \text{ in}^2$
    D. $A = 0.5h(a + b) = 0.5(7 \text{ ft})(8 \text{ ft} + 14 \text{ ft}) = 0.5(7 \text{ ft})(22 \text{ ft}) = 77 \text{ ft}^2$

10. A. $A = 0.5\,ab$ $A = 0.5(11 \text{ in})(60 \text{ in}) = 330 \text{ in}^2$
    B. $A = 0.5\,ab$ $A = 0.5(7 \text{ cm})(24 \text{ cm}) = 84 \text{ cm}^2$

11. Draw $FX \perp ED$, draw CF. $GF = 6$. $AB + CD = GX = 6 + 4 = 10$. $FX = 4$. $BC + DX = AG = 10$. $DX = 6$. $XE = 5$. Area $ABFG = 6(10) = 60$. Area $CDXF = 4(6) = 24$. Area $XFE = 0.5(7)(4) = 14$. Total area $ABCDEFG = 60 + 24 + 14 = 98$ square units.

▬ 11.17 CIRCLES

A **circle** is the set of all points in a plane that are at a fixed distance, or **radius**, from a given point, the **center**. We name a circle by its center.

- Two circles are **concentric** if they have the same center.
- An **arc** is part of a circle.
- A **semicircle** is an arc that is one-half of a circle.
- A **minor arc** is an arc of a circle that is less than a semicircle.
- A **major arc** is an arc of a circle that is greater than a semicircle.

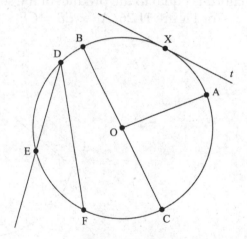

Figure 11.25

In circle O (Figure 11.25), AO is a radius, BC is a diameter, DF is a chord, $\angle AOB$ is a central angle, $\angle FDE$ is an inscribed angle, line t is a tangent to circle O at X, and line ED is a secant. Arc BAC is a semicircle, arc AC is a minor arc, and FDA is a major arc. In circle O, the region enclosed by arc AC, radius CO, and radius AO is called a sector of the circle.

➤ The **circumference** of a circle is the distance around the circle. $C = \pi d$. Pi (π) is the ratio of the circumference of a circle to its diameter. π is approximately 3.14 or $3\frac{1}{7}$.

➤ A **chord** of a circle is a line segment joining two points on the circle.
➤ A **diameter** is a chord that goes through the center of the circle.

➤ A **radius** is a line segment from the center of a circle to a point on the circle. A diameter is the same length as two radii.
➤ A **secant** is a line that intersects a circle in two points.
➤ A **tangent** is a line that intersects a circle in exactly one point.
➤ An **inscribed angle** is an angle whose sides are chords of a circle and with its vertex on the circle.
➤ A **central angle** is an angle with its vertex at the center of the circle, and with sides that are radii.

The measure of a central angle is the same as the degree measure of its arc. However, the degree measure of an inscribed angle is one-half the measure of its arc. $\angle AOC$ = arc AC and $\angle FDE$ = 0.5 arc EF. If an angle is inscribed in a semicircle, it is a right angle.

A diameter separates a circle into two semicircles. A semicircle is half of a circle, so it is half of 360° or 180°. A minor arc is a part of a circle that is less than a semicircle, so it has a measure that is less than 180°.

The circumference of circle C equals π multiplied by the diameter. $C = \pi d$, or $C = 2\pi r$.

The area A of a circle is equal to π multiplied by the square of the radius. $A = \pi r^2$.

The area A of a sector of a circle is to the area of the circle as the central angle of the sector is to 360°. That is, find the ratio of the central angle of the sector to 360° and multiply that ratio times the area of the circle.

When two chords intersect inside a circle, the product of the segments of one chord is equal to the product of the segments of the other chord.

For Figure 11.26, $AE \times EB = CE \times ED$. Also, $\angle AED = 0.5(\text{arc } AD + \text{arc } BC)$.

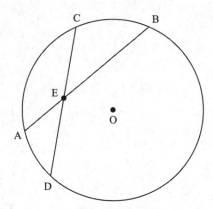

Figure 11.26

Example:

If a diameter of a circle is 30 cm, what is the circumference of the circle?

Solution:

$C = \pi d = 30\pi$ cm.

Example:

If a diameter of a circle is 40 ft, what is the area of the circle?

Solution:

Since d = 40 ft, then r = ½d = 20 ft. A = πr² = π(20 ft)² = 400π ft².

Example:

In Figure 11.26, if arc AD = 42°, arc AC = 72°, and arc BC = 46°, what is the measure of arc BD?

Solution:

arc AD + arc AC + arc BC + arc BD = 360°
42° + 72° + 46° + arc BD = 360°
160° + arc BD = 360°
arc BD = 200°

▬ 11.18   PRACTICE PROBLEMS

1. In Figure 11.27, if $AC = 12$ and $BC = 35$, what is the radius of circle O?

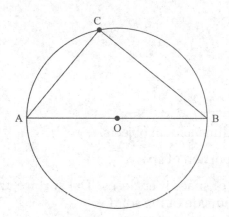

Figure 11.27

2. In Figure 11.27, if $AC = BC = 10$ cm, what is the length of the diameter of circle O?

3. In Figure 11.27, if $AC = AO = 5$, what is the area of triangle $\triangle ABC$?

4. In Figure 11.26, if $AE = 12$, $BE = 7$, and $CE = 6$, what is the length of CD?

5. In Figure 11.26, if $CE = ED$, $AE = 2$, and $EB = 6$, what is the length of CD?

6. In Figure 11.26, if arc $CB = 35°$ and arc $AD = 95°$, what is the measure of $\angle AED$?

7. In Figure 11.26, if arc $BC = 125°$ and arc $AD = 105°$, what is the measure of $\angle BEC$?

8. What is the circumference of a circle whose radius is 10 cm?

9. What is the area of a circle whose diameter is 10 in?

10. What is the circumference of a circle whose area is 49π cm²?

11. What is the area of a sector of a circle whose central angle is 60° and whose radius is 12 ft?

12. What is the area of a circle whose radius is 12.2 cm?

13. Find the circumference of a circle whose radius is 18 cm.

▬ 11.19   SOLUTIONS

1. Triangle ACB is a right triangle. Line segment AB is the hypotenuse of the right triangle and the diameter of the circle. $12² + 35² = 144 + 1225 = 1369 = 37²$, so $AB = 37$. The radius is one-half of the diameter. So, $r = 0.5(37) = 18.5$.

2. $AC = BC = 10$ cm in right triangle ABC. $AC² + BC² = AB²$, $10² + 10² = AB²$, $100 + 100 = AB²$ $200 = AB²$, $\sqrt{200} = AB$, $AB = \sqrt{100}\sqrt{2} = 10\sqrt{2}$

Note: Right triangle ABC is an isosceles right triangle, so it is a $45° - 45° - 90°$ right triangle and $AB = AC\sqrt{2}$.

3. $AC = AO$, so $2AC = AB$, so it is a $30° - 60° - 90°$ right triangle, so $BC = AC\sqrt{3} = 5\sqrt{3}$.
   $A = 0.5\,ab = 0.5(5)(5\sqrt{3}) = 12.5\sqrt{3}$.

4. $AE \times EB = CE \times ED$
   $AE = 12, EB = 7, CE = 6,$
   $ED = x$
   $12 \times 7 = 6x$
   $84 = 6x$
   $14 = x$
   $CD = CE + ED = 6 + 14 = 20$

5. $AE \times EB = CE \times ED$
   $2 \times 6 = y \times y$
   $12 = y^2$
   $\sqrt{12} = y$
   $\sqrt{4}\sqrt{3} = y$
   $2\sqrt{3} = y$
   $CD = 2y = 4\sqrt{3}$

6. $\angle AED = 0.5(35° + 95°) = 0.5(130°) = 65°$

7. $\angle BEC = 0.5(125° + 105°) = 0.5(230°) = 115°$

8. $C = 2\pi r = 2(\pi)(10 \text{ cm}) = 20\pi$ cm

9. $A = \pi r^2\ d = 10$ in, so $r = 5$ in
   $A = \pi(5 \text{ in})^2 = 25\pi$ in^2

10. $A = \pi r^2 = 49\pi$ cm^2 so $r^2 = 49$ cm^2 $r = 7$ cm
    $C = 2\pi r = 2\pi(7 \text{ cm}) = 14\pi$ cm

11. $A = \pi r^2 = \pi(12 \text{ ft})^2 = 144\pi$ ft^2. $360 \div 60° = 6$
    area of sector = Area of circle $\div 6 =$
    144π ft$^2 \div 6 = 24\pi$ ft^2

12. $A = \pi r^2 = \pi(12.2 \text{ cm})^2 = 148.84\pi$ cm^2

13. $C = 2\pi r = 2(\pi)(18 \text{ cm}) = 36\pi$ cm

▩ 11.20 SOLID GEOMETRY

Solid geometry has to do with three-dimensional objects.

➤ A **solid** is a figure that encloses a portion of space.

The **volume** of a solid is the amount of space it encloses. The **surface area** of a solid is the area needed to cover the outside of the solid.

➤ A **polyhedron** is a solid made up of parts of planes.

The sides of a polyhedron are called **faces**. Two faces intersect in an **edge**. Two edges intersect in a **vertex**.

A **rectangular solid** is a polyhedron with all faces being rectangles. See Figure 11.28. Most rooms and boxes are rectangular solids. The dimensions of a rectangle solid are its length l, width w, and height h. The volume $V = lwh$. The surface area $S = 2lw + 2lh + 2wh$.

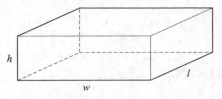

Figure 11.28

A **cube** is a rectangular solid with all faces being identical squares. See Figure 11.29. The edge e of any one of the squares is the length, width, and height for the cube.

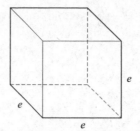

Figure 11.29

The volume $V = e^3$ and the surface area $S = 6e^2$.

A **right circular cylinder** has bases that are circles and the curved surface is perpendicular to the bases. See Figure 11.30. The volume $V = \pi r^2 h$ and the surface area $S = 2\pi r h + 2\pi r^2$, where h is the height of the cylinder and r is the radius of the base.

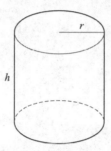

Figure 11.30

A **pyramid** is a solid with one base. See Figure 11.31. The vertices of the base are joined to the apex or vertex of the pyramid. The volume $V = \frac{1}{3}Bh$, where B is the area of the base of the pyramid and h is the height from the base to the apex.

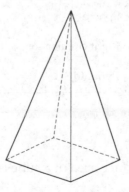

Figure 11.31

A **right circular cone** is a solid that has one circular base. See Figure 11.32. The vertex of the cone is a point on a line perpendicular to the circle at its center. The volume V of the cone $= \frac{1}{3}\pi r^2 h$, where r is the radius of the base and h is the height of the cone.

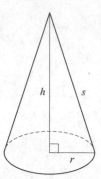

Figure 11.32

The surface area $S = \pi r^2 + \pi r s$, where s is the slant height and $s^2 = r^2 + h^2$.

A **sphere** is a ball-shaped solid. See Figure 11.33. The volume $V = \dfrac{4}{3}\pi r^3$ and the surface area $S = 4\pi r^2$, where r is the radius of the sphere.

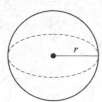

Figure 11.33

Volume and Surface Areas of Solids

Solid	Volume	Surface Area
Rectangular solid	$V = lwh$	$S = 2lw + 2lh + 2wh$
Cube	$V = e^3$	$S = 6e^2$
Right circular cylinder	$V = \pi r^2 h$	$S = 2\pi r h + 2\pi r^2$
Sphere	$V = \dfrac{4}{3}\pi r^3$	$S = 4\pi r^2$
Right circular cone	$V = \dfrac{1}{3}\pi r^2 h$	$S = \pi r^2 + \pi r s$
Pyramid	$V = \dfrac{1}{3}Bh$	--

Example:

If the surface area of a cube is 1,176 in^2, what is the length of an edge of the cube?

Solution:

$S = 6e^2$
$1{,}176 \text{ in}^2 = 6e^2$
$196 \text{ in}^2 = e^2$
$\sqrt{196 \text{ in}^2} = e$. Because e is the length of an edge of a cube, only the positive square root is used. $14 \text{ in} = e$.

Example:

If the radius and height of a right circular cylinder are 8 cm and 10 cm, respectively, what is the surface area of the cylinder?

Solution:

$S = 2\pi rh + 2\pi r^2$
$S = 2\pi(8 \text{ cm})(10 \text{ cm}) + 2\pi(8 \text{ cm})^2$
$S = 160\pi \text{ cm}^2 + 128\pi \text{ cm}^2$
$S = 288\pi \text{ cm}^2$

▆▆ 11.21 PRACTICE PROBLEMS

1. A rectangular solid has length l, width w, and height h. Find the volume of the solid.
   A. $l = 10$ ft, $w = 8$ ft, $h = 7$ ft
   B. $l = 30$ cm, $w = 24$ cm, $h = 12$ cm

2. A rectangular solid has length l, width w, and height h. Find the surface area of the solid.
   A. $l = 10$ in, $w = 6$ in, $h = 12$ in
   B. $l = 8$ m, $w = 7$ m, $h = 10$ m

3. A cube has edge e. Find the volume of the cube.
   A. $e = 0.4$ m   B. $e = 10$ in
   C. $e = 12$ cm   D. $e = 5$ m

4. A cube has edge e. Find the surface area of the cube.
   A. $e = 2$ in   B. $e = 7$ ft
   C. $e = 1.2$ cm   D. $e = 0.4$ m

5. A right circular cylinder has a radius r and height h. Find the volume of the cylinder.
   A. $r = 4$ in, $h = 8$ in   B. $r = 7$ cm, $h = 3.5$ cm

6. A right circular cylinder has radius r and height h. Find the surface area of the cylinder.
   A. $r = 8.4$ m, $h = 10.2$ m   B. $r = 8$ ft, $h = 7$ ft

7. A sphere has radius r. Find the volume of the sphere.
   A. $r = 6$ in   B. $r = 8$ cm   C. $r = 10$ m

8. A sphere has a radius r. Find the surface area of the sphere.
   A. $r = 5$ cm   B. $r = 1.4$ m   C. $r = 9$ in

9. A right circular cone has a radius r and height h. Find the volume of the cone.
   A. $r = 6$ in, $h = 10$ in   B. $r = 4$ cm, $h = 12$ cm

10. A right circular cone has r and slant height s. Find the surface area of the cone.
    A. $r = 6$ cm, $s = 12$ cm   B. $r = 10$ ft, $s = 8$ ft

11. Find the volume of a pyramid that has a rectangular base with length l and width w and the height of the pyramid is h.
    A. $l = 8$ in, $w = 6$ in, $h = 12$ in
    B. $l = 7$ cm, $w = 4$ cm, $h = 6$ cm

12. Find the volume of a pyramid that has a square base with side e. The height of the pyramid is h.
    A. $e = 6$ cm, $h = 10$ cm
    B. $e = 9$ in, $h = 14$ in

13. The dimensions of a classroom are 24 ft by 17 ft by 12 ft. Find the number of cubic feet of air space that it contains.

14. Find the volume of a cube whose surface area is 54 in^2.

15. A cylindrical can is 6 in tall and the radius of the base is 3 in. What is the surface area of the can?

16. A cylindrical tank holds 490π ft^3 of water. The base is a circle with a 7 ft diameter. What is the height of the cylinder?

17. Find the volume of a pyramid when the area of the base is 25 in^2 and the height is 6 in.

18. If the volume of a sphere is 288π cm^3, what is the radius?

19. What is the surface area of a rectangular solid whose dimensions are 17 cm, 19 cm, and 20 cm?

20. What is the surface area of a cube whose edge is 14 mm?

▬▬ 11.22 SOLUTIONS

1. A. $V = lwh = (10 \text{ ft})(8 \text{ ft})(7 \text{ ft}) = 560 \text{ ft}^3$
   B. $V = lwh = (30 \text{ cm})(24 \text{ cm})(12 \text{ cm}) = 8,640 \text{ cm}^3$

2. A. $S = 2lw + 2lh + 2wh = 2(10 \text{ in})(6 \text{ in}) + 2(10 \text{ in})(12 \text{ in}) + 2(6 \text{ in})(12 \text{ in})$
   $S = 120 \text{ in}^2 + 240 \text{ in}^2 + 144 \text{ in}^2 = 504 \text{ in}^2$
   B. $S = 2lw + 2lh + 2wh = 2(8 \text{ m})(7 \text{ m}) + 2(8 \text{ m})(10 \text{ m}) + 2(7 \text{ m})(10 \text{ m})$
   $S = 112 \text{ m}^2 + 160 \text{ m}^2 + 140 \text{ m}^2 = 412 \text{ m}^2$

3. A. $V = e^3 = (0.4 \text{ m})^3 = 0.064 \text{ m}^3$
   B. $V = e^3 = (10 \text{ in})^3 = 1,000 \text{ in}^3$
   C. $V = e^3 = (12 \text{ cm})^3 = 1,728 \text{ cm}^3$
   D. $V = e^3 = (5 \text{ m})^3 = 125 \text{ m}^3$

4. A. $S = 6e^2 = 6(2\text{in})^2 = 6(4\text{in}^2) = 24\text{in}^2$
   B. $S = 6e^2 = 6(7\text{ft})^2 = 6(49\text{ft}^2) = 294\text{ft}^2$
   C. $S = 6e^2 = 6(1.2\text{cm}) = 6(1.44\text{cm}^2) = 8.64\text{cm}^2$
   D. $S = 6e^2 = 6(0.4\text{m}) = 6(0.16\text{m}^2) = 0.96\text{m}^2$

5. A. $V = \pi r^2 h = \pi(4\text{in})^2(8\text{in}) = \pi(16\text{in}^2)(8\text{in}) = 128\pi \text{ in}^3$
   B. $V = \pi r^2 h = \pi(7 \text{ cm})^2(3.5 \text{ cm}) = \pi(49 \text{ cm}^2)(3.5 \text{ cm}) = 171.5\pi \text{ cm}^3$

6. A. $S = 2\pi r^2 + 2\pi rh = 2(\pi)(8.4\text{m})^2 + 2(\pi)(8.4 \text{ m})(10.2 \text{ m}) = 2\pi(70.56\text{m}^2) + 2\pi(85.68\text{m}^2) = 141.12\pi\text{m}^2 + 171.36\pi \text{ m}^2 = 312.48\pi \text{ m}^2$
   B. $S = 2\pi r^2 + 2\pi rh = 2(\pi)(8 \text{ ft})^2 + 2(\pi)(8 \text{ ft})(7 \text{ ft}) = 2\pi(64 \text{ ft}^2) + 2\pi(56 \text{ ft}^2) = 128\pi \text{ ft}^2 + 112\pi \text{ ft}^2 = 240\pi \text{ ft}^2$

7. A. $V = \frac{4}{3}\pi r^3 = \frac{4}{3}(\pi)(6 \text{ in})^3 = \frac{4}{3}\pi(216) \text{ in}^3 = 288\pi \text{ in}^3$
   B. $V = \frac{4}{3}\pi r^3 = \frac{4}{3}(\pi)(8 \text{ cm})^3 = \frac{4}{3}\pi(512) \text{ cm}^3 = 682.7\pi \text{ cm}^3$
   C. $V = \frac{4}{3}\pi r^3 = \frac{4}{3}(\pi)(10\text{m})^3 = \frac{4}{3}\pi(1,000)\text{m}^3 = 1,333\pi\text{m}^3$

8. A. $S = 4\pi r^2 = 4(\pi)(5 \text{ cm})^2 = 100\pi \text{ cm}^2$
   B. $S = 4\pi r^2 = 4(\pi)(1.4 \text{ m})^2 = 7.84\pi \text{ m}^2$
   C. $S = 4\pi r^2 = 4(\pi)(9 \text{ in})^2 = 324\pi \text{ in}^2$

9. A. $V = \frac{1}{3}\pi r^2 h = \frac{1}{3}(\pi)(6 \text{ in})^2(10 \text{ in}) = 120\pi \text{ in}^3$
   B. $V = \frac{1}{3}\pi r^2 h = \frac{1}{3}(\pi)(4 \text{ cm})^2(12 \text{ cm}) = 64\pi \text{ cm}^3$

10. A. $S = \pi r^2 + \pi rs = \pi(6\text{cm})^2 + \pi(6\text{cm})(12\text{cm}) = 36\pi\text{cm}^2 + 72\pi\text{cm}^2 = 108\pi \text{ cm}^2$
    B. $S = \pi r^2 + \pi rs = \pi(10 \text{ ft})^2 + \pi(10 \text{ ft})(8 \text{ ft}) = 100\pi \text{ ft}^2 + 80\pi \text{ ft}^2 = 180\pi \text{ ft}^2$

11. A. $V = \frac{1}{3}Bh = \frac{1}{3}(lw)h = \frac{1}{3}(8 \text{ in})(6 \text{ in})(12 \text{ in}) = 192 \text{ in}^3$
    B. $V = \frac{1}{3}Bh = \frac{1}{3}(lw)h = \frac{1}{3}(7 \text{ cm})(4 \text{ cm})(6 \text{ cm}) = 56 \text{ cm}^3$

12. A. $V = \frac{1}{3}Bh = \frac{1}{3}(e^2)h = \frac{1}{3}(6 \text{ cm})^2(10 \text{ cm}) = 120 \text{ cm}^3$
    B. $V = \frac{1}{3}Bh = \frac{1}{3}(e^2)h = \frac{1}{3}(9 \text{ in})^2(14 \text{ in}) = 378 \text{ in}^3$

13. A room is a rectangular solid. You need to find the volume.
    $V = lwh = (24 \text{ ft})(17 \text{ ft})(12 \text{ ft}) = 4,896 \text{ ft}^3$
    There are $4,896 \text{ ft}^3$ of air space in the room.

14. Surface area of the cube is 54 in^2.
    $S = 6e^2 \qquad V = e^3$
    $54 \text{ in}^2 = 6e^2 \qquad V = (3 \text{ in})^3$
    $9 \text{ in}^2 = e^2 \qquad V = 27 \text{ in}^3$
    $3 \text{ in} = e$
    The volume of the cube is 27 in^3.

15. $S = 2\pi r^2 + 2\pi rh = 2\pi(3 \text{ in})^2 + 2\pi(3 \text{ in})(6 \text{ in}) = 18\pi \text{ in}^2 + 36\pi \text{ in}^2 = 54\pi \text{ in}^2$
    The surface area of the can is $54\pi \text{ in}^2$.

16. Cylinder with circular base, volume is $490\pi \text{ ft}^3$.
    Diameter = 7 ft, so radius is 3.5 ft.
    $V = \pi r^2 h$
    $490\pi \text{ ft}^3 = \pi(3.5 \text{ ft})^2 h$
    $490 \text{ ft}^3 = 12.25 \text{ ft}^2 h$
    $40 \text{ ft} = h$
    The height of the cylinder is 40 ft.

17. $V = \frac{1}{3}Bh = \frac{1}{3}(25 \text{ in}^2)(6 \text{ in}) = 50 \text{ in}^3$
    The volume of the pyramid is 50 in^3.

18. $V = \dfrac{4}{3}\pi r^3 = 288\pi \text{ cm}^3$

$4\pi r^3 = 864\pi \text{ cm}^3$
$r^3 = 216 \text{ cm}^3$
$r = 6 \text{ cm}$
The radius of the sphere is 6 cm.

19. $S = 2lw + 2lh + 2wh = 2(17 \text{ cm})(19 \text{ cm}) + 2(17 \text{ cm})(20 \text{ cm}) + 2(19 \text{ cm})(20 \text{ cm}) = 646 \text{ cm}^2 + 680 \text{ cm}^2 + 760 \text{ cm}^2 = 2{,}086 \text{ cm}^2$
The surface area of the rectangular solid is 2,086 cm².

20. $S = 6e^2 = 6(14 \text{ mm})^2 = 6(196 \text{ mm}^2) = 1{,}176 \text{ mm}^2$
The surface area of the cube is 1,176 mm².

11.23 COORDINATE GEOMETRY

Coordinate geometry is the branch of geometry that deals with planes on a coordinate system. Each point in the plane has a unique **coordinate** (x, y). The horizontal number line is the **x axis**, and the vertical number line is the **y axis**. The zeros on each number line match up, and this point is called the **origin** for the coordinate system. The plane is divided into four parts, which are called **quadrants** and are numbered I, II, III, and IV in counterclockwise order from the upper right. See Figure 11.34.

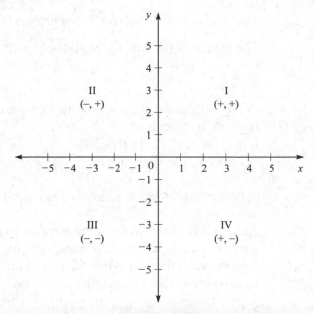

Figure 11.34

Points in Quadrant I have both coordinates positive, (pos., pos.). In Quadrant II, the x coordinate is negative, but the y coordinate is positive (neg., pos.). Points in Quadrant III have both coordinates negative (neg., neg.). Points in Quadrant IV have a positive x coordinate and a negative y coordinate (pos., neg.).

To locate a point, such as $(5, -3)$, in the plane, start at the origin and move 5 units in the positive x direction (right). Then move from that location 3 units in the negative y direction (down). The point $(-3, 5)$ is three units left from the origin on the x axis and then 5 units up on the y axis. See Figure 11.35.

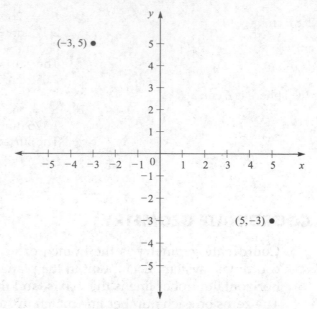

Figure 11.35

Let $P = (x_1, y_1)$ and $Q = (x_2, y_2)$. The **distance** from P to Q, denoted d, can be found by the distance formula: $d = \sqrt{(x_2 - x_1)^2 + (y_2 - y_1)^2}$

The **midpoint**, M, of the line segment PQ is: $M = \left(\dfrac{x_1 + x_2}{2}, \dfrac{y_1 + y_2}{2}\right)$

The **slope** of the line through the points P and Q is:

$m = \dfrac{y_2 - y_1}{x_2 - x_1}$, when $x_1 \neq x_2$

Let l_1 with slope m_1 and l_2 with slope m_2 be any two nonvertical lines.
If $m_1 = m_2$ then $l_1 \parallel l_2$.
If $m_1 \times m_2 = -1$, then $l_1 \perp l_2$.
A vertical line does not have slope and has the form $x = k$, where k is a real number.
A horizontal line has a slope of zero and has the form $y = h$, where h is a real number.
Any two horizontal lines are parallel to each other.
Any two vertical lines are parallel to each other.
Any vertical line is perpendicular to any horizontal line.
If a line crosses the x axis, the point at which it crosses is called the **x intercept**. If a line crosses the y axis, the point at which it crosses is called the **y intercept**. The x intercept is the value of x when $y = 0$ for the line. The y intercept is the value of y when $x = 0$.
Let the slope of a line be m and the y intercept of the line be b. The **slope-- intercept equation of a line** is: $y = mx + b$.

Example:

What is the slope of the line containing the points $(6, -4)$ and $(-8, -6)$?

Solution:

$m = \dfrac{y_2 - y_1}{x_2 - x_1} = \dfrac{-6 - (-4)}{-8 - 6} = \dfrac{-6 + 4}{-14} = \dfrac{-2}{-14} = \dfrac{1}{7}$. The slope of the line is $\dfrac{1}{7}$.

Example:

What is midpoint of the line segment with endpoints $(6, -4)$ and $(-8, -6)$?

Solution:

$$M = \left(\frac{x_1 + x_2}{2}, \frac{y_1 + y_2}{2} \right) = \left(\frac{6 - 8}{2}, \frac{-4 - 6}{2} \right) = \left(\frac{-2}{2}, \frac{-10}{2} \right) = (-1, -5).$$

The midpoint of the line segment is $(-1, -5)$.

Example:

What is the distance between the points $(6, -4)$ and $(-8, -6)$?

Solution:

$$d = \sqrt{(x_2 - x_1)^2 + (y_2 - y_1)^2} = \sqrt{(-8 - 6)^2 + (-6 - (-4))^2} =$$
$$\sqrt{(-14)^2 + (-2)^2} = \sqrt{196 + 4} = \sqrt{200} = \sqrt{100} \cdot \sqrt{2} = 10\sqrt{2}$$

The distance between the points is $10\sqrt{2}$.

Example:

If a line has a slope of -3 and a y-intercept of $(0, 4)$, what is the equation of the line?

Solution:

$m = -3$ and $b = 4$. $y = mx + b = -3x + 4$. The equation of the line is $y = -3x + 4$.

11.24 PRACTICE PROBLEMS

1. Name the quadrant or axis where each point is located.
   A. $(6, 1)$ B. $(-2, -4)$ C. $(-10, -2)$
   D. $(-8, 4)$ E. $(3, -6)$

2. Locate each point on a coordinate grid.
   A. $A = (3, 2)$ B. $B = (-2, -3)$ C. $C = (5, 0)$
   D. $D = (4, -2)$ E. $E = (0, -2)$ F. $F = (-3, 3)$
   G. $G = (-3, 0)$ H. $H = (-1, 4)$

3. Find the slope of the line through the given points.
   A. $(-2, -3)$ and $(-1, 5)$ B. $(8, 1)$ and $(2, 6)$
   C. $(2, 4)$ and $(-4, 4)$ D. $(-3, 4)$ and $(-4, 3)$
   E. $(3, -13)$ and $(-6, -5)$

4. Write the equation of the line with the given slope m and given y-intercept b.
   A. $m = 1.5, b = -4$ B. $m = -4, b = 1.8$
   C. $m = 3, b = 0$ D. $m = -1.6, b = 4$

5. Write the equation of the line with the given slope m and intersection through the given point.

   A. $m = 0.5; (-3, 2)$ B. $m = -2; (0, 1)$
   C. $m = 1.25; (-2, -1)$ D. $m = -1.5; (2, 4)$

6. Determine if L_1 and L_2 are parallel, perpendicular, or neither.
   A. L_1 goes through $(4, 6)$ and $(-8, 7)$; L_2 goes through $(7, 4)$ and $(-5, 5)$
   B. L_1 goes through $(9, 15)$ and $(-7, 12)$; L_2 goes through $(-4, 8)$ and $(-20, 5)$
   C. L_1 goes through $(2, 0)$ and $(5, 4)$; L_2 goes through $(6, 1)$ and $(2, 4)$
   D. L_1 goes through $(0, -7)$ and $(2, 3)$; L_2 goes through $(0, -3)$ and $(1, -2)$

7. Find the distance between the given points.
   A. $(3, 4)$ and $(-2, 1)$ B. $(-2, 1)$ and $(3, 2)$
   C. $(-2, 4)$ and $(3, -2)$

8. Find the midpoint of the line segment between the points.
   A. $(7, -3)$ and $(-4, 2)$
   B. $(0, -5)$ and $(4, -12)$
   C. $(5, -1)$ and $(-3, -7)$

■■■ 11.25 SOLUTIONS

1.
A. I   B. III   C. III
D. II   E. IV

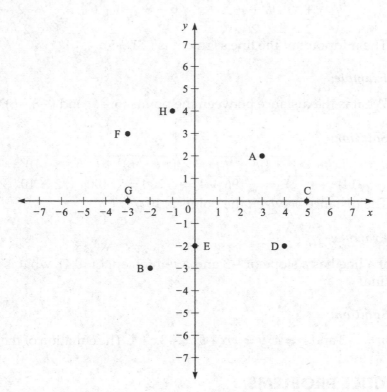

Figure 11.36

2. Notes: $A = (3, 2)$ $B = (-2, -3)$ $C = (5, 0)$
   $D = (4, -2)$ $E = (0, -2)$ $F = (-3, 3)$
   $G = (-3, 0)$ $H = (-1, 4)$

3. A. $m = \dfrac{y_2 - y_1}{x_2 - x_1} = \dfrac{5 + 3}{-1 + 2} = \dfrac{8}{1} = 8$

   B. $m = \dfrac{y_2 - y_1}{x_2 - x_1} = \dfrac{6 - 1}{2 - 8} = \dfrac{5}{-6} = -\dfrac{5}{6}$

   C. $m = \dfrac{y_2 - y_1}{x_2 - x_1} = \dfrac{4 - 4}{-4 - 2} = \dfrac{0}{-6} = 0$

   D. $m = \dfrac{y_2 - y_1}{x_2 - x_1} = \dfrac{3 - 4}{-4 + 3} = \dfrac{-1}{-1} = 1$

   E. $m = \dfrac{y_2 - y_1}{x_2 - x_1} = \dfrac{-5 + 13}{-6 - 3} = \dfrac{8}{-9} = -\dfrac{8}{9}$

4. $y = mx + b$

   A. $y = 1.5x - 4$
   B. $y = -4x + 1.8$
   C. $y = 3x$
   D. $y = -1.6x + 4$

5. A. $y = mx + b, (x, y) = (-3, 2), m = 0.5$
   $2 = 0.5(-3) + b$
   $2 = -1.5 + b$
   $3.5 = b$
   $y = 0.5x + 3.5$

   B. $y = mx + b, (x, y) = (0, 1), m = -2$
   $1 = -2(0) + b$
   $1 = b$
   $y = -2x + 1$

   C. $y = mx + b, (x, y) = (-2, -1), m = 1.25$
   $-1 = 1.25(-2) + b$
   $-1 = -2.5 + b$
   $1.5 = b$
   $y = 1.25x + 1.5$

   D. $y = mx + b, (x, y) = (2, 4), m = -1.5$
   $4 = -1.5(2) + b$
   $4 = -3 + b$
   $7 = b$
   $y = -1.5x + 7$

6. A. L_1 has slope $m = \dfrac{y_2 - y_1}{x_2 - x_1} = \dfrac{7-6}{-8-4} = \dfrac{1}{-12}$

so $m_1 = -\dfrac{1}{12}$

L_2 has slope $m = \dfrac{y_2 - y_1}{x_2 - x_1} = \dfrac{5-4}{-5-7} = \dfrac{1}{-12}$

so $m_2 = -\dfrac{1}{12}$

Since $m_1 = m_2$, $L_1 \parallel L_2$.

B. $L_1 : m_1 = \dfrac{12-15}{-7-9} = \dfrac{-3}{-16} = \dfrac{3}{16}$

$L_2 : m_2 = \dfrac{5-8}{4-20} = \dfrac{-3}{-16} = \dfrac{3}{16}$

Since $m_1 = m_2$, $L_1 \parallel L_2$.

C. $L_1 : m_1 = \dfrac{4-0}{5-2} = \dfrac{4}{3}$,

$L_2 : m_2 = \dfrac{4-1}{2-6} = \dfrac{3}{-4} = -\dfrac{3}{4}$

$m_1 \times m_2 = \dfrac{4}{3} \times -\dfrac{3}{4} = -1$

Since $m_1 \times m_2 = -1$, $L_1 \perp L_2$.

D. $L_1 : m_1 = \dfrac{3+7}{2-0} = \dfrac{10}{2} = 5$,

$L_2 : m_2 = \dfrac{-2+3}{1-0} = \dfrac{1}{1} = 1$

$m_1 \neq m_2$, and $m_1 \times m_2 = 5 \times 1 = 5 \neq -1$
Since $m_1 \neq m_2$, and $m_1 \times m_2 \neq -1$, L_1 and L_2 are neither parallel nor perpendicular.

7. A. $d = \sqrt{(x_2 - x_1)^2 + (y_2 - y_1)^2}$

$= \sqrt{(-2-3)^2 + (1-4)^2} = \sqrt{(-5)^2 + (-3)^2}$

$= \sqrt{25+9} = \sqrt{34}$

$d = \sqrt{34}$

B. $d = \sqrt{(x_2 - x_1)^2 + (y_2 - y_1)^2}$

$= \sqrt{(3+2)^2 + (2-1)^2}$

$= \sqrt{(5)^2 + (1)^2} = \sqrt{25+1} = \sqrt{26}$

$d = \sqrt{26}$

C. $d = \sqrt{(x_2 - x_1)^2 + (y_2 - y_1)^2}$

$= \sqrt{(3+2)^2 + (-2-4)^2}$

$= \sqrt{(5)^2 + (-6)^2} = \sqrt{25+36} = \sqrt{61}$

$d = \sqrt{61}$

8. A. $M = \left(\dfrac{x_1 + x_2}{2}, \dfrac{y_1 + y_2}{2} \right)$

$= \left(\dfrac{7-4}{2}, \dfrac{-3+2}{2} \right) = \left(\dfrac{3}{2}, -\dfrac{1}{2} \right)$

B. $M = \left(\dfrac{x_1 + x_2}{2}, \dfrac{y_1 + y_2}{2} \right)$

$= \left(\dfrac{0+4}{2}, \dfrac{-5-12}{2} \right)$

$= \left(\dfrac{4}{2}, \dfrac{-17}{2} \right) = \left(2, -\dfrac{17}{2} \right)$

C. $M = \left(\dfrac{x_1 + x_2}{2}, \dfrac{y_1 + y_2}{2} \right)$

$= \left(\dfrac{5-3}{2}, \dfrac{-1-7}{2} \right) = \left(\dfrac{2}{2}, \dfrac{-8}{2} \right) = (1, -4)$

11.26 GEOMETRY TEST

Use the following test to assess how well you have mastered the material in this chapter. Mark your answers by blackening the corresponding answer oval in each question. An answer key and solutions are provided at the end of the test.

1. Which is an angle with a measure of 180°?

   (A) Acute
   (B) Obtuse
   (C) Reflex
   (D) Right
   (E) Straight

2. Which is the supplement of a 72° angle?

   (A) 18°
   (B) 36°
   (C) 72°
   (D) 108°
   (E) 288°

3. What is the name for the opposite angles formed when two lines intersect?

   (A) Complementary
   (B) Supplementary
   (C) Vertical
   (D) Interior
   (E) Exterior

4. If two lines are cut by a transversal, which angles will always be equal?

   (A) Adjacent angles
   (B) Alternate exterior angles
   (C) Alternate interior angles
   (D) Corresponding angles
   (E) Vertical angles

5. If an angle is five times its complement, what is the measure of the angle?

   (A) 15°
   (B) 18°
   (C) 36°
   (D) 75°
   (E) 144°

6. When two parallel lines are cut by a transversal, which pair of angles will always be supplementary?

   (A) Alternate interior angles
   (B) Consecutive interior angles
   (C) Corresponding angles
   (D) Alternate exterior angles
   (E) Vertical angles

7. Which is a quadrilateral that is equilateral but not always equiangular?

   (A) Trapezoid
   (B) Parallelogram
   (C) Rectangle
   (D) Square
   (E) Rhombus

8. Which polygon always has diagonals of the same length?

   (A) Pentagon
   (B) Rectangle
   (C) Parallelogram
   (D) Rhombus
   (E) Hexagon

9. Which is the fewest number of sides a polygon can have?

   (A) 0
   (B) 1
   (C) 2
   (D) 3
   (E) 4

10. In which figure do the diagonals NOT bisect each other?

   (A) Trapezoid
   (B) Parallelogram
   (C) Square
   (D) Rhombus
   (E) Rectangle

11. Which is a part of a line with one endpoint?

   (A) Line
   (B) Ray
   (C) Line segment
   (D) Angle
   (E) Point

12. Which is NOT a requirement for adjacent angles?

   (A) Two angles
   (B) Common vertex
   (C) Sum of angles is 180°
   (D) No interior points in common
   (E) Common side

13. Which are the angles of a right triangle?

   (A) 11°, 79°, 90°
   (B) 33°, 67°, 90°
   (C) 90°, 90°, 90°
   (D) 10°, 20°, 60°, 90°
   (E) 90°, 20°, 70°, 90°

14. Which lengths could be the sides of a triangle?

    (A) 4, 9, 16
    (B) 3, 4, 7
    (C) 4, 6, 11
    (D) 5, 7, 9
    (E) 5, 5, 10

15. In right $\triangle ABC$, C is the right angle. If side $a = 5$ and side $c = 13$, what is the length of side b?

    (A) 5
    (B) 12
    (C) 13
    (D) 25
    (E) 144

16. In a $30° - 60° - 90°$ triangle, if the hypotenuse is 10, what is the length of the side opposite the 60° angle?

    (A) $10\sqrt{3}$
    (B) $5\sqrt{3}$
    (C) $20\sqrt{3}$
    (D) 5
    (E) 20

17. Which term is used to indicate the perimeter of a circle?

    (A) Diameter
    (B) Radius
    (C) Secant
    (D) Chord
    (E) Circumference

18. Which is a line that has exactly one point in common with a circle?

    (A) Radius
    (B) Chord
    (C) Diameter
    (D) Tangent
    (E) Secant

19. Two chords intersect inside a circle. The lengths of the two parts of one chord are 9 and 16. The lengths of the two parts of the second chord are 6 and x. What is the value of x? (See Figure 11.26.)

    (A) 24
    (B) 19
    (C) 13
    (D) 8
    (E) 2

20. A room is 20 ft by 30 ft by 8 ft. What is the surface area of the room?

    (A) 4,800 ft²
    (B) 800 ft²
    (C) 1,000 ft²
    (D) 1,520 ft²
    (E) 2,000 ft²

21. If the altitude of a cylinder is 12 cm and the radius of the base is 9 cm, what is the volume of the cylinder?

    (A) 108π cm³
    (B) 216π cm³
    (C) 378π cm³
    (D) 972π cm³
    (E) $1,296\pi$ cm³

22. If the area of the base of a pyramid is 24 cm² and the altitude of the pyramid is 18 cm, what is the volume of the pyramid?

    (A) 144 cm³
    (B) 432 cm³
    (C) 1,296 cm³
    (D) 1,764 cm³
    (E) 20,736 cm³

23. If the radius of the base of a cone is 15 m and the slant height of the cone is 6 m, what is the surface area of the cone?

    (A) 225π m²
    (B) 90π m²
    (C) 630π m²
    (D) 126π m²
    (E) 315π m²

24. What is the surface area of a sphere whose radius is 14 in?

    (A) 56π in²
    (B) 196π in²
    (C) 784π in²
    (D) $2,744\pi$ in²
    (E) $10,976\pi$ in²

25. A cylinder has a height of 12 ft and the radius of the base is 9 ft. What is the surface area of the cylinder?

    (A) 189π ft²
    (B) 252π ft²
    (C) 378π ft²
    (D) 504π ft²
    (E) 540π ft²

GEOMETRY TEST
Answer Key

1. E	6. B	11. B	16. B	21. D
2. D	7. E	12. C	17. E	22. A
3. C	8. B	13. A	18. D	23. E
4. E	9. D	14. D	19. A	24. C
5. D	10. A	15. B	20. E	25. C

▬▬ 11.27 SOLUTIONS

1. **E** Straight
   straight $= 180°$

2. **D** 108°
   $180° - 72° = 108°$

3. **C** Vertical

4. **E** Vertical angles
   You are not told that the lines are parallel, so only vertical angles must always be equal.

5. **D** 75°
   $n + 5n = 90, 6n = 90, n = 15, 5n = 75$. The angle is 75°.

6. **B** Consecutive interior angles
   The adjacent interior angles are always supplementary. The interior angles on the same side are transversal supplementary when the lines are parallel. So, when the lines are parallel, consecutive interior angles are supplementary.

7. **E** Rhombus
   A square is both equilateral and equiangular. A rectangle is always equiangular, but only squares are equilateral. A rhombus is always equilateral, but only squares are equiangular.

8. **B** Rectangle
   The diagonals of a rectangle are always equal.

9. **D** 3
   A triangle has the fewest sides of any polygon, 3.

10. **A** Trapezoid
    The diagonals of a parallelogram always bisect each other. The trapezoid is the only non-parallelogram listed.

11. **B** Ray
A ray is a part of a line with one endpoint.

12. **C** Sum of angles is 180°
Adjacent angles are two angles in the same plane that have a common vertex and a common side, and the common side separates the angles.

13. **A** 11°, 79°, 90°
Every right triangle has a right angle, 90°, so the other two angles must have a sum of 90° or $11° + 79° = 90°$. Note: D and E are not triangles.

14. **D** 5, 7, 9
In a triangle, any two sides must exceed the third side. Since $5 + 7 = 12 > 9$, and 9 is the longest side, these can be the sides of a triangle.

15. **B** 12
$a^2 + b^2 = c^2$ and $a = 5$ and $c = 13$
$b^2 = c^2 - a^2 = 13^2 - 52 = 169 - 25 = 144$
$b^2 = 144$
$b = \sqrt{144} = 12$

16. **B** $5\sqrt{3}$
In a $30° - 60° - 90°$ triangle, the sides are a, $a\sqrt{3}$, and $2a$. Side a is opposite the 30° angle and $2a$ is the hypotenuse. $2a = 10$, $a = 5$, $a\sqrt{3} = 5\sqrt{3}$.

17. **E** Circumference

18. **D** Tangent
A tangent is a line that intersects the circle in exactly one point.

19. **A** 24
When two chords intersect inside a circle, the product of the segments, or parts, of each chord is the same.
So, $9(16) = 6x$ $6x = 144$ and $x = 24$

20. **E** 2,000 ft²
A room is a rectangular solid, so $S = 2lw + 2lh + 2wh$.
$S = 2(20 \text{ ft})(30 \text{ ft}) + 2(20 \text{ ft})(8 \text{ ft}) + 2(30 \text{ ft})(8 \text{ ft})$
$= 1,200 \text{ ft}^2 + 320 \text{ ft}^2 + 480 \text{ ft}^2$
$S = 2,000 \text{ ft}^2$

21. **D** 972π cm³
$V = \pi r^2 h$
$V = (\pi)(9 \text{ cm})^2(12 \text{ cm}) = (\pi)(81 \text{ cm}^2)(12 \text{ cm}) = 972\pi \text{ cm}^3$

22. **A** 144 cm³
$V = \frac{1}{3}Bh = \frac{1}{3}(24 \text{ cm}^2)(18 \text{ cm}) = 8 \text{ cm}^2(18 \text{ cm}) = 144 \text{ cm}^3$

23. **E** 315π m²
$S = \pi r^2 + \pi rs = \pi(15 \text{ m})^2 + \pi(15 \text{ m})(6 \text{ m}) = 225\pi \text{ m}^2 + 90\pi \text{ m}^2$
$S = 315\pi \text{ m}^2$

24. **C** 784π in^2
$S = 4\pi r^2 = 4\pi(14 \text{ in})^2 = 4\pi(196 \text{ in}^2) = 784\pi \text{ in}^2$

25. **C** 378π ft^2
$S = 2\pi r^2 + 2\pi rh = 2\pi(r^2 + rh) = 2\pi[(9 \text{ ft})^2 + (9 \text{ ft})(12 \text{ ft})] = 2\pi(81 \text{ ft}^2 + 108 \text{ ft}^2) = 2\pi(189 \text{ ft}^2) = 378\pi \text{ ft}^2$

11.28 SOLVED GRE PROBLEMS

For each question, select the best answer unless otherwise instructed.

Quantity A	Quantity B

1. Number of sides on a hexagon   Number of sides on a pentagon

   (A) Quantity A is greater.
   (B) Quantity B is greater.
   (C) The two quantities are equal.
   (D) The relationship cannot be determined from the given information.

2. **Which polygon has exactly eight sides?**

   (A) Hexagon
   (B) Quadrilateral
   (C) Octagon
   (D) Pentagon
   (E) Triangle

For this question, enter your answer in the box.

3. **If the sides of a right triangle are 60, 91, and 109, what is the area of the right triangle?**

4. **In $\triangle$ ABC, $\angle A > \angle B > \angle C$, AB = 31 cm, and AC = 50 cm. Which could be the length of BC?**

   (A) 25 cm
   (B) 30 cm
   (C) 40 cm
   (D) 45 cm
   (E) 80 cm

Quantity A	Quantity B
5. The measure of a right angle	The measure of the supplement of a 31° angle

(A) Quantity A is greater.
(B) Quantity B is greater.
(C) The two quantities are equal.
(D) The relationship cannot be determined from the given information.

11.29 SOLUTIONS

1. **A** Quantity A is greater.
   A hexagon has six sides and a pentagon has five sides.
2. **C** Octagon
   An octagon is an eight-sided polygon.
3. 2,730
   In a right triangle, A = 0.5 ab where a and b are the legs of the right triangle. Because the hypotenuse is the longest side in a right triangle, the legs have lengths of 60 and 91. A = 0.5(60)(91) = 2,730.
4. **E** 80 cm
   Because $\angle A > \angle B > \angle C$, BC > AC > AB, BC > AC > 50. Also, BC < AB + AC = 31 + 50 = 81. Thus, 81 > BC > 50, and 80 is the only choice in the interval.
5. **B** Quantity B is greater.
   The measure of a right angle is 90°. The supplement of a 31° angle is 180° − 31° = 149°.

11.30 GRE PRACTICE PROBLEMS

For each question, select the best answer unless otherwise instructed.

1. **Which is NOT a property of parallelogram?**

   (A) The opposite sides are parallel.
   (B) The opposite sides are equal.
   (C) The consecutive angles are supplementary.
   (D) The opposite angles are equal.
   (E) The diagonals bisect the angles.

2. **Which is NOT a type of triangle?**

   (A) Scalene
   (B) Equiangular
   (C) Isosceles
   (D) Reflex
   (E) Regular

The perimeter of quadrilateral Q is 64 cm.

Quantity A	Quantity B

3. The area of Q if Q is a rectangle.   The area of Q if Q is a square.

Ⓐ Quantity A is greater.
Ⓑ Quantity B is greater.
Ⓒ The two quantities are equal.
Ⓓ The relationship cannot be determined from the given information.

For this question, enter your answer in the box.

4. **What is the smallest number of acute angles a triangle can have?**

5. **Which set of lengths could be the sides of a right triangle?**

Ⓐ 8, 15, 17
Ⓑ 4, 5, 6
Ⓒ 9, 12, 14
Ⓓ 1, 1, 2
Ⓔ $2, 2\sqrt{2}, 4$

Quantity A	Quantity B

6. Circumference of circle O   Radius of circle O

Ⓐ Quantity A is greater.
Ⓑ Quantity B is greater.
Ⓒ The two quantities are equal.
Ⓓ The relationship cannot be determined from the given information.

Quantity A	Quantity B

7. The volume of cube T   The surface area of cube T

Ⓐ Quantity A is greater.
Ⓑ Quantity B is greater.
Ⓒ The two quantities are equal.
Ⓓ The relationship cannot be determined from the given information.

8. **In a $45° - 45° - 90°$ triangle, if the hypotenuse is $4\sqrt{6}$, what is the length of a leg?**

Ⓐ $2\sqrt{6}$
Ⓑ $4\sqrt{3}$
Ⓒ $4\sqrt{2}$
Ⓓ $8\sqrt{6}$
Ⓔ $8\sqrt{3}$

T is a right circular cylinder with radius of the base x cm and height y cm. R is a right circular cone with radius of the base x cm and height y cm.

Quantity A	**Quantity B**
9. Volume of T	Volume of R

Ⓐ Quantity A is greater.
Ⓑ Quantity B is greater.
Ⓒ The two quantities are equal.
Ⓓ The relationship cannot be determined from the given information.

10. **If the sides of a polygon are 2.5 cm, 3.8 cm, 11 cm, 4.9 cm, and 5.28 cm, what is the perimeter of the polygon?**

Ⓐ 1.758 cm
Ⓑ 27.48 cm
Ⓒ 65.1 cm
Ⓓ 274.8 cm
Ⓔ 651 cm

ANSWER KEY

1.	E	6.	A
2.	D	7.	D
3.	D	8.	B
4.	2	9.	A
5.	A	10.	B

PART IV

GRE MATH PRACTICE SECTIONS

The questions on the following practice sections are designed to be just like questions that have appeared on the mathematics sections of the GRE. These questions were written specifically for these practice sections and are not endorsed by the GRE. Each practice section has a balance of questions over the content areas of arithmetic, algebra, and geometry, along with information presented in graphs.

GRE MATH PRACTICE SECTIONS

Like the actual GRE Quantitative sections, each GRE math practice section has 20 questions and a 35-minute time limit. The questions are divided into the following categories:

    8 quantitative comparison questions
    12 general problem-solving questions, including graphs and table
      questions

Each practice section will be an accurate refection of how well you'll do on test day if you treat it like the actual examination. Here is how to take each section under conditions similar to the actual exam:

➤ Find a place where you can work comfortably and without interruption.
➤ Complete the section in one sitting.
➤ Tear out the answer sheet and mark your answers by blackening the corresponding answer oval. (Note that on the real test, you will mark your answer by clicking on an answer choice oval on the computer screen.)
➤ Time yourself and observe the given time limits. Note how many questions remain, if any, when time runs out.
➤ Become familiar with the directions to the test and the reference information provided to test-takers. You will save time on the actual test day if you are already familiar with this information.

Once you have completed a practice section, check your answers against the answer key provided. Then review the solutions to each problem, paying particular attention to the problems you missed. For those problems, you may want to go back and reread the corresponding topic review section in this book.

Note: On the real GRE you will enter your answers by clicking on answer ovals on the computer screen.

(1)   Ⓐ   Ⓑ   Ⓒ   Ⓓ   Ⓔ
(2)   Ⓐ   Ⓑ   Ⓒ   Ⓓ   Ⓔ
(3)   Ⓐ   Ⓑ   Ⓒ   Ⓓ   Ⓔ
(4)   Ⓐ   Ⓑ   Ⓒ   Ⓓ   Ⓔ
(5)   Ⓐ   Ⓑ   Ⓒ   Ⓓ   Ⓔ

(6)   Ⓐ   Ⓑ   Ⓒ   Ⓓ   Ⓔ
(7)   Ⓐ   Ⓑ   Ⓒ   Ⓓ   Ⓔ
(8)   Ⓐ   Ⓑ   Ⓒ   Ⓓ   Ⓔ
(9)   Ⓐ   Ⓑ   Ⓒ   Ⓓ   Ⓔ
(10)   Ⓐ   Ⓑ   Ⓒ   Ⓓ   Ⓔ

(11)   Ⓐ   Ⓑ   Ⓒ   Ⓓ   Ⓔ
(12)   Ⓐ   Ⓡ   Ⓒ   Ⓓ   Ⓔ
(13)   Ⓐ   Ⓑ   Ⓒ   Ⓓ   Ⓔ

(14)   [   ]
(15)   Ⓐ   Ⓑ   Ⓒ   Ⓓ   Ⓔ

(16)   Ⓐ   Ⓑ   Ⓒ   Ⓓ   Ⓔ
(17)   Ⓐ   Ⓑ   Ⓒ   Ⓓ   Ⓔ
(18)   Ⓐ   Ⓑ   Ⓒ   Ⓓ   Ⓔ
(19)   Ⓐ   Ⓑ   Ⓒ   Ⓓ   Ⓔ
(20)   Ⓐ   Ⓑ   Ⓒ   Ⓓ   Ⓔ

GRE MATH PRACTICE SECTION 1

20 Questions

Time—35 Minutes

➤ The numbers on this test are real numbers.

➤ You may assume that positions of points, lines, and angles are in the order shown.

➤ You can NOT assume measures of line segments or angles unless the figure is drawn to scale.

➤ Figures are in a plane unless stated otherwise.

➤ You may assume that lines that appear to be straight lines are straight lines.

➤ Lines are NOT parallel or perpendicular unless you are told they are or you can deduce it from information given.

DIRECTIONS: Quantitative comparison questions provide two quantities, Quantity A and Quantity B. You are to compare the two quantities. Then choose

A if Quantity A is greater
B if Quantity B is greater
C if the two quantities are equal
D if the relationship can not be determined from the given information

NOTE: There is no answer E for these questions, so marking E is always wrong.

Common

Information: In a question, information concerning one or both of the quantities to be compared is centered above the two columns. A symbol that appears in both columns represents the same thing for both Quantity A and Quantity B.

	Quantity A	Quantity B	Sample Answers
Example 1:	$3 + 6$	$3 \div 2$	● Ⓑ Ⓒ Ⓓ
Example 2:	n^2	$2n$	Ⓐ Ⓑ Ⓒ ●

(If $n = 3$, $n^2 = 9$, $2n = 6$ and $n^2 > 2n$; if $n = 0.5$, $n^2 = 0.25$, $2n = 1$, and $n^2 < 2n$.)

GO ON TO THE NEXT PAGE.

	Quantity A	Quantity B	Sample Answers
Example 3:	3×2	2×3	Ⓐ Ⓑ ● Ⓓ
Example 4:	$5 - 3$	$5 - (-3)$	Ⓐ ● Ⓒ Ⓓ

	Quantity A	Quantity B
1.	25^2	50

The ratio of two adult brothers' ages is 5 to 8.

	Quantity A	Quantity B
2.	The ratio of the brothers' ages 8 years ago.	The ratio of the brothers' ages in 6 years.

James bought 16 ounces of dried fruit P for 89¢ and 24 ounces of dried fruit Q for $1.29.

	Quantity A	Quantity B
3.	per ounce cost of dried fruit P	per ounce cost of dried fruit Q

	Quantity A	Quantity B
4.	$(3x - 7y)^2$	$9x^2 - 42xy + 49y^2$

X is a square with sides of length s.

	Quantity A	Quantity B
5.	units in the perimeter of X	units in the area of X

Salary S reduced by 10% = salary T

Salary T increased by 10% = salary W.

	Quantity A	Quantity B
6.	Salary S	Salary W

parallelogram ABCD

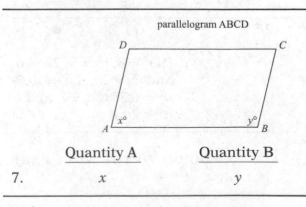

	Quantity A	Quantity B
7.	x	y

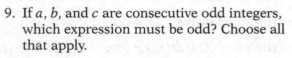

	Quantity A	Quantity B
	$0 < t < 1$	
8.	$1 - t$	$\dfrac{1}{2} + t$

9. If a, b, and c are consecutive odd integers, which expression must be odd? Choose all that apply.

Ⓐ $a(b + c)$

Ⓑ $(a + b) - c$

Ⓒ abc

Ⓓ $2(c - a) + b$

Ⓔ $a + b + c$

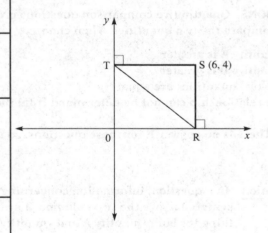

Figure 1

10. What is the area of $\triangle RST$? (Use Figure 1.)

Ⓐ 6

Ⓑ $2\sqrt{13}$

Ⓒ 10

Ⓓ 12

Ⓔ 24

GO ON TO THE NEXT PAGE.

11. What is the value of $\left(3 + \dfrac{1}{2}\right)\left(2 - \dfrac{1}{4}\right)$?

(A) $-\dfrac{3}{4}$

(B) $5\dfrac{7}{8}$

(C) $6\dfrac{1}{8}$

(D) $6\dfrac{1}{4}$

(E) $7\dfrac{7}{8}$

Questions 12–14 refer to the following graph.

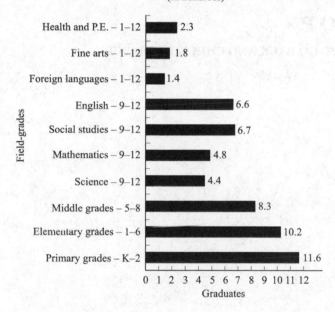

Number of Graduates from State University
with Degrees in Education, 1980–2005
(in hundreds)

Field-grades

Field-grades	Graduates
Health and P.E. – 1–12	2.3
Fine arts – 1–12	1.8
Foreign languages – 1–12	1.4
English – 9–12	6.6
Social studies – 9–12	6.7
Mathematics – 9–12	4.8
Science – 9–12	4.4
Middle grades – 5–8	8.3
Elementary grades – 1–6	10.2
Primary grades – K–2	11.6

0 1 2 3 4 5 6 7 8 9 10 11 12
Graduates

12. Approximately how many education degrees in grades 1–12 fields were earned at State University in 1980–2005?

(A) 5.5
(B) 41
(C) 55
(D) 410
(E) 550

13. What percent of the 5,810 graduates had degrees in primary grades?

(A) 2%
(B) 12%
(C) 18%
(D) 20%
(E) 37%

For this question, enter your answer in the box.

14. How many graduates earned degrees in mathematics?

15. If $\dfrac{x}{2} = 5$ and $\dfrac{3}{y} = 6$, what is the value of $\dfrac{2 - y}{x + 4}$?

(A) $-\dfrac{4}{9}$

(B) 0

(C) $\dfrac{1}{14}$

(D) $\dfrac{3}{28}$

(E) $\dfrac{5}{28}$

16. If a right triangle has legs of 20 cm and 21 cm, what is the length of the hypotenuse?

(A) $\sqrt{41}$ cm
(B) 20.5 cm
(C) 29 cm
(D) 41 cm
(E) 45 cm

17. What is the average (arithmetic mean) of the values 41, 42, 41, 47, 44, 37, and 49?

(A) 41
(B) 43
(C) 44
(D) 47
(E) 49

18. If the sale price of an air conditioner is $306 after a 10% reduction was made, what was the regular price of the air conditioner?

(A) $275.40
(B) $316.00
(C) $336.60
(D) $340.00
(E) $374.00

GO ON TO THE NEXT PAGE.

19. Last week, Jane worked 40 hours of regular time and 8 hours of overtime. If she earns $11.50 per hour regularly and gets $1\frac{1}{2}$ times that for overtime, how much did Jane earn last week?

    (A) $391
    (B) $414
    (C) $460
    (D) $575
    (E) $598

20. Laura's bag contains 14 red marbles and 16 green marbles. If Laura selects one marble at random from the bag, what is the probability that the marble she selects will be green?

    (A) $\dfrac{1}{30}$

    (B) $\dfrac{7}{15}$

    (C) $\dfrac{8}{15}$

    (D) $\dfrac{7}{8}$

    (E) $\dfrac{8}{7}$

STOP

IF YOU FINISH BEFORE TIME RUNS OUT, GO BACK AND CHECK YOUR WORK.

GRE MATH PRACTICE SECTION 1

Answer Key

1. A
2. B
3. A
4. C
5. D
6. A
7. D
8. D
9. B,C,D,E
10. D
11. C
12. E
13. D
14. 480
15. D
16. C
17. B
18. D
19. E
20. C

GRE MATH PRACTICE SECTION 1

SOLUTIONS

1. **A** Quantity A is greater.
   $A = 25^2 = 625$, $B = 50$, $625 > 50$, $A > B$

2. **B** Quantity B is greater.
   Currently the ratio is 5 to 8 and both brothers are adults, so use the ages of $5 \times 4 = 20$ and $4 \times 8 = 32$ as an example. Eight years ago, the ages would have been 12 and 24 for a ratio of $\dfrac{1}{2}$.
   In 6 years, the ages will be 26 and 38 for a ratio of $\dfrac{13}{19}$.
   $$\dfrac{13}{19} > \dfrac{1}{2}$$
   The answer is B.

3. **A** Quantity A is greater.
   P is $\$0.89 \div 16 = \$0.0556$ per ounce
   Q is $\$1.29 \div 24 = \$0.0538$ per ounce
   Since $P > Q$, the answer is A.

4. **C** The two quantities are equal.
   $(3x - 7y)^2 = 9x^2 - 42xy - y^2$
   The answer is C.

5. **D** The relationship can not be determined from the given information.
   perimeter of X $= 4s$ area of X $= s^2$
   $s = 3$, $P = 12$, $A = 9$ $s = 10$, $P = 40$, $A = 100$
   Since the relationship depends on s, the answer is D.

6. **A** Quantity A is greater.
   S is the original salary. Salary $T = 90\%$ of Salary $S = 0.9S$.
   Salary W = Salary T + 10% of Salary T = $0.9S + 0.1(0.9S) = 0.99S$
   Since $S > T$, the answer is A.

7. **D** The relationship cannot be determined from the given information.
   You cannot tell which is greater from the figure. In a parallelogram (given), you know that $x° + y° = 180°$. $45° + 135° = 180°$, so x could be less than y. But $120° + 60° = 180°$, so x could also be greater than y. The answer is D.

8. **D** The relationship cannot be determined from the given information.
   $0 < t < 1$
   If $t = 0.1$, then $1 - t = 1 - 0.1 = 0.9$ and $0.5 + t = 0.5 + 0.1 = 0.6$. $1 - t > 0.5 + t$
   If $t = 0.5$, then $1 - t = 1 - 0.5 = 0.5$ and $0.5 + t = 0.5 + 0.5 = 1$. $1 - t < 0.5 + t$
   The answer is D.

9. **B, C, D, E** Only choice.
   A. $a(b + c)$ is even
   $b + c = \text{odd} + \text{odd} = \text{even}$
   $a(b + c) = \text{odd} \times \text{even} = \text{even}$

10. **D** 12
    Since two sides of the triangle are perpendicular to the axes and go through the point (6, 4), the angle at S is a right angle and $TS = 6$ and $RS = 4$.
    Area of $\triangle RST = 0.5(6)(4) = 12$.

11. **C** $6\dfrac{1}{8}$
    $$\left(3 + \frac{1}{2}\right)\left(2 - \frac{1}{4}\right) = (3.5)(1.75) = 6.125 = 6\frac{1}{8}$$

12. **E** 550
    Foreign languages $= 1.4 \times 100 = 140$
    Fine arts $= 1.8 \times 100 = 180$
    Health and PE $= 2.3 \times 100 = 230$
    The total is 550.

13. **D** 20%
    Primary grades: $11.6 \times 100 = 1,160$;
    $1,160 \div 5,810 = 0.1990$ or 20%

14. $\boxed{480}$

    From the graph, there are $4.8(100) = 480$
    mathematics graduates.

15. **D** $\dfrac{3}{28}$

    Since $\dfrac{x}{2} = 5$, $x = 10$ and since $\dfrac{3}{y} = 6$,

    $y = \dfrac{1}{2}$. Then, $\dfrac{2-y}{x+4} = \dfrac{2 - \dfrac{1}{2}}{10+4} = \dfrac{1.5}{14} = \dfrac{3}{28}$.

16. **C** 29 cm
    In a right triangle, $a^2 + b^2 = c^2$, so
    $20^2 + 21^2 = 400 + 441 = 841 = 29^2$. The
    hypotenuse is 29 cm.

17. **B** 43
    $\text{AVE} = \dfrac{SUM}{SUM}$
    $= \dfrac{41 + 42 + 41 + 47 + 44 + 37 + 49}{7}$
    $= \dfrac{301}{7} = 43$.

18. **D** $340.00
    $(1.00 - 0.10)P = \$306$, $0.9P = \$306$,
    $P = \$340$.

19. **E** $598
    $\text{Pay} = 40(\$11.50) + 8[1.5(\$11.50)]$
    $= \$460 + \$138 = \$598$.

20. **C** $\dfrac{8}{15}$

    $P(green) = \dfrac{16}{14 + 16} = \dfrac{16}{30} = \dfrac{8}{15}$.

Note: On the real GRE you will enter your answers by clicking
on answer ovals on the computer screen.

(1)   Ⓐ   Ⓑ   Ⓒ   Ⓓ   Ⓔ
(2)   Ⓐ   Ⓑ   Ⓒ   Ⓓ   Ⓔ
(3)   Ⓐ   Ⓑ   Ⓒ   Ⓓ   Ⓔ
(4)   Ⓐ   Ⓑ   Ⓒ   Ⓓ   Ⓔ
(5)   Ⓐ   Ⓑ   Ⓒ   Ⓓ   Ⓔ

(6)   Ⓐ   Ⓑ   Ⓒ   Ⓓ   Ⓔ
(7)   Ⓐ   Ⓑ   Ⓒ   Ⓓ   Ⓔ
(8)   Ⓐ   Ⓑ   Ⓒ   Ⓓ   Ⓔ
(9)   Ⓐ   Ⓑ   Ⓒ   Ⓓ   Ⓔ
(10) Ⓐ   Ⓑ   Ⓒ   Ⓓ   Ⓔ

(11) Ⓐ   Ⓑ   Ⓒ   Ⓓ   Ⓔ
(12) Ⓐ   Ⓑ   Ⓒ   Ⓓ   Ⓔ
(13) Ⓐ   Ⓑ   Ⓒ   Ⓓ   Ⓔ
(14) Ⓐ   Ⓑ   Ⓒ   Ⓓ   Ⓔ
(15) Ⓐ   Ⓑ   Ⓒ   Ⓓ   Ⓔ

(16) [   ]

(17) Ⓐ   Ⓑ   Ⓒ   Ⓓ   Ⓔ
(18) Ⓐ   Ⓑ   Ⓒ   Ⓓ   Ⓔ
(19) Ⓐ   Ⓑ   Ⓒ   Ⓓ   Ⓔ
(20) Ⓐ   Ⓑ   Ⓒ   Ⓓ   Ⓔ

GRE MATH PRACTICE SECTION 2

20 Questions
Time—35 Minutes

➤ The numbers on this test are real numbers.

➤ You may assume that positions of points, lines, and angles are in the order shown.

➤ You can NOT assume measures of line segments or angles unless the figure is drawn to scale.

➤ Figures are in a plane unless stated otherwise.

➤ You may assume that lines that appear to be straight lines are straight lines.

➤ Lines are NOT parallel or perpendicular unless you are told they are or you can deduce it from information given.

DIRECTIONS: Quantitative comparison questions provide two quantities, Quantity A and Quantity B. You are to compare the two quantities. Then choose

A if Quantity A is greater
B if Quantity B is greater
C if the two quantities are equal
D if the relationship can not be determined from the given information

NOTE: There is no answer E for these questions, so marking E is always wrong.

Common

Information: In a question, information concerning one or both of the quantities to be compared is centered above the two columns. A symbol that appears in both columns represents the same thing for both Quantity A and in Quantity B.

	Quantity A	Quantity B	Sample Answers
Example 1:	$3 + 6$	$3 \div 2$	● Ⓑ Ⓒ Ⓓ
Example 2:	n^2	$2n$	Ⓐ Ⓑ Ⓒ ●
			(If $n = 3$, $n^2 = 9$, $2n = 6$ and $n^2 > 2n$; if $n = 0.5$, $n^2 = 0.25$, $2n = 1$, and $n^2 < 2n$.)
Example 3:	3×2	2×3	Ⓐ Ⓑ ● Ⓓ
Example 4:	$5 - 3$	$5 - (-3)$	Ⓐ ● Ⓒ Ⓓ

GO ON TO THE NEXT PAGE.

	Quantity A	Quantity B
1.	Largest prime factor of 1,728	7

A ribbon 40 inches long is cut into three parts in the ratio 2 : 3 : 5.

	Quantity A	Quantity B
2.	the length of the longest part of the ribbon	20 inches

	Quantity A	Quantity B
3.	$(-5)^4$	$(-4)^5$

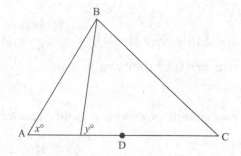

D is the midpoint of *AC*

	Quantity A	Quantity B
4.	x	y

	Quantity A	Quantity B
5.	number of positive integer divisors of 30	number of positive integer divisors of 18

The price of a coat is *P* dollars

	Quantity A	Quantity B
6.	Sale price of the coat after 3 separate discounts of 10%.	Sale price of the coat after a single discount of 30%

	Quantity A	Quantity B
7.	$4 - x^2$	5

	Quantity A	Quantity B
8.	$\sqrt{75} - \sqrt{27}$	$\sqrt{10}$

Questions 9–10 refer to the following graph.

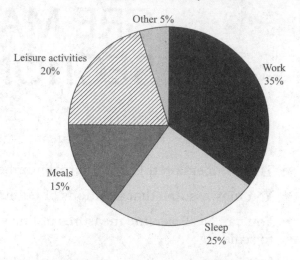

9. Based on the graph, what is the ratio of work time to leisure time?

   (A) 4 to 7
   (B) 7 to 4
   (C) 5 to 4
   (D) 4 to 5
   (E) 3 to 2

10. Based on the graph, if leisure activities were reduced by 25% and the time saved were added to sleep time, how many hours would then be spent on sleep?

   (A) 1.2%
   (B) 2.4%
   (C) 6 hours
   (D) 7.2 hours
   (E) 8 hours

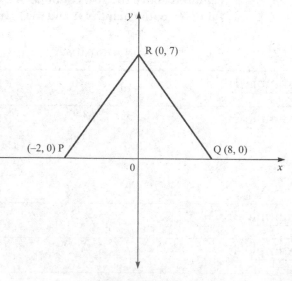

Figure 2

11. What is the area of △*PQR*? (Use Figure 2.)

    (A) 56
    (B) 35
    (C) 28
    (D) 14
    (E) 7

12. Three slices of pizza and a salad cost $6.45. Six slices of the same pizza and an identical salad cost $10.95. What is the cost of one slice of the pizza?

    (A) $1.50
    (B) $1.95
    (C) $4.50
    (D) $8.70
    (E) $12.80

13. If 60% of *P* is 40% of *S*, what is the ratio of *P* to *S*?

    (A) 1 to 5
    (B) 2 to 3
    (C) 1 to 3
    (D) 3 to 2
    (E) 3 to 1

14. If the radius of a right circular cylinder is doubled and the height is tripled, how does the volume of the cylinder change?

    (A) It is 2 times as much as before.
    (B) It is 4 times as much as before.
    (C) It is 3 times as much as before.
    (D) It is 6 times as much as before.
    (E) It is 12 times as much as before.

15. Based on the graph, what is the percentage change in the number of teachers from 1970 to 1990?

    (A) 18%
    (B) 20%
    (C) 27%
    (D) 36%
    (E) 13%

For this question, enter your answer in the box.

16. Based on the graph, about how many teachers were employed in 1995?

17. If the average of three numbers is $3x$ and two of the numbers are $2x - y$ and $x + y$, what is the third number in terms of x and y?

    (A) $x + y$
    (B) $x - y$
    (C) $6x - 2y$
    (D) $9x$
    (E) $6x$

18. If a family has three children, what is the probability that the family has at least one girl?

    (A) 1
    (B) 0.875
    (C) 0.75
    (D) 0.50
    (E) 0.375

Questions 15 and 16 are based on the following graph.

Number of Teachers Employed by Springfield Schools

GO ON TO THE NEXT PAGE.

19. If a restaurant has 4 appetizers, 7 entrees, 10 sides, and 3 desserts, how many different meals, each including one of each type of food, can it serve?

    (A) 24
    (B) 58
    (C) 82
    (D) 576
    (E) 840

20. Which are the roots of the equation $7x + 3x^2 = -2$? Choose all that apply.

    [A] -3
    [B] -2,
    [C] $-\dfrac{1}{2}$
    [D] $-\dfrac{1}{3}$
    [E] 2

STOP

IF YOU FINISH BEFORE TIME RUNS OUT, GO BACK AND CHECK YOUR WORK.

GRE MATH PRACTICE SECTION 2

Answer Key

1. B
2. C
3. A
4. B
5. A
6. A
7. B
8. A
9. B
10. D
11. B
12. A
13. B
14. E
15. D
16. 700
17. E
18. B
19. E
20. B,D

GRE MATH PRACTICE SECTION 2

SOLUTIONS

1. **B** Quantity B is greater.
   $1728 = 12 \times 12 \times 12 = (12)^3 = (2^2 \times 3)^3 = 2^6 \times 3^3; 7 > 3.$
   The answer is B.

2. **C** The two quantities are equal.
   $2 : 3 : 5$ means $2x + 3x + 5x = 40$ and $x = 4$
   $8 : 12 : 20$ The longest piece is 20 inches long.
   The answer is C.

3. **A** Quantity A is greater.
   $(-5)^4 = 625$ $(-4)^5 = -1024.$
   Since $625 > -1024$, the answer is A.

4. **B** Quantity B is greater.
   y is the measure of an exterior angle of $\triangle ABE$ and x is the measure of remote interior angle, so $y > x$.
   The answer is B.

5. **A** Quantity A is greater.
   30: 1, 2, 3, 5, 6, 10, 15, 30; 8 divisors
   18: 1, 2, 3, 6, 9, 18; 6 divisors
   Since $8 > 6$, the answer is A.

6. **A** Quantity A is greater.
   A: $100 item, 10% discount = $90, 10% discount = $81, 10% discount = $72.90.
   B: $100 item, 30% discount = $70.
   Since $A > B$, the answer is A.

7. **B** Quantity B is greater.
   $x^2 \geq 0$ for all real numbers, so $4 - x^2 \leq 4$ for all real numbers. Thus, $4 - x^2 < 5$.
   The answer is B.

8. **A** Quantity A is greater.
   $\sqrt{75} - \sqrt{27} = \sqrt{25}\sqrt{3} - \sqrt{9}\sqrt{3} =$
   $5\sqrt{3} - 3\sqrt{3} = 2\sqrt{3} = \sqrt{4}\sqrt{3} = \sqrt{12} > \sqrt{10}$
   The answer is A.

9. **B** 7 to 4
   35% to 20 % = 7 to 4

10. **D** 7.2 hours
    25% of 20% = 5%
    25% + 5% = 30%
    30% of 24 hours = 7.2 hours

11. **B**
    PQ is 10 units long and RQ is 7 units
    The area of $\triangle PQR$ is $0.5(10)(7) = 35$.

12. **A** $1.50
    6 slices and a salad = $10.95
    -3 slices and a salad = $6.45
    = 3 slices for $4.50
    Three slices cost $4.50, so one slice costs $1.50.

13. **B** 2 to 3
    $0.6\,P = 0.4S$, so $6P = 4S$ $\dfrac{P}{S} = \dfrac{4}{6} = \dfrac{2}{3}$
    P to S = 2 to 3

14. **E** It is 12 times as much as before.
    $V = \pi r^2 h$
    V after changes $= \pi(2r)^2(3h) = \pi(4r^2)(3h) = 12\pi r^2 h = 12V$

15. **D** 36%
    In 1970, 550 teachers were employed. In 1990, 750 teachers were employed.
    $750 - 550 =$ increase of 200
    $200 \div 550 \times 100\% = 36.36\%$ or 36%

16. **D** $\boxed{700}$

    In 1990, 750 teachers were employed. In 2000, 650 teachers were employed. Since 1995 is halfway between 1990 and 2000, the number of teachers employed in 1995 is about 700.

17. **E** $6x$

Let T be the third number.

$$\frac{2x - y + x + y + T}{3} = 3x \quad 3x + T = 9x \text{ and}$$

$T = 6x$.

18. **B** 0.875

All possible outcomes for a 3-child family are as follows: BBB, BBG, BGB, BGG, GBB, GBG, GGB, GGG

Seven of the eight outcomes have at least one girl, so $7 \div 8 = 0.875$.

19. **E** 840.

Number of different meals $= (4)(7)(10)(3) = 840$.

20. **B** and **D** $-2, -\dfrac{1}{3}$

$7x + 3x^2 = -2$

$3x^2 + 7x + 2 = 0$

$(3x + 1)(x + 2) = 0$

$3x + 1 = 0$ or $x + 2 = 0$

$x = -\dfrac{1}{3}$ or $x = -2$. The answers are B and D.

Note: On the real GRE you will enter your answers by clicking on answer ovals on the computer screen.

(1) Ⓐ Ⓑ Ⓒ Ⓓ Ⓔ
(2) Ⓐ Ⓑ Ⓒ Ⓓ Ⓔ
(3) Ⓐ Ⓑ Ⓒ Ⓓ Ⓔ
(4) Ⓐ Ⓑ Ⓒ Ⓓ Ⓔ
(5) Ⓐ Ⓑ Ⓒ Ⓓ Ⓔ

(6) Ⓐ Ⓑ Ⓒ Ⓓ Ⓔ
(7) Ⓐ Ⓑ Ⓒ Ⓓ Ⓔ
(8) Ⓐ Ⓑ Ⓒ Ⓓ Ⓔ
(9) Ⓐ Ⓑ Ⓒ Ⓓ Ⓔ
(10) Ⓐ Ⓑ Ⓒ Ⓓ Ⓔ

(11) Ⓐ Ⓑ Ⓒ Ⓓ Ⓔ
(12) Ⓐ Ⓑ Ⓒ Ⓓ Ⓔ
(13) Ⓐ Ⓑ Ⓒ Ⓓ Ⓔ
(14) Ⓐ Ⓑ Ⓒ Ⓓ Ⓔ

(15) [   ]

(16) Ⓐ Ⓑ Ⓒ Ⓓ Ⓔ
(17) Ⓐ Ⓑ Ⓒ Ⓓ Ⓔ
(18) Ⓐ Ⓑ Ⓒ Ⓓ Ⓔ
(19) Ⓐ Ⓑ Ⓒ Ⓓ Ⓔ
(20) Ⓐ Ⓑ Ⓒ Ⓓ Ⓔ

GRE MATH PRACTICE SECTION 3

20 Questions
Time—35 Minutes

➤ The numbers on this test are real numbers.

➤ You may assume that positions of points, lines, and angles are in the order shown.

➤ You can NOT assume measures of line segments or angles unless the figure is drawn to scale.

➤ Figures are in a plane unless stated otherwise.

➤ You may assume that lines that appear to be straight lines are straight lines.

➤ Lines are NOT parallel or perpendicular unless you are told they are or you can deduce it from information given.

DIRECTIONS: Quantitative comparison questions provide two quantities, Quantity A and Quantity B. You are to compare the two quantities. Then choose

A if Quantity A is greater
B if Quantity B is greater
C if the two quantities are equal
D if the relationship can not be determined from the given information

NOTE: There is no answer E for these questions, so marking E is always wrong.

Common

Information: In a question, information concerning one or both of the quantities to be compared is centered above the two columns. A symbol that appears in both columns represents the same thing for both Quantity A and in Quantity B.

	Quantity A	Quantity B	Sample Answers
Example 1:	$3 + 6$	$3 \div 2$	● Ⓑ Ⓒ Ⓓ
Example 2:	n^2	$2n$	Ⓐ Ⓑ Ⓒ ●
			(If $n = 3$, $n^2 = 9$, $2n = 6$ and $n^2 > 2n$; if $n = 0.5$, $n^2 = 0.25$, $2n = 1$, and $n^2 < 2n$.)
Example 3:	3×2	2×3	Ⓐ Ⓑ ● Ⓓ
Example 4:	$5 - 3$	$5 - (-3)$	Ⓐ ● Ⓒ Ⓓ

GO ON TO THE NEXT PAGE.

317

	Quantity A	Quantity B
1.	number of divisors of 121	number of divisors of 24

For all counting numbers x, $x^* = \dfrac{2}{x}$

	Quantity A	Quantity B
2.	x	$(x^*)^*$

P and Q are the roots of $x^2 - 9x = 36$

	Quantity A	Quantity B
3.	$P + Q$	$P \times Q$

$r > 0$

	Quantity A	Quantity B
4.	$\dfrac{7}{r+8}$	$\dfrac{7}{r}$

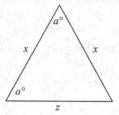

	Quantity A	Quantity B
5.	x	$\dfrac{1}{2}(x + z)$

r is the radius of sphere S and of the circle C.

	Quantity A	Quantity B
6.	The ratio of the surface area of sphere S to the area of circle C	4

7. If $\dfrac{x}{8} = \dfrac{3}{5}$, what is the value of x?

(A) 6
(B) 5.5
(C) 4.8
(D) 2.2
(E) 0.075

8. What is the selling price of a lamp if a store bought it for $118 and marked it up 30%?

(A) $35.40
(B) $82.60
(C) $148
(D) $153.40
(E) $200.60

9. What is the value of $\dfrac{3}{5}\left(\dfrac{5}{8} - 4\right)$?

(A) $-\dfrac{81}{40}$

(B) $-\dfrac{3}{20}$

(C) $\dfrac{3}{20}$

(D) $\dfrac{4}{9}$

(E) $\dfrac{81}{40}$

10. If the diameter (d) of a circle is an integer and 6 cm $\leq d \leq$ 10 cm, which could be the area of the circle?

(A) 36π cm²
(B) 25π cm²
(C) 14π cm²
(D) 6π cm²
(E) 3π cm²

Questions 11–15 refer to the following graph.

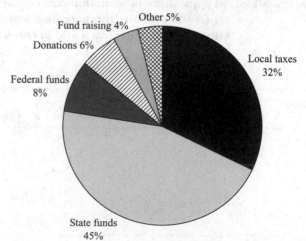

Funding Sources for Springfield School in 2006

GO ON TO THE NEXT PAGE.

11. What is the ratio of local taxes to donations for Springfield Schools in 2006?

    (A) 8 to 1
    (B) 32 to 1
    (C) 4 to 1
    (D) 16 to 3
    (E) 16 to 9

12. What percentage of the funds for Springfield Schools came from state, federal, and local government agencies?

    (A) 40%
    (B) 53%
    (C) 77%
    (D) 85%
    (E) 90%

DIRECTIONS: Questions 13 and 14 are quantitative comparison questions. Choose

A if Quantity A is greater
B if Quantity B is greater
C if the two quantities are equal
D if the relationship can not be determined from the given information

	Quantity A	Quantity B
13.	Local taxes plus federal funds	50%

	Quantity A	Quantity B
14.	Donations	Other

For this question, enter your answer in the box.

15. According to the graph, if Springfield Schools had a budget of $7,200,000 in 2006, how many dollars were raised from fund raising?

16. Which is the supplement of a 52° angle?

    (A) 38°
    (B) 52°
    (C) 68°
    (D) 78°
    (E) 128°

17. If the length of a rectangle is 16 cm and its width is 10 cm, what is the perimeter of the rectangle?

    (A) 26 cm
    (B) 42 cm
    (C) 52 cm
    (D) 80 cm
    (E) 160 cm

18. If a bag contains 18 yellow tiles and 22 black tiles, what is the probability that a tile selected at random from the bag will be yellow?

    (A) $\dfrac{1}{10}$
    (B) $\dfrac{9}{20}$
    (C) $\dfrac{11}{20}$
    (D) $\dfrac{9}{11}$
    (E) $\dfrac{11}{9}$

19. After a 5% raise, Kim has a salary of $40,530. What was Kim's salary before the raise?

    (A) $2,026.50
    (B) $38,503.50
    (C) $38,600.00
    (D) $42,556.50
    (E) $42,663.16

20. Which is equal to 0.0001? Choose all that apply.

    A $\dfrac{1}{10000}$
    B $(0.1)^3$
    C $(0.1)^4$
    D $\dfrac{1}{1000}$
    E 1×10^{-4}

GRE MATH PRACTICE SECTION 3

Answer Key

1. B
2. C
3. A
4. B
5. C
6. C
7. C
8. D
9. A
10. B
11. D
12. D
13. B
14. A
15. 288,000
16. E
17. C
18. B
19. C
20. A,C,E

GRE MATH PRACTICE SECTION 3

SOLUTIONS

1. **B** Quantity B is greater.
   121 : 1, 11, 121, so 3 divisors   24 : 1, 2, 3, 4,
   6, 8 12, 24, so 8 divisors
   $B > A$ The answer is B.

2. **C** The two quantities are equal.
   $x^* = \dfrac{2}{x}$ $\left(\dfrac{2}{x}\right)^* = \dfrac{2}{(2/x)} = x$, so $(x^*)^* = x$
   The answer is C.

3. **A** Quantity A is greater.
   $x^2 - 9x = 36$
   $x^2 - 9x - 36 = 0$
   $(x - 12)(x + 3) = 0$
   $x = 12$ or $x = -3$
   $P + Q = 12 + (-3) = 9$
   $PQ = 12(-3) = -36$
   $P + Q > PQ$
   The answer is A.

4. **B** Quantity B is greater.
   $r > 0$ $\dfrac{7}{r + 8}, \dfrac{7}{r}$ Cross-multiply.
   $7r < 7r + 56$ so $\dfrac{7}{r + 8} < \dfrac{7}{r}$
   The answer is B.

5. **C** The two quantities are equal.
   Since two sides are given as equal, the
   angles opposite these sides are equal. Thus,
   all three of the angles are equal because two
   of the angles were given to be $a°$. Since all
   angles of the triangle are equal, all sides of
   the triangle are equal. Thus $z = x$ and
   $\frac{1}{2}(x + z) = x$.
   The answer is C.

6. **C** The quantities are equal.
   The surface area of the sphere is $4\pi r^2$. The
   area of the circle is πr^2. $4\pi r^2 : \pi r^2 = 4 : 1$.
   The answer is C.

7. **C** 4.8
   $\dfrac{x}{8} = \dfrac{3}{5}$, $5x = 24$, $x = 24 \div 5$, $x = 4.8$
   The answer is C.

8. **D** $153.40
   Original price + mark-up = Selling price
   $118 + 0.3($118) = $118 + $35.40 = $153.40

9. **A** $-\dfrac{81}{40}$
   $\dfrac{3}{5}\left(\dfrac{5}{8} - 4\right) = \dfrac{3}{5} \times \dfrac{5}{8} - \dfrac{3}{5} \times \dfrac{4}{1} = \dfrac{3}{8} - \dfrac{12}{5} =$
   $\dfrac{15 - 96}{40} = -\dfrac{81}{40}$

10. **B** 25π cm^2
    d is an integer and 6 cm $\leq d \leq$ 10 cm.
    $r = \dfrac{1}{2}d$ so $3 \leq r \leq 5$ cm. Thus, $r = 3$ cm,
    4 cm, or 5 cm, and $A = 9\pi$ cm^2, 16π cm^2, or
    25π cm^2. Therefore of the areas listed, only
    25π cm^2 is a possible area. Note: Since r^2 is
    an integer in all the answers, you can ignore
    values of r when d is odd.

11. **D** 16 to 3
    Donations 6%   Local taxes 32%
    32% to 6% — 16 to 3.

12. **D** 85%
    State 45%   Federal 8%   Local taxes 32%
    45% + 8% + 32% = 85%

13. **B** Quantity B is greater.
    Local taxes + federal funds
    = 32% + 8% = 40%.

14. **A** Quantity A is greater.
    Donations = 6%. Other = 5%.

15. $\boxed{288{,}000}$

Because 4% of funds are raised by fund raising, fund raising = $0.04(7{,}200{,}000) = 288{,}000$.

16. **E** 128°

Supplementary angles have a sum of 180°. The supplement of a 52° angle $= 180° - 52° = 128°$.

17. **C** 52 cm

$P = 2l + 2w = 2(16 \text{ cm}) + 2(10 \text{ cm}) = 32 \text{ cm} + 20 \text{ cm} = 52 \text{ cm}$.

18. **B** $\dfrac{9}{20}$

$P(yellow) = \dfrac{18}{18 + 22} = \dfrac{18}{40} = \dfrac{9}{20}$.

19. **C** $38,600

S = previous salary; $S + 0.05S = \$40{,}530$; $1.05S = \$40{,}530$; $S = \$38{,}600$.

20. **A, C,** and **E**

$$\dfrac{1}{10000} = 0.0001$$

$$(0.1)^4 = (0.1)(0.1)(0.1)(0.1) = 0.0001$$

$$1 \times 10^{-4} = \dfrac{1}{10000} = 0.0001$$

Choices A, C, and E are all correct.